Physiology and Technology of Reproduction in Female Domestic Animals

Physiology and Technology of Reproduction in Female Domestic Animals

R. H. F. Hunter

School of Agriculture,
University of Edinburgh

1980

ACADEMIC PRESS
A Subsidiary of Harcourt Brace Jovanovich, Publishers
London New York Toronto Sydney San Francisco

ACADEMIC PRESS INC. (LONDON) LTD.
24/28 Oval Road
London NW1

United States Edition published by
ACADEMIC PRESS INC.
111 Fifth Avenue
New York, New York 10003

British Library Cataloguing in Publication Data

Hunter, R H F
Physiology and technology of reproduction in
female domestic animals
1. Livestock — Reproduction 2. Generative organs,
Female
I. Title
636.089'26'2 SF105 80–40897

ISBN 0–12–361950–5

Printed in Great Britain
by W & J Mackay Limited, Chatham

Physiology and Technology of Reproduction in Female Domestic Animals

R. H. F. Hunter

School of Agriculture,
University of Edinburgh

1980

ACADEMIC PRESS

A Subsidiary of Harcourt Brace Jovanovich, Publishers

London New York Toronto Sydney San Francisco

ACADEMIC PRESS INC. (LONDON) LTD.
24/28 Oval Road
London NW1

United States Edition published by
ACADEMIC PRESS INC.
111 Fifth Avenue
New York, New York 10003

British Library Cataloguing in Publication Data

Hunter, R H F
Physiology and technology of reproduction in female domestic animals
1. Livestock — Reproduction 2. Generative organs, Female
I. Title
636.089'26'2 SF105 80–40897

ISBN 0–12–361950–5

Printed in Great Britain
by W & J Mackay Limited, Chatham

Preface

This book has been written for advanced undergraduates in Animal Science, Veterinary Science and Applied Biology and, in the context of the University of Edinburgh, for students taking Applied Physiology in the Faculty of Science and the MSc course in Animal Breeding. The material should also prove useful to graduate students commencing research in mammalian reproductive physiology, and for those engaged in animal improvement programmes. Indeed, in recent years, we have offered refresher courses on the subject matter of this volume to staff of the Food and Agriculture Organisation and to candidates on British Council courses for livestock experts, and the response from all concerned has been encouraging.

My initial approach was to write a comprehensive volume of 18 chapters, dealing first with basic physiology in both the male and female and then with the various means of exploiting this knowledge. However, as the manuscript took shape, the treatment seemed too lengthy and the section on male reproductive physiology in particular was repetitive of material found in other texts. Accordingly, the manuscript was shortened to 12 chapters and its substance concentrated principally on those aspects of female reproductive physiology amenable to modification or more direct manipulation, not least since these might reasonably be expected to contribute to improved animal productivity. This approach led to some reduction in the discussion of equine reproduction and an increase in emphasis on the major meat-producing species of domestic animal found in many countries — cattle, sheep and pigs. Other domesticated species such as goats, camels or buffalo have not been mentioned, in part to avoid the text taking on the appearance of a catalogue. If strong feelings are raised by these omissions, it may become necessary to reconsider the scope of the book in a future edition.

As already stated, this is not an introductory text and a reasonable understanding of animal physiology and biological terminology is therefore assumed. With this in mind, the chapters have been written so that by and large they are self sufficient and can be perused as individual entities. A logical sequence of presentation of material has been attempted, but the use of an introductory section to each chapter has inevitably led to occasional repetition. Concerning a policy on references cited in the text, the objective has certainly not been to produce an exhaustive list but rather to present a sufficient selection to enable the reader to trace development of a given topic and to consult the established literature. Some attention has been paid to historical aspects in the hope that original contributions would be highlighted, but the author is only too well aware how easy it is to overlook key papers. In this regard, he would be pleased to hear from those who feel that such references have been omitted.

Finally, no apology is offered for frequent use of the term *reproductive technology*, even though this may provoke shudders in many and remind addicts of the Orwellian saga that the day of reckoning is fast approaching. Rather than criticise the development of this branch of animal science, a more responsible attitude is surely to appreciate the contribution that available techniques can make to augmenting the output of animal products in both developed and developing countries. Progress tends to be stimulated by those who recognise opportunities: one of these is the ability to regulate reproductive processes in female farm animals.

R. H. F. Hunter

Edinburgh
December 1979

Acknowledgements

The gestation length of a textbook is undoubtedly longer than many authors would like to claim, and in the present instance a reasonable estimate of timespan might well extend from my commencement of postgraduate studies in 1961. During the intervening period, a number of distinguished reproductive physiologists have given me the privilege of working in their laboratories, and of these I am especially grateful to: Professor T. R. R. Mann, FRS, and Mr L. E. A. Rowson, FRS, in Cambridge; Professors P. J. Dziuk and A. V. Nalbandov of the University of Illinois; Dr M. C. Chang of the Worcester Foundation for Experimental Biology, Massachusetts; and Professor C. Thibault and Dr F. Du Mesnil du Buisson of the Station de Physiologie Animale, Jouy-en-Josas, France. My warmest words of thanks go to Dr E. J. C. Polge, Director of the Animal Research Station in Cambridge, who has been a source of advice, assistance and encouragement for some 20 years.

Concerning preparation of the manuscript, help has been received from diverse quarters. Colleagues in Edinburgh who kindly commented on first drafts of individual chapters included Drs T. G. Baker, K. P. Bland, R. B. Land, I. Wilmut and D. G. M. Wood-Gush. Mr M. S. Smith of the Agricultural Development and Advisory Service, Wolverhampton, also provided numerous thoughtful comments on an early draft of Chapter IX. I am very appreciative of the time and trouble these specialists took on my behalf.

The task of illustrating the text was much eased by the line-drawings prepared by Dr T. A. Bramley and Mr R. Nichol, and photographs prepared by Mr G. Finnie. Further drawings and photographs were generously provided by the following friends and colleagues: Drs L. L. Anderson, D. T. Baird, T. G. Baker, J. M. Bedford, B. Cook, H. M. Dott, M. Dym, P. J. Dziuk, D. W. Fawcett, J. E. Fléchon, O. J. Ginther, G. J. King, J. A. Laing, R. Leiser, M. J. Meredith, D. Mitchell, R. M. Moor, J. S. Perry, E. J. C. Polge, N. O. Rasbech, L. E. A. Rowson, D. Szollosi, C. Thibault and D. F. Wishart. Others gave me permission to reproduce graphical data and to quote from their published work, and here I wish to thank: Drs C. E. Adams, C. R. Austin, N. Ayalon, D. M. Baldwin, M. J. Bosc, L. E. Casida, J. R. G. Challis, W. Y. Chan, M. C. Chang, I. J. Christiansen, G. Colas, R. S. Comline, M. J. Cooper, P. J. Dziuk, B. H. Erickson, R. H. Foote, T.

D. Glover, I. Gordon, R. B. L. Gwatkin, H. D. Hafs, W. Hansel, W. Haresign, R. B. Heap, Y. Heyman, L. B. Jeffcott, M. J. B. Joyce, G. E. Lamming, G. C. Liggins, G. M. Lindner, J. A. McCracken, A. McLaren, P. Mauléon, R. M. Moor, R. Newcomb, J. W. Overstreet, W. M. Palmer, P. D. Penning, J. S. Perry, E. J. C. Polge, T. D. Quinlivan, H. A. Robertson, T. J. Robinson, J. F. Roche, I. W. Rowlands, L. E. A. Rowson, M. Silver, J. Sreenan, G. H. Stabenfeldt, T. Sugie, H. R. Tervit, J. Testart, C. Thibault, D. Walker, R. A. S. Welch, H. L. Whitmore, D. G. Whittingham and A. H. Willemse. In connection with citations from published work, my gratitude is extended to the Editors of the following journals: American Journal of Veterinary Research, Anatomical Record, Animal Production, Animal Reproduction Science, Annales de Biologie Animale, Biochimie et Biophysique, Biology of Reproduction, British Veterinary Journal, Developmental Biology, Irish Veterinary Journal, Journal of Agricultural Science, Journal of Anatomy, Journal of Animal Science, Journal of Dairy Science, Journal of Embryology and Experimental Morphology, Journal of Endocrinology, Journal of Physiology, Journal of Reproduction and Fertility, Nature, New Zealand Veterinary Journal, Veterinary Record and the World Review of Animal Production; and to the following publishing houses: Baillière Tindall, Butterworths, Cambridge University Press, Elsevier-North Holland Inc., Karger, Martinus Nijhoff, Paul Parey, Plenum Press, Prentice Hall Inc. and the W. B. Saunders Company.

Typing of the manuscript was undertaken by Mrs Pat Gallie of the School of Agriculture and Hilary Prout of the Department of Genetics in Edinburgh; to both I express my best thanks for their meticulous attention to the text and for the trouble and care they took in cross-checking various figures, tables and references. The skills of Hilary Prout were kindly made available by Dr R. A. Beatty and sponsored by the Ford Foundation. I should also like to record my gratitude to the University of Edinburgh for facilities placed at my disposal during writing of this book; to the late Professor F. W. H. Elsley for encouragement when the book was in early gestation; and to the staff of Academic Press for much help during the process of delivery.

Finally, I must express my very best thanks to my wife for continuous help with preparation and checking of the manuscript, and to my whole family for tolerating my idiosyncrasies during the period of writing.

Contents

Glossary of abbreviations

ACTH	Adrenocorticotrophic hormone
ACTH-RH	Corticotrophin releasing hormone
BSA	Bovine serum albumin
CL	Corpus luteum (singular); corpora lutea (plural)
DMSO	Dimethyl sulphoxide
FGA	Fluorogestone acetate
FSH	Follicle stimulating hormone
FSH-RH	Follicle stimulating hormone releasing hormone
Gn-RH	Gonadotrophin releasing hormone
HAP	Horse anterior pituitary (preparation or extract)
HCG	Human chorionic gonadotrophin
i.d.	Inside (internal) diameter
i.u.	International units
LH	Luteinising hormone
LH-RH	Luteinising hormone releasing hormone
o.d.	Outside diameter
PGE	Prostaglandin E
$PGF_2\alpha$	Prostaglandin $F_2\alpha$
PIF	Prolactin inhibiting factor
PLH	Placental lactogenic hormone
PMSG	Pregnant mare serum gonadotrophin
RF	Releasing factor
RH	Releasing hormone
STH	Somatotrophin (growth hormone)
TCM	Tissue culture medium
TDN	Total digestible nutrients
TSH	Thyroid stimulating hormone (thyrotrophin)
w/v	Ratio of weight/volume

Units

mg	milligram (10^{-3}g)
μg	microgram (10^{-6}g)
ng	nanogram (10^{-9}g)
pg	pikogram (10^{-12}g)

Chapter I

Differentiation, Puberty, and the Oestrous Cycle

Introduction

In order to set the stage for subsequent chapters, these opening sections must present fairly concise descriptions of differentiation and morphology of the female reproductive structures, the functional relationship between the hypothalamus and pituitary gland, and the phenomenon of puberty. These sections are followed by a general description of the nature of the oestrous cycle and then an analysis of the underlying pattern of secretion of gonadotrophic and gonadal hormones for individual species. Apart from restricting itself to a consideration of events in the female, this introductory chapter also avoids describing such fundamental topics as primary and secondary sex ratios, sexual differentiation of the brain, and growth of specific endocrine organs. Rather, it places considerable emphasis on events after puberty since most of the information in the text refers to sexually mature animals. The aim has been to introduce some of the concepts that are central to the discipline of reproductive physiology, and to pave the way for the more extensive material concerned with manipulation of reproductive processes in female domestic animals that forms the body of the book.

Formation of functional ovaries

The genetic sex of a mammal is fixed at the time of fertilisation according to whether the cytoplasm of the ovum is penetrated by an X- or a Y-bearing spermatozoon. Under the influence of its sex chromosome complement, the mammalian embryo undergoes anatomical modifications that lead to the differentiation of male or female gonads (testes and ovaries, respectively) and male or female duct systems. The very early embryo possesses the potential for sexual differentiation in either direction, and at this stage the presumptive gonads or genital ridges are colonised by primordial germ cells. These cells migrate to the gonad from the yolk sac endoderm, passing through the region of the hind gut (Fig. I.1). Active amoeboid movement is involved in their migration, and possibly also passive transport within the developing bloodstream.

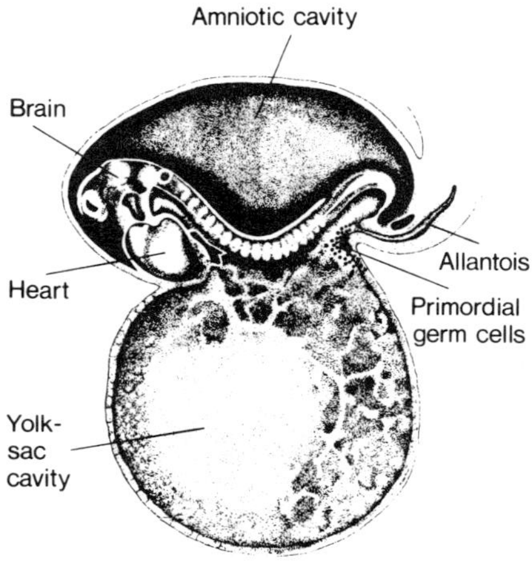

Fig. I.1. Diagram of a mammalian embryo to illustrate the migration of primordial germ cells from the yolk sac into the region of the hind gut. (Redrawn after E. Witschi, *Contrib. Embryol.* **32**, 69, Fig. 1, Plate 1 (1948). Courtesy of Dr T. G. Baker.)

In the later part of their migration, the germ cells move from the developing kidney (mesonephros) into the adjoining genital ridges, taking up either a cortical (outer) or medullary (central) position. If the gonad is to differentiate into an ovary, as will occur when the sex chromosome constitution is XX, the germ cells become concentrated in the cortex to form oogonia, whereas in the formation of a testis such cells generally pass into the medulla. Even so, cortical or medullary dominance is established before colonisation with germ cells, for gonadal tissue apparently commences to differentiate according to its own genetic constitution.

In the male, the cortex becomes obscured by the membranous tunica albuginea and the gonads remain relatively undifferentiated until shortly before puberty, consisting principally of tubules and interstitial tissue. In the female, by contrast, development of the cortex proceeds at the expense of involution of the medulla and the embryonic ovaries undergo development and differentiation so that production of potential gametes—the primary oocytes—is essentially complete by the time of birth. This is due to waves of mitosis causing enormous proliferation in the stock of oogonia, followed by the first phase of meiosis and extensive atresia even before the time of birth (review by Baker, 1972). Nonetheless, the foetus at term will contain a population of oocytes numbered in at least tens or hundreds of thousands rather than in units of a hundred. Whether the potential for growth and development is equally distributed amongst the population of oocytes remains an unresolved question. Certainly, only a minute proportion of the original stock is ever released at ovulation, and a smaller proportion still has the opportunity to be fertilised.

An oocyte can only grow and undergo further maturation within the ovary if it becomes surrounded by somatic cells. This process commences before the time of birth, when cells of epithelial origin arrange themselves around the limiting membrane of the oocyte. The structure formed when a complete layer of cells has evolved is termed a primary follicle, and this in turn becomes a secondary follicle when the cells divide to give several layers (Fig. I.2). In due course, with further cell division, fluid secretion and accumulation commence within the intercellular spaces, and gradual coalescence of these fluid islands leads to the formation of a cavity or antrum containing follicular fluid. Antrum formation signifies the stage of a tertiary or Graafian follicle, but only a small proportion of primary follicles achieves this stage.

During growth and development of an ovarian follicle, two processes occur that are essential to the production of a functional female gamete. First, an accessory membrane of mucopolysaccharide material is deposited around the oocyte due to the combined synthetic activities of the oocyte itself and the follicular cells (Fig. I.3). Secretion of this membrane—the zona pellucida—is initially as islands of fibrillar material, but formation of a complete and uniform zona is found long before the oocyte is encompassed within a Graafian follicle. Second, growth of the oocyte occurs, perhaps due to inherent synthetic capability but also because material is incorporated from the surrounding somatic cells. This latter process appears not to be impeded by the zona pellucida, and is thought to be facilitated by the interdigitations between the follicular cell processes and the microvilli of the egg surface. In any event, the oocyte acquires considerable cytoplasmic (vitelline) reserves of nutritive substances, of which vesicles of yolk and other lipid-like deposits are particularly prominent in the eggs of farm animals.

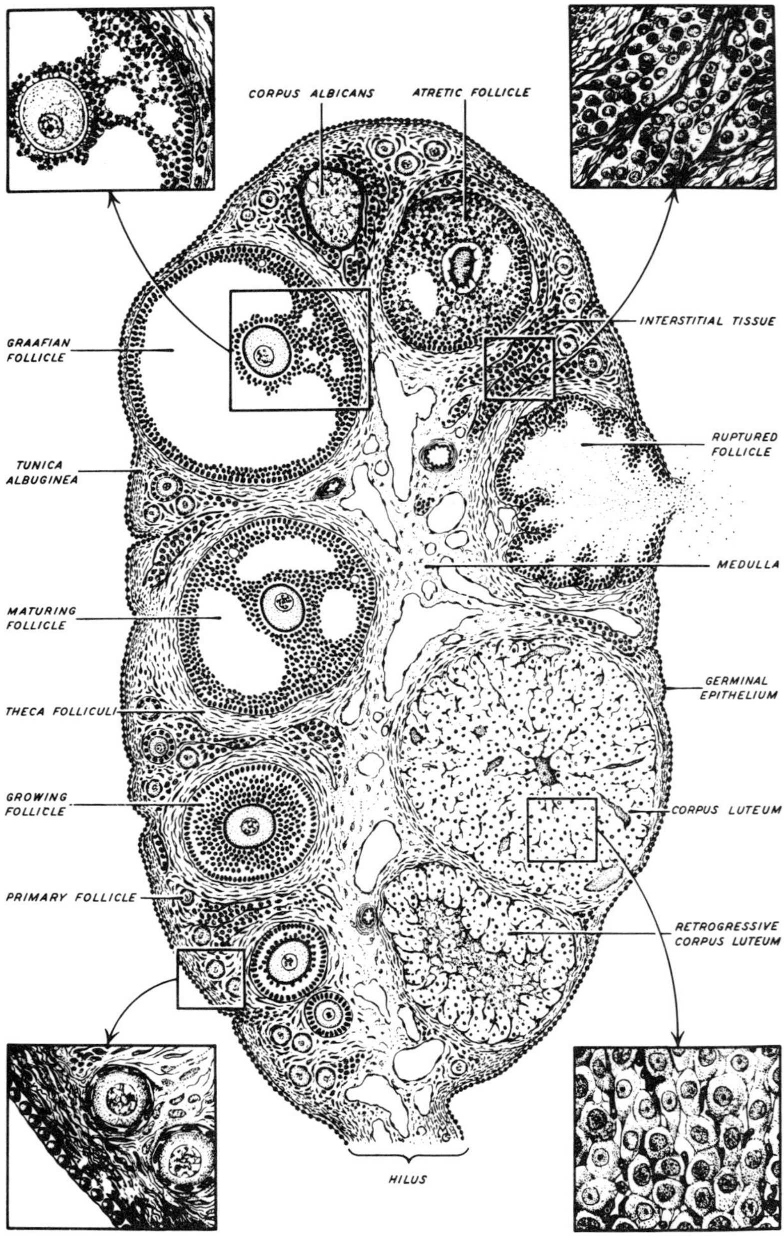

Fig. I.2. An idealised section through a mammalian ovary after puberty to illustrate the principal structural components. The section is composite in the sense that no distinction has been drawn between the luteal and follicular phases of the oestrous cycle. (Courtesy of C. D. Turner and J. T. Bagnara, *General Endocrinology*, 5th edition, Fig. 11–9, W. B. Saunders Company.)

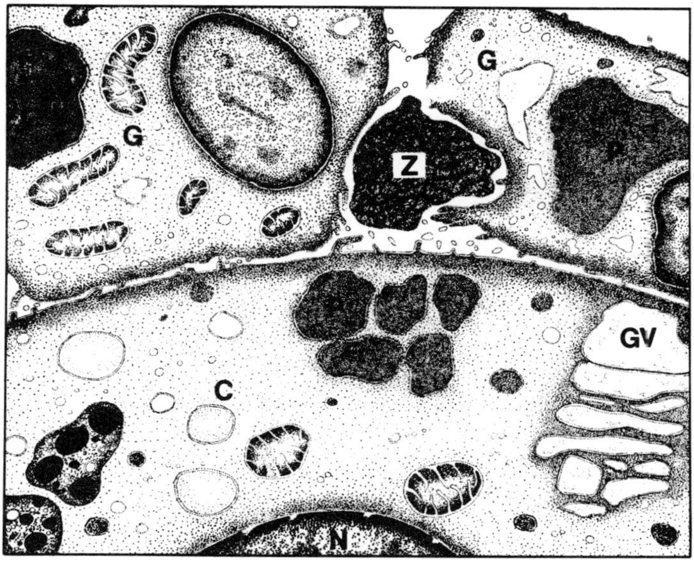

Fig. I.3. Partially diagrammatic electron micrograph depicting formation of the zona pellucida (Z) by means of a contribution from both the egg cytoplasm (C) and the granulosa cells (G). P represents precursor-like material and GV the Golgi vesicles. (From Austin, C. R. and Short, R. V. *Reproduction in Mammals*, Cambridge University Press, Volume 1, Fig. 2–8. Courtesy of Dr T. G. Baker.)

Concerning nuclear changes within the oocyte, the first meiotic prophase has been embarked upon in the foetal ovary, but proceeds only to pachytene or diplotene (Brambell, 1956; Zuckerman, 1960), at which stage events are arrested with the chromatin arranged around the nucleolus in a large resting nucleus or germinal vesicle (Fig. I.4). The nucleus of the primary oocyte remains in this arrested condition until overtaken either by the process of atresia or, when enclosed within a healthy Graafian follicle, meiosis resumes at diakinesis in response to the preovulatory surge of gonadotrophic hormones (see below). If and when the latter event occurs, meiotic changes proceed through metaphase, anaphase and telophase, with formation of a first polar body and arrangement of the chromosomes at metaphase on the second meiotic spindle (Fig. I.4). This stage is referred to as that of the secondary oocyte, and it is in this condition that the eggs of most species of mammal so far examined are ovulated.

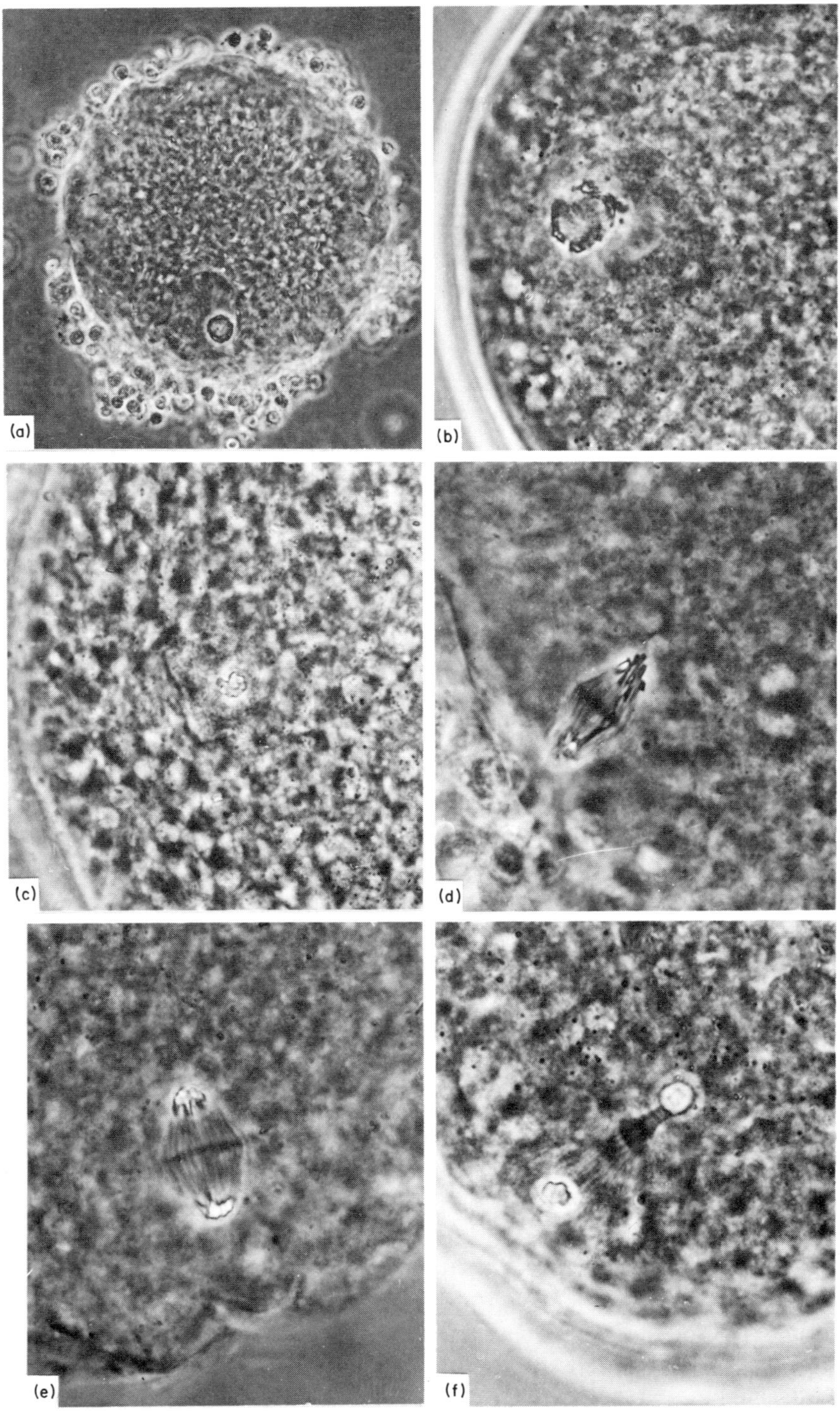

Fig. I.4. Events of the first meiotic division in pig oocytes from the germinal vesicle stage (a and b) through prometaphase (c), anaphase (d), telophase (e) and abstriction of the first polar body (f).

Structure and developmental fate of mature follicle

The structure of a Graafian follicle is presented in a simplified form in Fig. I.5. The cells that surround and support the oocyte are those of the cumulus oophorus, whose innermost layer (the corona radiata) is cuboidal and attached by processes to the surface of the zona pellucida.

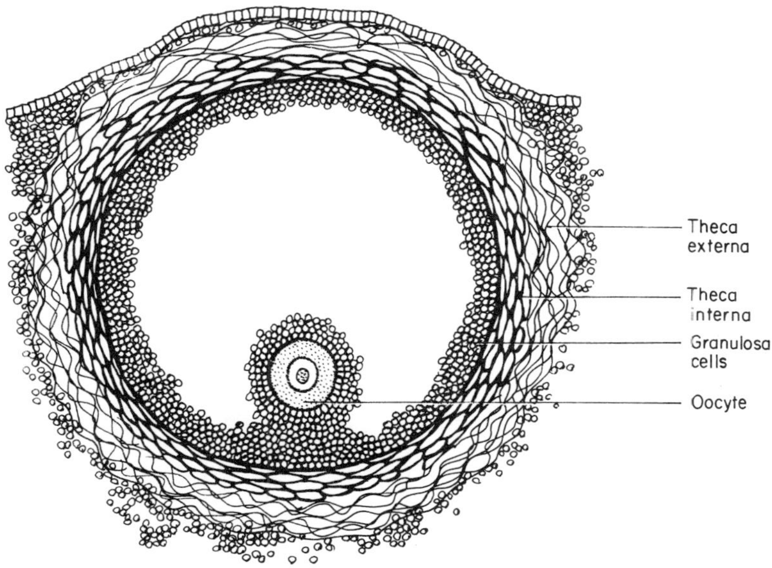

Fig. I.5. Semi-diagrammatic representation of a Graafian follicle to illustrate the theca externa, theca interna, and the oocyte supported within the follicular antrum by elements of the granulosa cell layers.

The cells of the cumulus oophorus represent only a portion of the granulosa cells that encircle the antrum and are arranged upon a basement membrane that separates them from the rest of the follicle. Except in the final few hours before ovulation, this basement membrane is not penetrated by the capillary network of the follicle, and so granulosa cells must obtain substrate and undertake gaseous exchange by diffusion across the membrane. Surrounding the basement membrane are the well-vascularised cell layers of the theca interna and theca externa, amongst which smooth muscle elements can be distinguished. Outermost of all, the germinal epithelium covers the follicular surface in the region where it protrudes from the ovary. Despite its name, this tissue is not germinal but continuous with the coelomic (visceral) peritoneum.

Cells of the theca interna predominate in the biosynthesis of oestrogens (Fig. I.6), whereas it is the granulosa cells that synthesise progesterone, largely after the time of ovulation when, as granulosa lutein cells, they are the principal component of the corpus luteum.

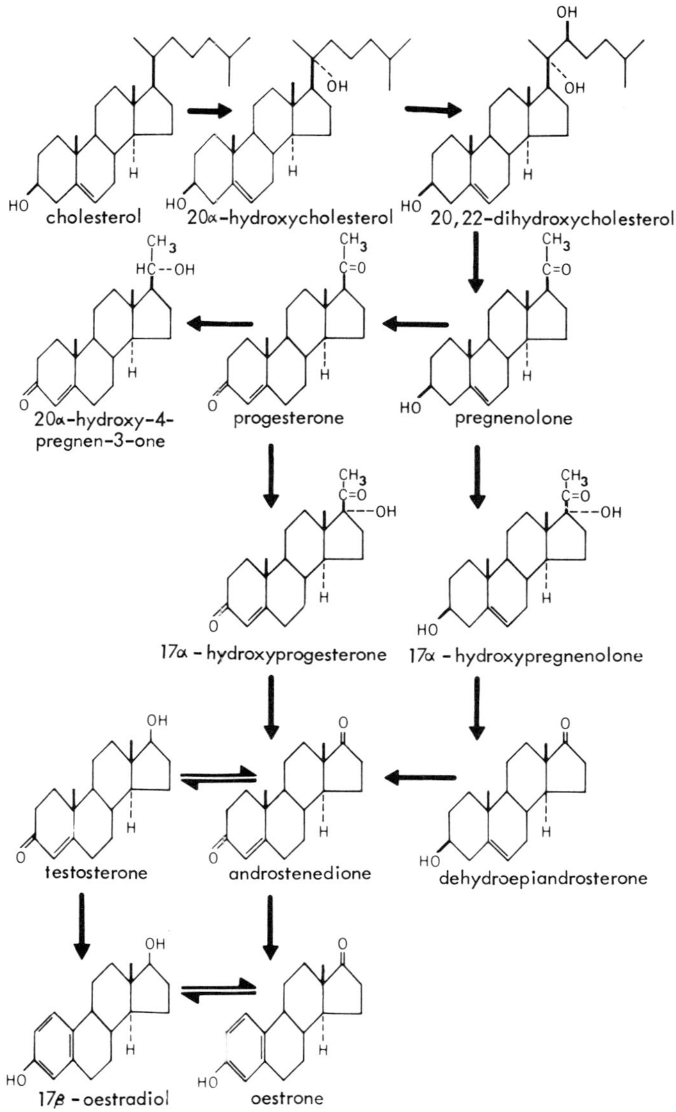

Fig. I.6. The biosynthetic pathway for gonadal steroid hormones illustrating the derivation of progestins, androgens and oestrogens by progressive enzymatic modification of the cholesterol molecule. (Courtesy of Dr B. Cook.)

After collapse of the follicle at the time of ovulation, the newly vascularised granulosa cells hypertrophy and proliferate rapidly to form this structure which soon becomes a spherical body achieving approximately the same diameter as a mature Graafian follicle, although of a solid cellular nature rather than fluid-filled. In species such as the pig in which cells of the theca interna also contribute to formation of the corpus luteum as theca lutein cells (Corner, 1919), oestrogen secretion can apparently occur simultaneously with that of progesterone, as in primates (see Baird *et al.*, 1975). A composite section through a mature ovary is presented in Fig. I.2, and shows the range of follicular structures that exist after puberty, as well as developing and regressing corpora lutea. The latter are known as corpora albicantia and are no longer steroidogenically active. Interstitial tissue is also an important structural component of the ovary, being derived principally from the theca of atretic follicles, and in some species shows considerable steroid biosynthetic potential, probably producing androgens.

Morphogenesis of the female genital tract

Developmental changes in the very young embryo furnish it with two sets of primitive genital tracts, the Wolffian and Mullerian ducts (Fig. I.7). In early embryonic life, there is thus the potential to develop either

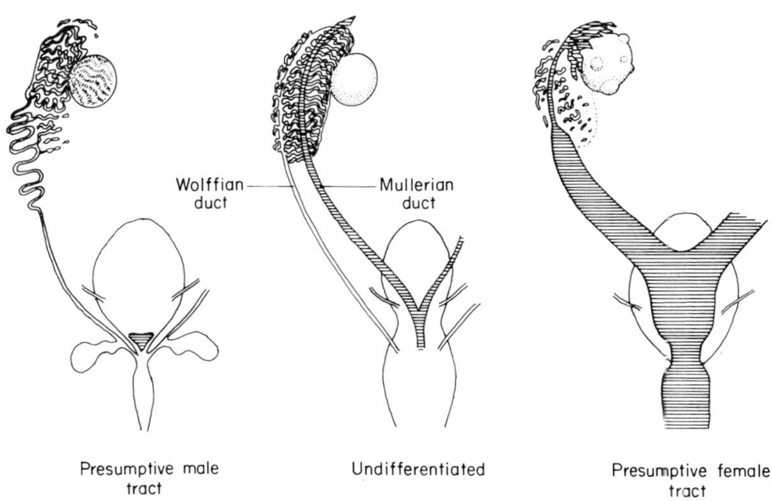

Fig. I.7. Semi-diagrammatic representation of the morphological derivation of male or female genital tracts from the Wolffian or Mullerian duct systems.

a male or a female reproductive system. In normal circumstances, only one set of ducts develops, the other becoming vestigial, although this may not be so in cases of chromosome or hormonal disturbance. The factor that dictates which of the duct systems will develop is the presence or absence of a Y chromosome, although the effect is mediated through secretion (or absence of secretion) of androgens from the embryonic testis. Maleness promotes stabilisation and development of the Wolffian ducts due to the influence of androgens, and regression of the Mullerian system due to the influence of some other testicular product, probably protein in nature. In the absence of male sex hormone, the Wolffian ducts regress and the Mullerian ducts develop. Thus, it is the absence of androgens rather than a positive influence of femaleness that permits development of the Mullerian ducts. These morphogenetic changes proceed according to an accurately-defined schedule of embryonic development (review by Burns, 1961).

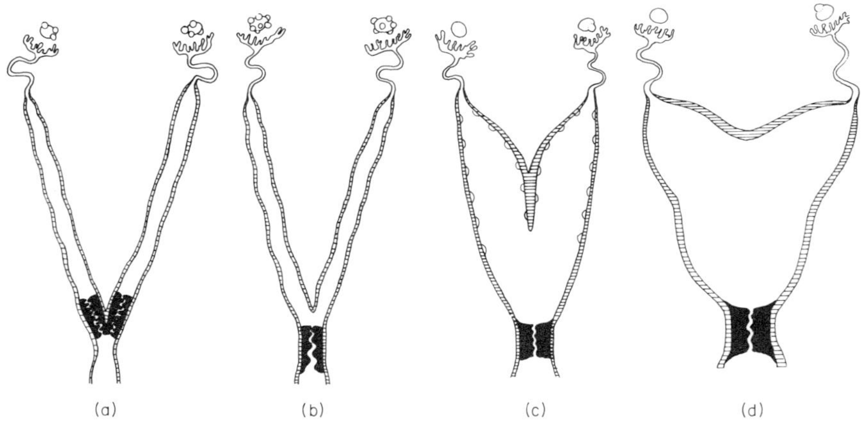

Fig. I.8. Semi-diagrammatic representation of the female genital tract showing the morphological transition from the duplex uterus of rodents (a) to the bicornuate uterus of the pig (b), the bipartite uterus of ruminants (c), and the simplex uterus of mares and primates (d).

Differentiation of the Mullerian ducts gives rise to four distinct regions of the female reproductive tract: the Fallopian tubes (oviducts), uterus, cervix and part of the vagina (Fig. I.8). Each of these regions is recognisable anatomically and histologically, and each performs specialised physiological functions. There are also a number of accessory glands associated with the duct system, especially in the region of the cervix and vagina, but these remain relatively small when compared with development of the seminal vesicles, prostate and bulbo-urethral glands in the male. However, the cervical secretion of mucus does play an important rôle in the mature animal both at mating and during pregnancy, as well as at the time of parturition.

The form of the female reproductive tract varies quite markedly between many of the common laboratory and farm species, even though all have arisen from the basic arrangement of the Mullerian system. Viewed from a quantitative standpoint, the form of the uterus and cervix appears most complex in the litter-bearing (polytocous) species, whereas animals giving rise predominantly to a single offspring (monotocous species) have a simpler arrangement. Thus, a duplex uterus is found in the rabbit and many small rodents, where each long uterine horn terminates caudally in a separate cervix before fusion of the duct system is achieved at the level of the vagina (Fig. I.8). The bicornuate uterus of the pig has a small region of uterine fusion, the uterine body, where the two sides of the tract are joined by a common lumen, and there is a single cervix. In some contrast, ruminants such as cattle and sheep have a more extensive uterine body, although partial fusion of the ducts leaves a prominent fold of tissue (or septum) in this region. Even so, this arrangement of a bipartite uterus permits the membranes of a single developing embryo to extend around the septum and utilise both horns of the uterus. Finally, the simplex uterus of the mare and indeed of primates shows a very extensive degree of fusion, so that the body of the uterus rather than the uterine horns predominates. Progressive fusion of the Mullerian ducts is associated with a reduction in the number of foetuses to be carried.

Hypothalamic-pituitary relationships

An overall integration of the nervous and endocrine systems is dependent upon the hypothalamus, a structure derived from the fore-brain. The hypothalamus exerts its influence over the anterior pituitary gland (adenohypophysis) by means of neurosecretion from specific groups of cells in the hypophysiotrophic area referred to as nuclei; these are arranged as bilaterally paired groups. Neurosecretion of releasing hormones (RHs) from the nerve endings into the network of capillary loops in the median eminence (Fig. I.9) and their local transport through the hypophyseal portal vessels permits control over the synthetic and secretory activities of the anterior pituitary gland (Table I.1). The

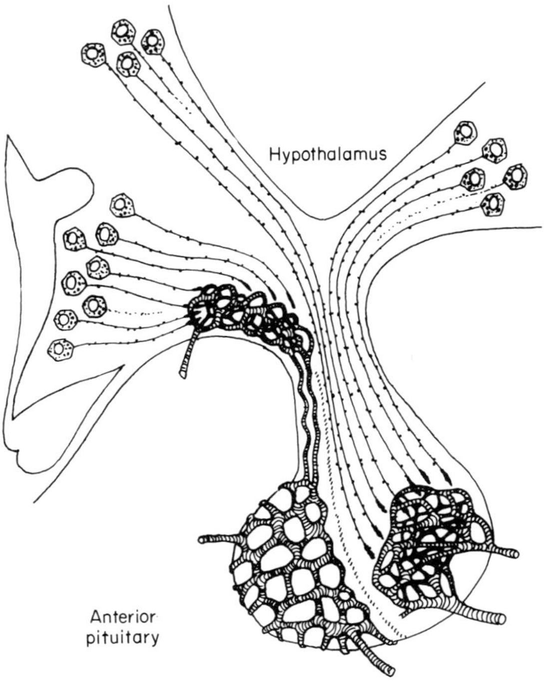

Fig. I.9. To illustrate the physiological relationships between the hypophysiotrophic area of the hypothalamus and the pituitary gland. Note particularly the capillary portal system to the anterior pituitary and direct innervation of the posterior pituitary.

Table I.1. Seven of the hypothalamic hormones, termed releasing hormones (RHs) or releasing factors (RFs), known to regulate the secretion of hormones from the anterior pituitary gland.

Hypothalamic hormone or factor	Abbreviation	
Follicle stimulating hormone (FSH)-releasing hormone[1]	FSH-RH or	FSH-RF
Luteinising hormone (LH)-releasing hormone[1]	LH-RH	LH-RF
Prolactin inhibiting hormone	PIH	PIF
Prolactin releasing hormone	PRH	PRF
Thyrotrophin (TSH)-releasing hormone	TSH-RH	TRF
Growth hormone or somatotrophin (STH)-releasing hormone	STH-RH	STH-RF
Corticotrophin (ACTH)-releasing hormone	ACTH-RH	CRF

[1] Strong evidence exists for a common hypothalamic factor (Gn-RH) that regulates release of the two gonadotrophic hormones. Injection of the synthetic decapeptide LH-RH at appropriate stages of the cycle will provoke increased concentrations of both LH and FSH in the blood of various species.

neuro-vascular concept of hypothalamic regulation of anterior pituitary function arose largely due to the meticulous studies of G. W. Harris in the late 1930s and 1940s. The experimental evidence leading to formulation of the concept is presented in his classical monograph (Harris, 1955) and has been summarised more recently by Everett (1972), Schally *et al.* (1973) and Sawyer (1975, 1978). Although the influence of releasing hormones is fundamental, it seems probable that the opioid pentapeptides termed endorphins and enkephalins are also involved in regulating hypothalamic and pituitary function, but experimental evidence for such a rôle is only just beginning to accrue.

The anterior pituitary gland is not of nervous tissue origin but has arisen embryologically as an ectodermal outgrowth from the roof of the buccal cavity or Rathke's pouch (Fig. I.10). Since there is no significant innervation of the gland, neurosecretion of releasing hormones into the capillary portal system is the only known means of integrating its function with that of the higher centres. By contrast, the posterior pituitary gland (neurohypophysis) is derived from nervous tissue, arising as a downgrowth from the floor of the third ventricle, and it retains its rich nerve supply from the ventral hypothalamus. In this manner, hormonal secretion from the posterior pituitary itself represents release of neurosecretory products, for oxytocin and vasopressin are synthesised in cells of the hypothalamus and simply stored in the posterior pituitary lobe.

It is no surprise, therefore, to note similarities between the neurosecretory releasing factors that regulate anterior pituitary function and the secretions of the posterior lobe. Both these groups of hypothalamic hormones (neurohormones) are small peptide molecules (octa or decapeptides) with molecular weights in the range of 1200–1500 or less. The secretions of the anterior pituitary, on the other hand, are distinguishable as proteins or glycoproteins with molecular weights in the general range of 4500–30 000 (Table I.2), even though the physiologically active moiety in these hormones may reside in a smaller peptide sequence within the molecule. The cells of the anterior pituitary that synthesise and secrete the trophic hormones are usually characterised microscopically on the basis of their staining reactions, although the elegant technique of immunofluorescence has also been used (Dubois *et al.*, 1971). The cells consist of acidophils synthesising prolactin and growth hormone, basophils synthesising the gonadotrophic hormones—follicle stimulating hormone (FSH) and luteinising hormone (LH), and chromophobes which are not fully differentiated for

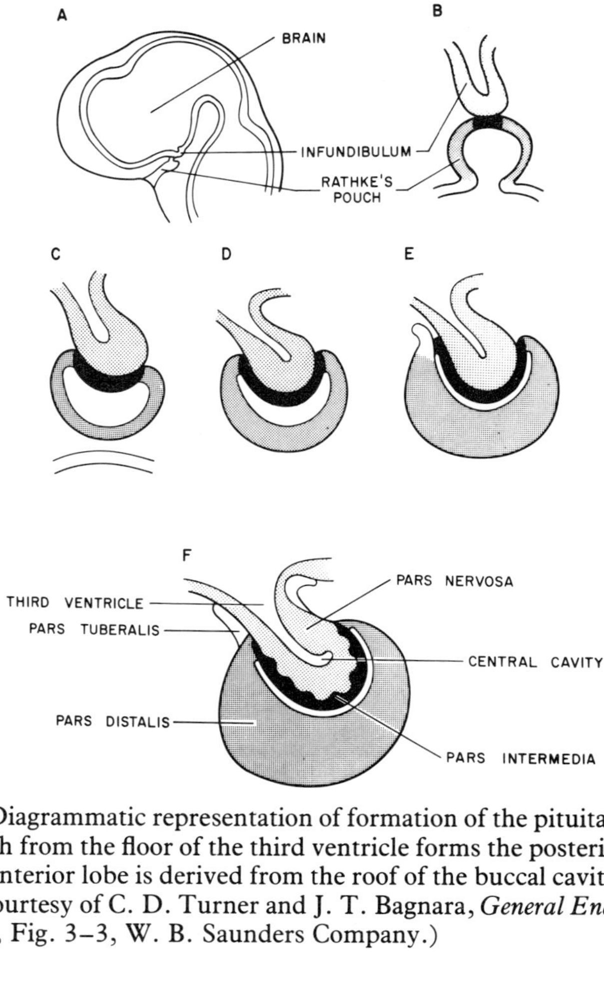

Fig. I.10. Diagrammatic representation of formation of the pituitary gland. A downgrowth from the floor of the third ventricle forms the posterior pituitary whilst the anterior lobe is derived from the roof of the buccal cavity (Rathke's pouch). (Courtesy of C. D. Turner and J. T. Bagnara, *General Endocrinology*, 5th edition, Fig. 3–3, W. B. Saunders Company.)

secretion and may act as precursor cells for replenishment of the acidophils and basophils. Two other trophic hormones, ACTH and TSH, are also thought to be synthesised by basophils.

Puberty

Puberty can be defined in several ways. For the purposes of the present text, it is considered as the age of the female at which oestrus is first detected and is followed by characteristic cyclic ovarian activity in the non-pregnant animal. By employing this definition, any controversy

Table I.2. The approximate molecular weight of six trophic hormones secreted by the anterior pituitary gland, with an indication of their chemical nature and characteristic cells of origin.

Hormone	Molecular wt (daltons)	Chemical nature	Secretory cells[1]
FSH	32 000	Glycoprotein	Basophils
LH	30 000	Glycoprotein	Basophils
Prolactin	23 000	Protein	Acidophils
TSH	26 000	Glycoprotein	Basophils
STH	22 000	Small protein	Acidophils
ACTH	4500	Polypeptide	Basophils

[1] This list represents a simplification of pituitary cytology since more than one cell type is distinguishable within each group.

over differences between the age at which reproduction is first possible and the time of sexual maturity is avoided. Nonetheless, when trying to explain mechanisms underlying the onset of puberty, it may not be fruitful to regard this event as discrete in time for an individual animal. On the contrary, Perry (1971) has suggested that puberty be viewed as a gradual and quantitative phenomenon rather than as a sudden and qualitative endocrinological event. In a similar vein, Donovan and van der Werff ten Bosch (1965) regarded puberty as the entire period during which the gonads secrete steroids in amounts sufficient to cause accelerated growth of the genital organs and the appearance of secondary sexual characteristics. This approach of considering puberty as a protracted period is emphasised by changes in gonadotrophic and gonadal hormone secretion as animals pass through the phase of sexual maturation. However, such endocrine estimations are largely the preserve of basic research, and the most meaningful criterion of puberty for animal breeders is likely to remain the age at which animals are competent to reproduce successfully.

The direct cause of sexual maturation at puberty is a rise in the output of pituitary hormones leading to an increase in size and activity of the ovaries. Underlying this process is some form of maturation of the hypothalamo-pituitary axis permitting changes in neurosecretion and thereby secretion of gonadotrophic hormones, but the nature of the modifications occurring at the higher levels is far from understood (Ramirez, 1973). Cellular and humoral events in the brain and hypothalamus associated with puberty are undoubtedly extremely complex, and need to monitor such factors as the animal's chronological age, body weight, organ growth, nutritional status, and a range of environmental factors. Reproduction is prevented before a certain stage of somatic development is attained, a stage which can usually be characterised as a proportion of the animal's mature body weight. Beyond these general statements, there remains the enigma of pineal and thymus involvement in sexual maturation. The secretions of the pineal gland have been implicated in the delay of the onset of puberty, perhaps by exerting an inhibitory action on gonadal development, although the route of a functional involvement with the anterior pituitary is uncertain. Concerning the thymus, there is the intriguing fact that involution of the gland begins close to the onset of puberty. Perhaps in some subtle manner, the activity of the thymus ensures that a female does not become sexually competent until she can tolerate the foreign (paternal) antigens that would be carried by the embryo.

Current interpretations of the phenomenon of puberty seem to fall into two principal schools of thought, although these are not mutually exclusive. First there is the view that emphasises the ability of very low concentrations of steroids (oestrogens) secreted by the prepuberal gonad to inhibit neurosecretion of gonadotrophin releasing factor(s) by the hypothalamus, an influence which is gradually removed as hypothalamic-pituitary sensitivity to this negative feedback diminishes towards the end of the prepuberal period. Factors underlying the changing threshold of hypothalamic sensitivity to oestrogen feedback remain unspecified but, by definition, the increase in gonadotrophin output at puberty is regarded as a release from inhibition. In lambs, this is seen as a change in the frequency rather than amplitude of the pulsatile LH releases (Foster *et al*., 1975). The second view is that a time-dependent maturation of the hypothalamo-pituitary axis is occurring more or less independently of gonadal activity but under the influence of inputs from the higher nervous structures of the brain. Competence is achieved in due course for releasing hormones to activate sufficient pituitary secretion of gonadotrophic hormones. In this model, involvement of steroid hormones from the gonads, or indeed inputs from any other peripheral endocrine organ, is clearly secondary to that of the developmental changes in the higher nervous centres, although a positive feedback of follicular oestrogens is of course essential for the first preovulatory surge of LH. Initiation of gonadotrophic function in this manner has been discussed in detail by Levasseur (1977), who considers the key event to be a change in the hypothalamic sensitivity to inhibition by the central nervous system.

The ability of the gonads of domestic animals to respond to the trophic influences of exogenous hormones some substantial interval before puberty implies a dearth of hypothalamic and/or pituitary activity as one of the limitations underlying the onset of puberty. Moreover, since animals stimulated in this way well before the onset of puberty do not then commence spontaneous oestrous cycles, it also suggests the need for a sufficient period of maturation of the higher centres before hypothalamo-pituitary activity can be triggered. That the deficiency is at the level of the hypothalamus or brain is shown by grafting pituitary tissue from immature rats beneath the median eminence of hypophysectomised adult females, enabling them to sustain full reproductive activity (Harris and Jacobsohn, 1952).

Age and weight at puberty

As in mammals generally, reproductive activity in the large farm species begins some time before growth is completed, and occurs at an earlier age in animals maintained under good conditions of nutrition. In developed systems of animal production, the *approximate* age and body weight at which oestrous cycles will commence can be predicted for a given breed within a species, but there is still considerable variation between individuals within breeds (Table I.3).

Cow

The age at first behavioural oestrus in heifers may vary from 8 to 18 months (more commonly 9–13 months), when animals such as Holsteins weigh about 260 kg (Dziuk, 1973). This range in ages is known to be influenced by breed and nutritional status (Hansel, 1959; Morrow, 1969), and under many farming systems is recorded earlier in dairy than in beef breeds. Summaries of studies concerned with puberty have been presented by Sorensen *et al.* (1959) and Joubert (1963). The former work concerned with the influence of level of feeding on a group of Holstein heifers also indicated that under the prevailing conditions, body size was more important than age in influencing the onset of the first oestrus. Heifers in the low feed-intake group at first oestrus (60% TDN compared with "standard") were nearly twice the age of those in the high-intake group (140% TDN compared with "standard") but they were only 18 kg lighter. As a general rule, a low plane of nutrition which may involve quantitative and/or qualitative elements will retard the age at first oestrus in cattle (Joubert, 1963). Puberty will be delayed until size is attained.

Sheep

Over and above the influence of genetic factors, the onset of puberty in lambs is determined by an interaction between age, day-length and nutritional status (review by Dyrmundsson, 1973). Because the ewe shows breeding seasons, and these are considerably influenced by the photoperiodicity of a given geographical latitude, the age at puberty will tend to be regulated by season of birth, especially at the higher latitudes. The breeding season of lambs starts later than that of adult ewes, but ewe lambs can achieve puberty at 6–7 months of age or even earlier. Later-born lambs show oestrus at progressively younger ages until the situation is reached in which those born very late in the year do not breed until the season of the following year at 15–16 months of age (Hammond, Jr., 1944). This means that animals born at the beginning of the lambing season will mature at a significantly younger age than those born towards the end of the season. Endocrine data relevant to the onset of puberty in lambs have been discussed by Land (1978), and some consideration given to the idea that puberty and seasonal breeding could be associated with changes in the pulse frequency of gonadotrophin release (see below). Certainly, the first behavioural oestrus is preceded by ovulation without overt oestrus (i.e. a silent heat) at both puberty and the onset of the breeding season.

Pig

The breeds of pig commonly found in the developed countries exhibit first oestrus at about 5–7 months of age when they usually weigh 85–90 kg. In some contrast to the situation in cattle and sheep, the onset of oestrous cycles in gilts appears dependent more on age than on weight: a marked variation in the latter measurement may be found at puberty. At least as important as the genetic background (i.e. breed and degree of inbreeding) in influencing the onset of oestrus seems to be the physical and social environment in which animals are kept. Gilts maintained in large groups or isolated from boars have delayed puberty (Dziuk, 1977). They seem to be particularly sensitive to certain forms of stress as they approach the time of puberty, as may be imposed, for example, by a high density of housing. Although the physiology underlying such a situation has not been analysed, movement of animals at the threshold of puberty to another set of pens will provoke the onset of oestrous cycles in many gilts within five or six days (Du Mesnil du Buisson and Signoret, 1962). This interval corresponds closely to that required for growth and development of a small Graafian follicle to mature size. Stimulation with an appropriate preparation of gonadotrophic hormone shortly before the expected onset of puberty will also provoke oestrus followed by the spontaneous occurrence of oestrous cycles.

Horse

Fillies usually reach puberty at an age of 15–18 months, except under conditions where mares show seasonal breeding activity. In this less common situation, puberty is reached at approximately 24–28 months of age.

The oestrous cycle

In the absence of successful mating, mature female mammals of many species undergo a repeated series of ovarian changes, particularly in secretion of steroid hormones, that influence both the reproductive tract and the animal's sexual behaviour. This cycle of ovarian endocrine events finds expression in the oestrous cycle of most eutherian mammals or the menstrual cycle of primates. Although histological changes can be identified in the genital tract as a response to the changing pattern of ovarian steroid secretion, a key stage of the oestrous cycle can be recognised more directly as the time when the animal is receptive to the male and can be mated. This is the period of oestrus when the female shows characteristic sexual behaviour in the presence of a mature male, such as immobility, raising the hindquarters or arching the back, pricking of the ears—features that are collectively termed lordosis in small laboratory animals; mounting and riding behaviour between females is also common. If the animal does not mate at this period of oestrus, she becomes refractory to the male and proceeds through a cycle of ovarian events that is not expressed clearly in terms of sexual behaviour until once more she returns to oestrus after an interval appropriate to the species. Cycle lengths range from 4–5 days in small laboratory species (e.g. rat, hamster, when there is not a true luteal phase) to approximately 21 days in pigs, cows and horses, with sheep having a somewhat shorter cycle of 16–17 days (Table I.3).

Table I.3. Some commonly observed measurements of the age at puberty, together with characteristics of the oestrous cycle in the large farm species.

	Cow	Sheep	Pig	Horse
Age at puberty (months)	9–13	5–10[1]	5–7	15–24[1]
Length of oestrous cycle (days)	20–21	16–17	20–22	21–22
Length of luteal phase (days)	17–18	14–15	15–16	14–15
Duration of oestrus (hours)	12–26	24–36	40–70	96–168
Timing of ovulation (hours)	10–12 after *end* of oestrus	24–26 after *onset* of oestrus	36–40 after *onset* of oestrus	24–36 before *end* of oestrus

[1] But see qualifying remarks in the text concerning the influence of season of the year.

Ovulation occurs during or shortly after the period of oestrus (see below), and is usually taken as a reference point for description of the cycle. After the opportunity to mate, the reproductive tract is specifically prepared under the influence of increasing titres of ovarian progesterone to nourish and support a developing embryo. This period of progesterone secretion by the corpus luteum comprises two thirds or more of the duration of the oestrous cycle in the large farm species (Table I.3), and is referred to as the *luteal phase* of the cycle. It thus provides a major interval towards the end of which the animal must respond to the presence of an embryo or return to oestrus. The lifespan of the corpus luteum of the oestrous cycle will be extended into that of pregnancy if one or more viable embryos are present in the uterus (see Chapter VI). If the animal has not conceived, the corpus luteum is caused to regress rapidly under the influence of a lytic factor from the uterus (prostaglandin $F_{2\alpha}$), secretion of progesterone diminishes as a consequence of this luteolytic activity, and the oestrous cycle continues with the objective of promptly restoring the condition of oestrus and the opportunity to mate. Removal of the negative feedback influence of progesterone permits active pituitary secretion of gonadotrophins, under the control of which one or more Graafian follicles will mature, secrete increasing amounts of oestrogen(s), and promote the changes that precede oestrus and ovulation. This much shorter phase of the cycle is referred to as the *follicular phase* and, viewed classically, continues until rupture or collapse of the Graafian follicle(s) at ovulation.

The oestrous cycle thus consists of an alternation between a relatively prolonged luteal phase during which progesterone secretion from the corpus luteum is the principal ovarian event, and a compact follicular phase when oestrogen secretion from one or more maturing follicles predominates. Nature's viewpoint appears to be that all mature females should be pregnant (or lactating) in order to propagate the species and, failing achievement of that state, the animal adopts a strategy in repeated pursuit of a successful mating. However, there are periods of ovarian inactivity in some species (see Thibault *et al.*, 1966), and in monoestrous and seasonally polyoestrous animals the period of sexual quiescence between seasons is termed anoestrus.

Many years ago, especially before the advent of a straightforward means of measuring hormones in blood plasma, the oestrous cycle was frequently referred to as the dioestrous cycle in which four stages were distinguished on the basis of vaginal histology. (Sampling simply involved taking a smear of the vaginal wall with a spatula or cotton wool swab.) The stages were pro-oestrus, oestrus, metoestrus and dioestrus, and the descriptions were used more with respect to small laboratory animals (Stockard and Papanicolaou, 1917; Long and Evans, 1922) than to the large farm species. Dioestrus constituted a major portion of the cycle, being the time when progesterone secretion from the corpus luteum predominated and when the female was therefore refractory to the male. Pro-oestrus represented the follicular phase of the cycle when the animal showed increasing sexual excitement under the influence of increasing secretion of follicular oestrogens. Oestrus corresponded to the period of lordosis, and metoestrus was the short, postovulatory phase when active sexual behaviour was waning under the influence of progesterone from the developing corpus luteum (Eckstein and Zuckerman, 1956). Characteristics of the vaginal smear in laboratory rodents have been summarised more recently by Perry (1971). Although changes in vaginal histology can be detected in farm species during the oestrous cycle, these are not precise and are sometimes difficult to distinguish. In any event, it is the period of oestrus itself that is of greatest concern to farmers and the insemination service, and recognition of the receptive period usually relies on overt, especially behavioural, signs in the presence of other animals.

Before discussing patterns of hormone secretion during the oestrous cycle, it is important to note that although ovulation is a "spontaneous" event during the cycle in cattle, sheep, pigs and horses, this is not so in all species. In animals such as the rabbit, cat and ferret, the surge of pituitary gonadotrophin secretion that provokes ovulation requires the stimulus of mating for its release, and in this way a time relationship is established between deposition of spermatozoa in the female genital tract and release of the eggs at ovulation. Thus, induced ovulators require an afferent neural component arising from vaginal and cervical stimulation before rupture or collapse of the Graafian follicles is triggered, whereas ovulation is spontaneous in farm animals to the extent that mating is *not* required to provoke the preovulatory surge of gonadotrophin: its release is brought about by a positive feedback influence of increasing oestrogen concentrations from the maturing follicle. Nonetheless, there have been repeated claims that in circumstances of heightened sexual excitement, the presence of a male and particularly the act of mating may accelerate or induce ovulation. Such an influence seems plausible since the preovulatory gonadotrophin surge in spontaneous ovulators may on occasions occur some time after the onset of behavioural oestrus.

Endocrinology of oestrous cycles

This section deals with individual species and covers the division of the oestrous cycle into luteal and follicular phases, the duration of oestrus, and the timing of ovulation. Measurements of the concentration of progesterone, oestrogens and LH in peripheral blood throughout the cycle are also referred to (Table I.4), and are drawn from studies using

Table I.4. Comparison of some common maximum concentrations of ovarian steroids and LH in peripheral blood during the oestrous cycle.

			LH	
	Oestrogens pg ml^{-1}	Progesterone ng ml^{-1}	Basal values ng ml^{-1}	Preovulatory peak ng ml^{-1}
Cow	15–25	6–8	<2·0	10–65
Sheep	30–50	<3·0	<5·0	30–100
Pig	60–90	25–35	<1·0	5–7
Horse	11–40	5–8	10–20	40–150

radioimmunoassay or competitive protein binding techniques. The concentration of FSH in blood plasma or serum during the cycle is not discussed since sensitive and specific assays for this gonadotrophin are not widespread or completely reliable at present. In fact, LH seems relatively more important than FSH in regulating the time of ovulation and, over and above a general facilitatory rôle of FSH, the same comment may apply to its involvement in development of a Graafian follicle during the later stages of the oestrous cycle. Even so, the preovulatory peak of LH found close to the onset of oestrus in cattle, sheep and pigs is invariably matched by a coincident but smaller peak of FSH. One interpretation of this finding is that a common hypothalamic releasing hormone exists for the two gonadotrophins.

Concerning the notation used for timing the oestrous cycle, it is customary to designate the first day of oestrus as Day 0. In this manner, ovulation will usually be occurring on Day 1 of the cycle in cows, sheep and pigs.

Cow

The duration of the oestrous cycle is commonly 20 or 21 days (Hammond, 1927; Robinson, 1977), but the range may extend from 18–24 days or more. The luteal phase of the cycle occupies approximately 17 days, followed by a compact follicular phase of 3–4 days. The duration of oestrus falls within the range of 12–26 hours, averaging 15–18 hours (Gomes, 1978), this period of receptivity tending to be shorter during the winter months or the heaviest phase of lactation. Ovulation occurs 10–12 hours after the end of oestrus, or some 25–30 hours after the preovulatory surge of gonadotrophin (Henricks *et al.*, 1970; Schams *et al.*, 1977), when the fully developed follicle may have reached a diameter of 16–18 mm. Some of the principal endocrine changes during the bovine oestrous cycle are summarised diagrammatically in Fig. I.11. Detailed curves have been presented by Hansel and Echternkamp (1972) and Chenault *et al.* (1975), and cyclic changes in the concentration of steroids and gonadotrophins in blood have been reviewed by Hansel *et al.* (1973) and more fully by Hunter (1976).

The concentration of progesterone in peripheral blood is less than 1·0 ng ml^{-1} at around oestrus, and does not rise appreciably until Day 5. Thenceforward, it increases steadily until Day 16 or 17, having a mean value of about 5·4 ng ml^{-1} during the luteal phase of the cycle and a mean peak value of about 6–7 ng ml^{-1} at the end of the luteal phase. At

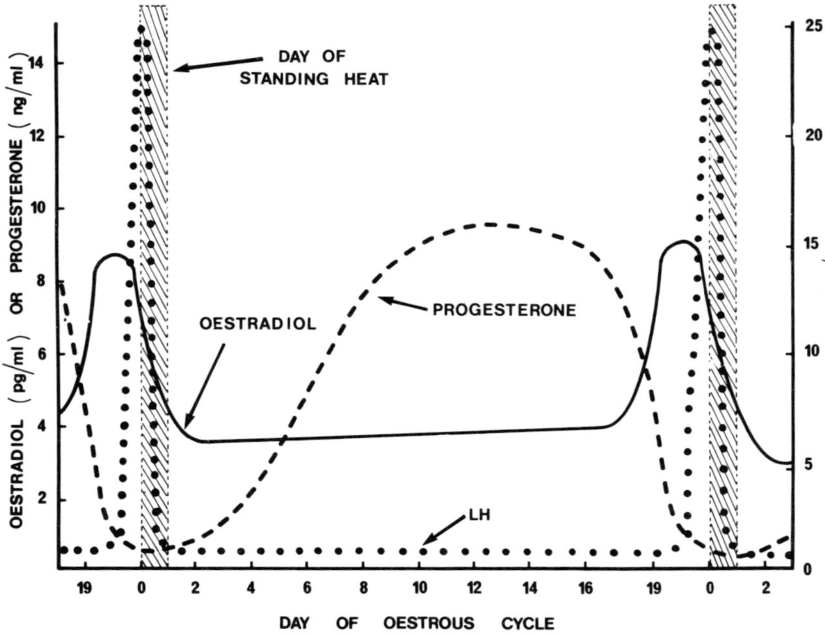

Fig. I.11. Simplified presentation of the changing concentrations of major ovarian steroids and luteinising hormone in peripheral blood during the oestrous cycle of cows. (Redrawn from Lamming *et al.*, 1975.)

this time, when the corpus luteum may be 2 cm or more in diameter, a steep decline in progesterone concentration is found in the cyclic animal. The onset of the drop is somewhat variable between Days 16 and 19 but, once it occurs, the decline in progesterone concentration is always precipitous. The time interval between this event and the occurrence of oestrus may vary from one to five days (Stabenfeldt *et al.*, 1969).

By contrast, concentrations of oestrogens in the peripheral plasma are in the pikogram range, a mean figure of 3·6 pg ml^{-1} being noted for the luteal phase of the cycle (Wettemann *et al.*, 1972). Oestrogen concentrations remained less than 10 pg ml^{-1} until the day prior to oestrus, when they rose to a peak of 15–25 pg ml^{-1}, but they decreased within two to five hours of the onset of oestrus. When oestrogen levels were measured in the peripheral plasma of a group of mated cows that returned to oestrus, Henricks *et al.* (1972) recorded 4·1 pg ml^{-1} on Day 3 after oestrus, 6·8 pg ml^{-1} on Day 6, 13·3 pg ml^{-1} on Day 9, and 6·0 pg ml^{-1} on Days 12 and 15. The Day 9 value represented a minor but statistically significant peak that may reflect a wave of follicular activity and/or be concerned with subsequent regression of the corpus luteum.

A peak of oestradiol has been noted 4 to 6 days after oestrus in some animals (Dobson *et al.*, 1973; Glencross *et al.*, 1973), and is probably associated with growth of Graafian follicles in the middle of the luteal phase.

With regard to pituitary hormones, the cyclicity of gonadotrophin secretion is a characteristic property of the female, and the most conspicuous gonadotrophin release in the mature animal is the preovulatory LH peak. During the luteal phase, the basal level of LH in peripheral plasma rarely exceeds 2–3 ng ml^{-1}, whereas high levels are found in the blood of all animals at oestrus, although the rise is not significant until plasma progesterone concentrations decline below 1 ng ml^{-1} (Hansel *et al.*, 1973). The preovulatory peak corresponds fairly closely with the onset of oestrus (± 8 hours), and does not usually last for more than 6–8 h at which time LH titres may rise to 10–15 ng ml^{-1} or considerably more (i.e. 19–35 ng ml^{-1}: Swanson and Hafs, 1971; 50–65 ng ml^{-1}: Black and Hansel, 1972).

Sheep

As already noted, the ewe is a seasonal breeder, but she is polyoestrous for four or five months of the year (in the absence of mating) with cycles of 16 or 17 days' duration (range 14–19). The luteal phase lasts for a little over 14 days, after which the preovulatory follicle(s) matures rapidly to reach a terminal diameter of 10–12 mm. The duration of oestrus is approximately 24–30 hours but sometimes longer, with ovulation occurring at the end of this time or 24–26 hours after the preovulatory surge of gonadotrophin (Dziuk, 1973). These facts and the supporting literature have been reviewed repeatedly, perhaps most recently by Robertson (1977). In this literature, one frequently notes the suggestion that mating of ewes at the onset of oestrus reduces the anticipated duration of receptivity to the male: in fact, exposure to the ram during pro-oestrus may hasten the preovulatory surge of gonadotrophin and thereby shorten the period of oestrus.

Because the general pattern of endocrine changes underlying the oestrous cycle in sheep is in many ways similar to that in the cow, only a concise description is given here with supplementary detail in Fig. I.12. Progesterone concentrations in peripheral plasma are low during the first few days of the cycle, rise rapidly from Days 3–11, and achieve maximum levels of about 3 ng ml^{-1}—that is approximately half the concentration found at a comparable stage in cattle (Table I.4). Trophic support by both LH and prolactin is necessary for full luteal function.

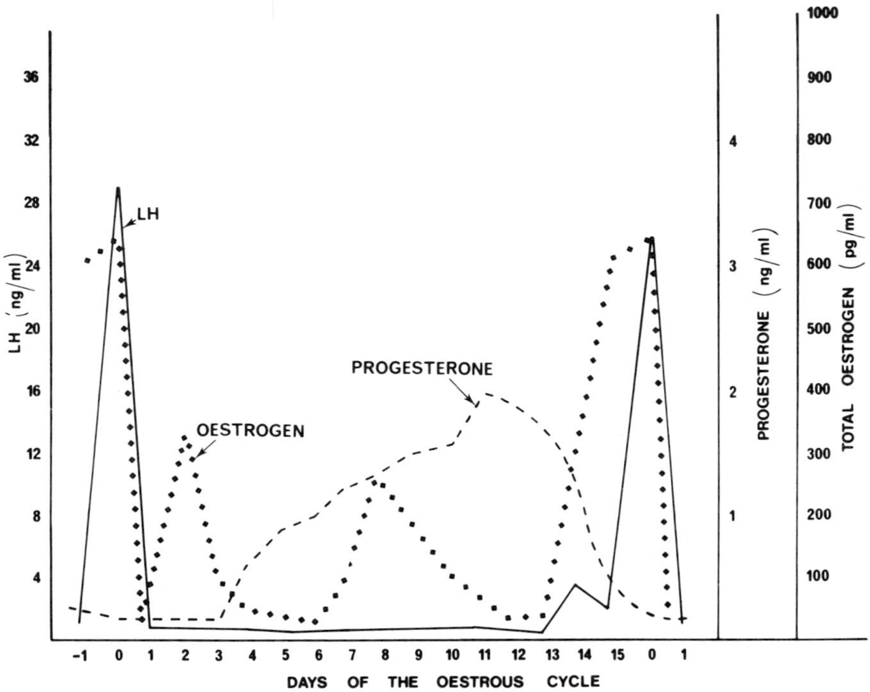

Fig. I.12. Generalised representation of peripheral plasma levels of luteinising hormone and progesterone, together with the concentration of total oestrogens in ovarian venous plasma, during the oestrous cycle of sheep. (Redrawn from Hansel *et al.*, 1973.)

Once again, there appear to be at least two waves of follicular growth during the luteal phase (Smeaton and Robertson, 1971), this activity being reflected in the curve of oestrogen secretion—principally oestradiol-17β—into the ovarian vein which commences its characteristic follicular phase rise on about Day 14. A further point of detail that distinguishes the ewe from the cow is that peak levels of plasma LH seem to be slightly higher (20–100 ng ml^{-1}) in sheep although, once again, this gonadotrophin rise is not found until progesterone concentrations fall below 1 ng ml^{-1} (Hansel *et al.*, 1973).

Pig

The duration of the oestrous cycle in pigs is usually 20–22 days, but the range may extend from 18–24 days. As the domestic pig is polyoestrous, cycles will be repeated throughout the year in the absence of pregnancy, lactation, or severe forms of environmental or social stress. The period of oestrus itself lasts for 40–70 hours, tending to be longer in sows than gilts, and ovulation occurs on the second day of oestrus. The corpora lutea are functional for 15 or 16 days, and apparently require no hypophyseal support during the luteal phase apart from any residual effect of the preovulatory surge of gonadotrophin (Sammelwitz *et al.*, 1961; Brinkley *et al.*, 1964a, b). However, other studies have suggested that these structures are not fully autonomous during this period to the extent that development to full weight seemingly requires a trophic influence of pituitary hormones, as judged from measurements in hypophysectomised animals (Du Mesnil du Buisson and Léglise, 1963).

The follicular phase of the cycle lasts for 5–6 days, and is therefore longer than in cattle or sheep. During this period Graafian follicles increase in mean diameter from 4–5 mm on Day 15 to the preovulatory diameter of 9–11 mm on Day 21 (Corner, 1921; McKenzie, 1926; Burger, 1952). The actual number of follicles achieving this final diameter in any one cycle is reduced to an ovulation rate characteristic of the breed or strain, and is well defined by the time of the onset of oestrus. A selection of follicles therefore occurs during this latter phase of the cycle, although evidence from experiments involving superovulation (Hunter, 1964) suggests that this may be solely due to competition for circulating gonadotrophic hormones. The concentration of oestrogens in the peripheral plasma increases from 10–30 pg ml^{-1} to 60–90 pg ml^{-1} over this period of six days as the follicles mature (Shearer *et al.*, 1971; Henricks *et al.*, 1972; Hunter *et al.*, 1972).

The preovulatory peak of LH may reach serum concentrations of 4–5 ng ml^{-1} (Niswender *et al.*, 1970), with ovulation occurring 40–48 hours after this peak (Liptrap and Raeside, 1966) or 40–42 hours after injection of an LH-rich preparation given shortly before oestrus to simulate the endogenous release of gonadotrophin (Dziuk *et al.*, 1964; Hunter, 1967). Minor breed and strain differences in the timing of the response to such an injection can be demonstrated (see Dziuk and Baker, 1962; Hunter, 1972). Reports that ovulation occurs 35–36 hours after the onset of oestrus could be reconciled with the above figure of 40–42 hours if there was some error in detecting the onset of oestrus.

The concentration of progesterone increases from preovulatory values of $<1{\cdot}0$ ng ml^{-1} (Hunter *et al.*, 1972) to mid luteal phase values of 20–35 ng ml^{-1} of plasma (Stabenfeldt *et al.*, 1969; Edqvist and Lamm,

1971; Henricks *et al.*, 1972) before commencing a fairly abrupt decline during Days 15 and 16 of the cycle. All these endocrine changes are summarised graphically in Fig. I.13.

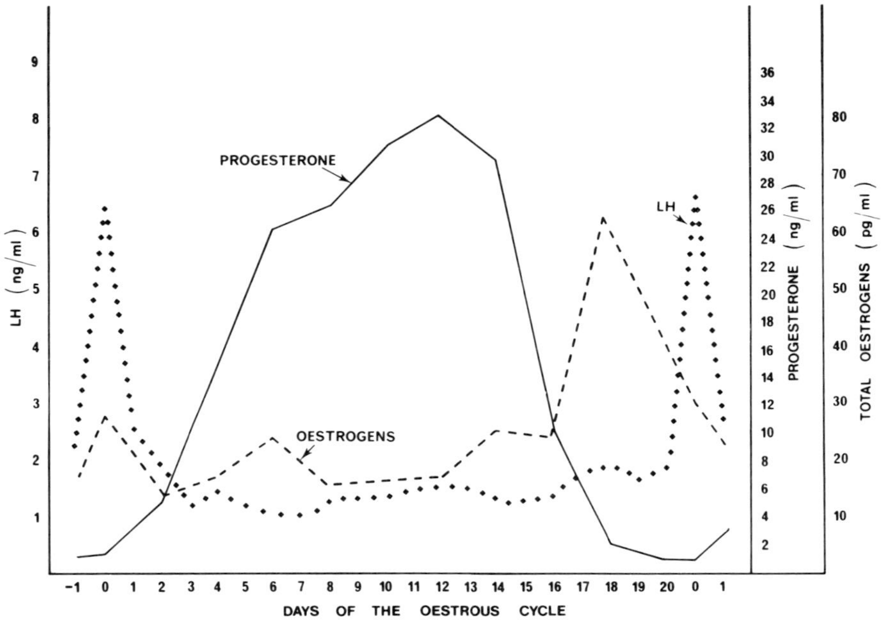

Fig. I.13. Generalised representation of peripheral plasma levels of luteinising hormone, progesterone and oestrogens during the oestrous cycle of pigs. (Redrawn from Hansel *et al.*, 1973.)

Horse

Although some mares may exhibit successive oestrous cycles throughout the year, a period of anoestrus is common during the winter months, especially at the higher latitudes. The cycle has a mean duration of 21 or 22 days, but differs significantly from that of cows or pigs in that the period of oestrus may last for 5–7 days or more. Longer periods of oestrus are associated with longer oestrous cycles, and are found shortly before and especially immediately after the winter anoestrus, but the dioestrous interval usually remains within the range of 14–19 days. Again, in contrast to the situation in other domestic livestock, behavioural oestrus in the mare shows a gradual increase in intensity over several days (Andrews and McKenzie, 1941). The time of ovulation is more closely related to the end of oestrus than the onset, occurring 24–48 hours before the former when the follicle achieves a diameter of 35–55 mm (Stabenfeldt and Hughes, 1977).

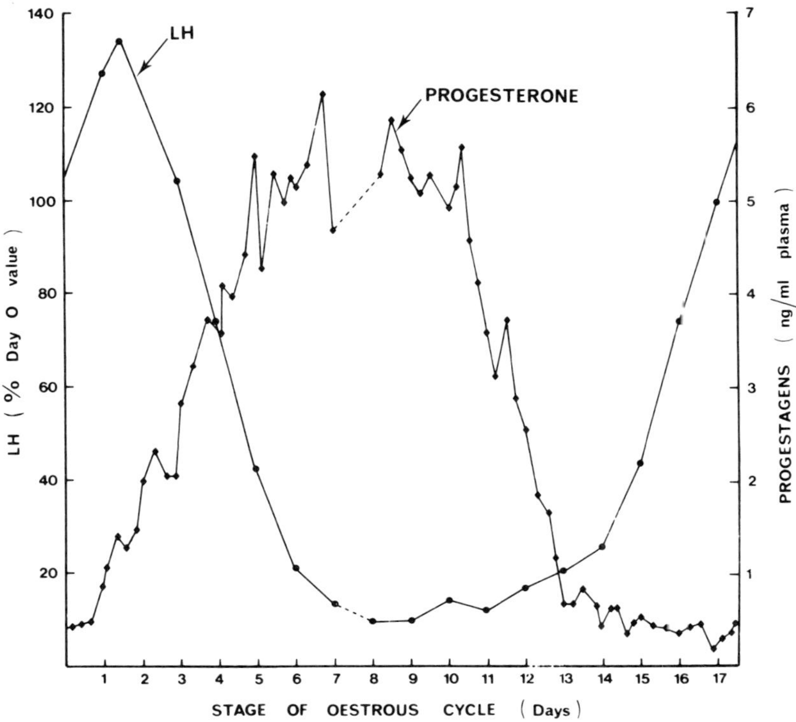

Fig. I.14. Concentration of luteinising hormone and progestagens in peripheral plasma of mares. (Redrawn from the data of Stabenfeldt *et al.*, 1975.)

The protracted phase of follicular development is to some extent mirrored by the pattern of secretion of LH. Instead of the sharp preovulatory peak seen near the onset of oestrus in the other farm species, the curve of LH secretion spans the time of ovulation and shows a slow rise and then decline over a period of approximately 10 days (Fig. I.14). The concentration of total oestrogens in peripheral plasma has increased at this time but remains relatively low, reaching values of 11·5 pg ml^{-1} for oestradiol and 12·5 pg ml^{-1} for oestrone one day before ovulation (Noden *et al.*, 1975). After ovulation, progesterone secretion by the corpus luteum reaches a mean concentration of

5–6 ng ml^{-1} or more within six days (Stabenfeldt *et al.*, 1975), the complete luteal phase lasting for some 14 days.

References

Andrews, F. N. and McKenzie, F. F. (1941). Estrus, ovulation and related phenomena in the mare. *Missouri Agric. Exp. Sta. Res. Bull.*, No. 329.

Baird, D. T., Baker, T. G., McNatty, K. P. and Neal, P. (1975). Relationship between the secretion of the corpus luteum and the length of the follicular phase of the ovarian cycle. *J. Reprod. Fert.* **45**, 611–619.

Baker, T. G. (1972). Oogenesis and ovarian development. *In* "Reproductive Biology" (H. Balin and S. Glasser, eds) pp. 398–437. Excerpta Medica, Amsterdam.

Black, D. L. and Hansel, W. (1972). Endocrine factors affecting reproduction in the bovine female. *Univ. Mass. Agric. Exp. Sta. Res. Bull.*, No. 596.

Brambell, F. W. R. (1956). Ovarian changes. I. Development of the ovary and oogenesis. *In* "Marshall's Physiology of Reproduction", 3rd edition (A. S. Parkes, ed.) Vol. 1, part 1, pp. 397–542. Longmans, Green and Co., London.

Brinkley, H. J., Norton, H. W. and Nalbandov, A. V. (1964a). Role of a hypophysial luteotrophic substance in the function of porcine corpora lutea. *Endocrinology* **74**, 9–13.

Brinkley, H. J., Norton, H. W. and Nalbandov, A. V. (1964b). Is ovulation alone sufficient to cause formation of corpora lutea? *Endocrinology* **74**, 14–20.

Burger, J. F. (1952). Sex physiology of pigs. *Onderstepoort J. vet. Res.* Suppl. No. 2.

Burns, R. K. (1961). Role of hormones in the differentiation of sex. *In* "Sex and Internal Secretions", 3rd edition (W. C. Young, ed.) Vol. 1, pp. 76–158. Williams and Wilkins, Baltimore.

Chenault, J. R., Thatcher, W. W., Kalra, P. S., Abrams, R. M. and Wilcox, C. J. (1975). Transitory changes in plasma progestins, estradiol, and luteinizing hormone approaching ovulation in the bovine. *J. Dairy Sci.* **58**, 709–717.

Corner, G. W. (1919). On the origin of the corpus luteum of the sow from both granulosa and theca interna. *Amer. J. Anat.* **26**, 117–183.

Corner, G. W. (1921). Cyclic changes in the ovaries and uterus of the sow, and their relation to the mechanism of implantation. *Contrib. Embryol. Carneg. Instn.* **13**, 119–146.

Dobson, H., Hopkinson, C. R. N. and Ward, W. R. (1973). Progesterone, 17-β-oestradiol and LH in relation to ovulation in cows. *Vet. Rec. 93*, **76** (Abstract).

Donovan, B. T. and van der Werff ten Bosch, J. J. (1965). "Physiology of Puberty". Edward Arnold, London.

Dubois, M. P., de Reviers, M. M. and Courot, M. (1971). Activité gonadotrope de l'éminence médiane après hypophysectomie chez le rat. Étude en immunofluorescence. *Expn anim.* **4**, 213–226.

Du Mesnil du Buisson, F. and Léglise, P. C. (1963). Effet de l'hypophysectomie sur les corps jaunes de la truie. Résultats préliminaires. *C.R. Acad. Sci. (Paris)* **257**, 261–263.

Du Mesnil du Buisson, F. and Signoret, J. P. (1962). Influences de facteurs externes sur le déclenchement de la puberté chez la truie. *Annls Zootech.* **11**, 53–59.

Dyrmundsson, O. R. (1973). Puberty and early reproductive performance in sheep. I. Ewe lambs. *Anim. Breed. Abstr.* **41**, 273–289.

Dziuk, P. J. (1973). Occurrence, control, and induction of ovulation in pigs, sheep and cows. *In* "Handbook of Physiology", Endocrinology, II, Part 1, pp. 143–152. American Physiological Society, Washington.

Dziuk, P. J. (1977). Reproduction in pigs. *In* "Reproduction in Domestic Animals", 3rd edition (H. H. Cole and P. T. Cupps, eds) pp. 455–474. Academic Press, New York and London.

Dziuk, P. J. and Baker, R. D. (1962). Induction and control of ovulation in swine. *J. Anim. Sci.* **21**, 697–699.

Dziuk, P. J., Polge, C. and Rowson, L. E. (1964). Intra-uterine migration and mixing of embryos in swine following egg transfer. *J. Anim. Sci.* **23**, 37–42.

Eckstein, P. and Zuckerman, S. (1956). The oestrous cycle in the mammalia. *In* "Marshall's Physiology of Reproduction", 3rd edition (A. S. Parkes, ed.) Vol. 1, part 1, pp. 226–396. Longmans, Green and Co., London.

Edqvist, L. E. and Lamm, A. M. (1971). Progesterone levels in plasma during the oestrous cycle of the sow measured by a rapid competitive protein binding technique. *J. Reprod. Fert.* **25**, 447–449.

Everett, J. W. (1972). The third annual Carl G. Hartman lecture. Brain, pituitary gland and the ovarian cycle. *Biol. Reprod.* **6**, 3–12.

Foster, D. L., Lemons, J. A., Jaffe, R. B. and Niswender, G. D. (1975). Sequential patterns of circulating luteinizing hormone and follicle stimulating hormone in female sheep from early postnatal life through the first estrous cycle. *Endocrinology* **97**, 985–994.

Glencross, R. G., Munro, I. B., Senior, B. E. and Pope, G. S. (1973). Concentrations of oestradiol-17β, oestrone and progesterone in jugular venous plasma of cows during the oestrous cycle and in early pregnancy. *Acta endocr.* **73**, 374–384.

Gomes, W. R. (1978). The estrous cycle. *In* "Physiology of Reproduction and Artificial Insemination of Cattle", 2nd edition (G. W. Salisbury, N. L. van Demark and J. R. Lodge, eds) pp. 52–90. W. H. Freeman and Co., San Francisco.

Hammond, J. (1927). "The Physiology of Reproduction in the Cow". University Press, Cambridge.

Hammond, J., Jr. (1944). On the breeding season in the sheep. *J. agric. Sci.* **34**, 97–105.

Hansel, W. (1959). The estrous cycle of the cow. *In* "Reproduction in Domestic Animals", 1st edition (H. H. Cole and P. T. Cupps, eds) pp. 223–265. Academic Press, New York and London.

Hansel, W., Concannon, P. W. and Lukaszewska, J. H. (1973). Corpora lutea of the large domestic animals. *Biol. Reprod.* **8**, 222–245.

Hansel, W. and Echternkamp, S. E. (1972). Control of ovarian function in domestic animals. *Amer. Zool.* **12**, 225–243.

Harris, G. W. (1955). "Neural Control of the Pituitary Gland". Edward Arnold, London.

Harris, G. W. and Jacobsohn, D. (1952). Functional grafts of the anterior pituitary gland. *Proc. R. Soc. Lond. B.* **139**, 263–276.

Henricks, D. M., Dickey, J. F., Hill, J. R. and Johnston, W. E. (1972). Plasma estrogen and progesterone levels after mating, and during late pregnancy and postpartum in cows. *Endocrinology* **90**, 1336–1342.

Henricks, D. M., Dickey, J. F. and Niswender, G. D. (1970). Serum luteinizing hormone and plasma progesterone levels during the estrous cycle and early pregnancy in cows. *Biol. Reprod.* **2**, 346–351.

Henricks, D. M., Guthrie, H. D. and Handlin, D. L. (1972). Plasma estrogen, progesterone and luteinizing hormone levels during the estrous cycle in pigs. *Biol. Reprod.* **6**, 210–218.

Hunter, R. H. F. (1964). Superovulation and fertility in the pig. *Anim. Prod.* **6**, 189–194.

Hunter, R. H. F. (1967). Porcine ovulation after injection of human chorionic gonadotrophin. *Vet. Rec.* **81**, 21–23.

Hunter, R. H. F. (1972). Ovulation in the pig: timing of the response to injection of human chorionic gonadotrophin. *Res. Vet. Sci.* **13**, 356–361.

Hunter, R. H. F. (1976). Reproductive physiology in the bovine female: a review. *In* "Beef Cattle Production in Developing Countries" (A. J. Smith, ed.) pp. 79–104. University Press, Edinburgh.

Hunter, R. H. F., Hall, J. P., Cook, B. and Taylor, P. D. (1972). Oestrogens and progesterone in porcine peripheral plasma before and after induced ovulation. *J. Reprod. Fert.* **31**, 499–501.

Joubert, D. M. (1963). Puberty in female farm animals. *Anim. Breed. Abstr.* **31**, 295–306.

Lamming, G. E., Hafs, H. D. and Manns, J. G. (1975). Hormonal control of reproduction in cattle. *Proc. Brit. Soc. Anim. Prod.* **4**, 71–78.

Land, R. B. (1978). Reproduction in young sheep: some genetic and environmental sources of variation. *J. Reprod. Fert.* **52**, 427–436.

Levasseur, M. C. (1977). Thoughts on puberty. Initiation of the gonadotrophic function. *Annls Biol. anim. Biochim. Biophys.* **17**, 345–361.

Liptrap, R. M. and Raeside, J. I. (1966). Luteinizing hormone activity in blood and urinary oestrogen excretion by the sow at oestrus and ovulation. *J. Reprod. Fert.* **11**, 439–446.

Long, J. A. and Evans, H. M. (1922). The oestrous cycle in the rat and its associated phenomena. *Mem. Univ. Calif.* **6**, 1–148.

McKenzie, F. F. (1926). The normal oestrous cycle in the sow. *Missouri Agric. Exp. Sta. Res. Bull.*, No. 86.

Morrow, D. A. (1969). Estrous behaviour and ovarian activity in prepuberal and postpuberal dairy heifers. *J. Dairy Sci.* **52**, 224–227.

Niswender, G. D., Reichert, L. E. and Zimmerman, D. R. (1970).

Radioimmunoassay of serum levels of luteinizing hormone throughout the estrous cycle in pigs. *Endocrinology* **87**, 576–580.

Noden, P. A., Oxender, W. D. and Hafs, H. D. (1975). The cycle of oestrus, ovulation and plasma levels of hormones in the mare. *J. Reprod. Fert., Suppl.* **23**, 189–192.

Perry, J. S. (1971). "The Ovarian Cycle of Mammals". Oliver and Boyd, Edinburgh.

Ramirez, V. D. (1973). Endocrinology of puberty. *In* "Handbook of Physiology", Endocrinology II, part 1, pp. 1–28. American Physiological Society, Washington.

Robertson, H. A. (1977). Reproduction in the ewe and the goat. *In* "Reproduction in Domestic Animals", 3rd edition (H. H. Cole and P. T. Cupps, eds) pp. 475–498. Academic Press, New York and London.

Robinson, T. J. (1977). Reproduction in cattle. *In* "Reproduction in Domestic Animals", 3rd edition (H. H. Cole and P. T. Cupps, eds) pp. 433–454. Academic Press, New York and London.

Sammelwitz, P. H., Aldred, J. P. and Nalbandov, A. V. (1961). Mechanisms of maintenance of corpora lutea in pigs and rats. *J. Reprod. Fert.* **2**, 387–393.

Sawyer, C. H. (1975). First Geoffrey Harris Memorial Lecture. Some recent developments in brain-pituitary-ovarian physiology. *Neuroendocrinology* **17**, 97–124.

Sawyer, C. H. (1978). History of the neurovascular concept of hypothalamo-hypophysial control. *Biol. Reprod.* **18**, 325–328.

Schally, A. V., Arimura, A. and Kastin, A. J. (1973). Hypothalamic regulatory hormones. *Science* **179**, 341–350.

Schams, D., Schallenberger, E., Hoffmann, B. and Karg, H. (1977). The oestrous cycle of the cow: hormonal parameters and time relationships concerning oestrus, ovulation, and electrical resistance of the vaginal mucus. *Acta endocr.* **86**, 180–192.

Shearer, I. J., Haynes, N. B. and Crighton, D. B. (1971). Peripheral steroid hormone levels and nitrogen retention in the gilt from prepuberty to parturition. *J. Reprod. Fert.* **27**, 491–492.

Smeaton, T. C. and Robertson, H. A. (1971). Studies on the growth and atresia of Graafian follicles in the ovary of the sheep. *J. Reprod. Fert.* **25**, 243–252.

Sorensen, A. M., Hansel, W., Hough, W. H., Armstrong, D. T., McEntee, K. and Bratton, R. W. (1959). Causes and prevention of reproductive failures in dairy cattle. *Cornell Univ. Agric. Exp. Sta. Bull.*, No. 936.

Stabenfeldt, G. H., Akins, E. L., Ewing, L. L. and Morrissette, M. C. (1969). Peripheral plasma progesterone levels in pigs during the oestrous cycle. *J. Reprod. Fert.* **20**, 443–449.

Stabenfeldt, G. H., Ewing, L. L. and McDonald, L. E. (1969). Peripheral plasma progesterone levels during the bovine oestrous cycle. *J. Reprod. Fert.* **19**, 433–442.

Stabenfeldt, G. H. and Hughes, J. P. (1977). Reproduction in horses. *In* "Reproduction in Domestic Animals", 3rd edition (H. H. Cole and P. T. Cupps, eds) pp. 401–431. Academic Press, New York and London.

Stabenfeldt, G. H., Hughes, J. P., Evans, J. W. and Geschwind, I. I. (1975). Unique aspects of the reproductive cycle of the mare. *J. Reprod. Fert., Suppl.* **23**, 155–160.

Stockard, C. R. and Papanicolaou, G. N. (1917). The existence of a typical oestrous cycle in the guinea pig—with a study of its histological and physiological changes. *Amer. J. Anat.* **22**, 225–283.

Swanson, L. V. and Hafs, H. D. (1971). LH and prolactin in blood serum from estrus to ovulation in Holstein heifers. *J. Anim. Sci.* **33**, 1038–1041.

Thibault, C., Courot, M., Martinet, L., Mauléon, P., Du Mesnil du Buisson, F., Ortavant, R., Pelletier, J. and Signoret, J. P. (1966). Regulation of breeding season and estrous cycles by light and external stimuli in some mammals. *J. Anim. Sci., Suppl.* **25**, 119–142.

Turner, C. D. and Bagnara, J. T. (1971). "General Endocrinology", 5th edition. W. B. Saunders, Philadelphia and London.

Wettemann, R. P., Hafs, H. D., Edgerton, L. A. and Swanson, L. V. (1972). Estradiol and progesterone in blood serum during the bovine estrous cycle. *J. Anim. Sci.* **34**, 1020–1024.

Zuckerman, S. (1960). Origin and development of oocytes in foetal and mature mammals. *Mem. Soc. Endocr.* No. 7, p. 63.

Chapter II

Techniques for Modification of the Oestrous Cycle: Synchronisation of Oestrus and Programmed Breeding

Introduction

Although collection and storage of semen from selected males enables a considerable degree of control to be imposed prior to the moment of breeding, methods of actually detecting oestrus for the efficient and timely use of artificial insemination leave much to be desired. Moreover, with the ever increasing scale of animal enterprises and likewise the steeply rising cost of labour inputs, it seems inappropriate in many agricultural situations to permit mature animals to impose their intrinsic reproductive rhythms on the livestock enterprise. In fact, the detection of oestrus by regular surveillance or by the use of vasectomised teaser males may not be warranted or possible and, in any event, neither of these approaches in itself enables insemination to be performed at the optimum time with regard to fertility. More logical and satisfactory by far would be a situation in which the oestrous cycles of animals to be bred could be controlled by a pharmacological or pseudo-physiological treatment such that the time of onset of oestrus could be predicted in the great majority of animals receiving the treatment. Furthermore, if the precise time of ovulation itself could be predetermined, it might then be possible to inseminate animals at a fixed time without reference to the state of behavioural oestrus, thereby avoiding the penalties associated with incorrectly timed insemination and the ageing of gametes (see Chapter V).

In essence, therefore, a system of oestrous cycle control would attempt to dictate the desired timetable of breeding, rather than permitting females to impose their own reproductive rhythms on the farming system. In practice, as one of the objectives of oestrous cycle synchronisation is to have a herd or large batch of animals exhibiting oestrus at the same time, it should be clear that artificial insemination rather than natural mating is the usual sequel to the techniques to be described. Using this approach, the limitations of frequency of ejaculation in individual males together with the problem of lowered sperm concentration in semen samples arising from repeated daily matings can be overcome in a simple and satisfactory way. The benefits that might accrue from synchronisation arise in several ways—managerial, veterinary and husbandry—and those relating principally to reproduction are summarised in Table II.1.

Table II.1. Some potential advantages associated with the synchronisation of oestrus in farm animals. The sequence of presentation is not intended to suggest a particular order of priority.

1. To reduce the time needed for detection of oestrus
2. To facilitate the use of artificial insemination, especially in herds of beef cattle, by treating animals in groups
3. In conjunction with a procedure for controlling the time of ovulation, to permit insemination on a predetermined schedule
4. To enable the feeding of animals in uniform groups, especially where this involves a change of diet according to the stage of pregnancy
5. As a sequel to synchronised breeding, to limit the overall period of parturition in a herd or flock
6. To permit supervision at birth, (perhaps in conjunction with induced parturition), in order to reduce neonatal mortality and arrange cross fostering
7. After successful control of breeding, to permit weaning, fattening and marketing of uniform groups of animals
8. To enable introduction of stricter measures for the control of disease, especially with regard to the use of buildings
9. To facilitate use of the technique of embryo transplantation
10. As a general management aid, to rationalise the use of labour, buildings and other resources

Arguments in support of programmed breeding, as control of oestrus is frequently termed, are by no means new: serious approaches to the subject, especially in cattle, were initiated in the United States and elsewhere as early as 1940. But recent developments in the livestock industry, such as use of the technique of embryo transplantation in North America, Europe and the Antipodes, together with a clearer understanding of the endocrine basis of the oestrous cycle, have acted as additional incentives to developmental work on a meaningful field scale.

Principles underlying synchronisation treatments

As already described, the oestrous cycle in all species of domestic livestock is composed of hormonally well-defined luteal and follicular phases which vary in overall length according to species. However, in contrast to the situation during the menstrual cycle of primates, the luteal phase is always significantly longer than the follicular phase, in cattle, sheep, pigs and horses being at least twice as long (Table I.3). In practical terms, synchronisation of oestrus in groups of animals involves two alternative approaches to manipulation of the oestrous cycle. The first is by removing or inducing the demise of the corpus luteum so that all animals in an appropriate group enter the follicular phase of the cycle at the same time, and hopefully are still closely synchronised at the ensuing oestrus. The second approach involves suppression of follicular development during an artificially extended luteal phase so that upon removal of the pharmacological blockade after a sufficient period of treatment, all animals should enter the follicular phase approximately synchronously (Fig. II.1). The more compact follicular phase will therefore be working to experimental advantage during the rebound period after cessation of treatment, in the sense that it restricts the time available for any loss of synchrony due to expression of intrinsic reproductive rhythms. In contrast to manipulations of the luteal phase, it should be noted that the follicular phase can be shortened by only a day or two at the most in cattle, sheep and pigs. Attempts to prolong an active follicular phase inevitably result in ovulation or atresia, or in follicular abnormalities such as cysts.

Bearing in mind that attempts will usually be made to synchronise the breeding of animals in a large batch or herd, it is necessary to assume a random distribution of females with respect to stages of the oestrous cycle found at any given time. This would almost certainly be the situation in a large herd of cattle or flock of sheep maintained under extensive conditions. A satisfactory treatment must therefore aim to regulate the cycle of all animals such that on cessation or withdrawal of treatment, a very high proportion, if not all, of the animals subsequently exhibit oestrus on the same day. Assuming a random distribution of stages of the oestrous cycle at the onset of treatment, the *duration*

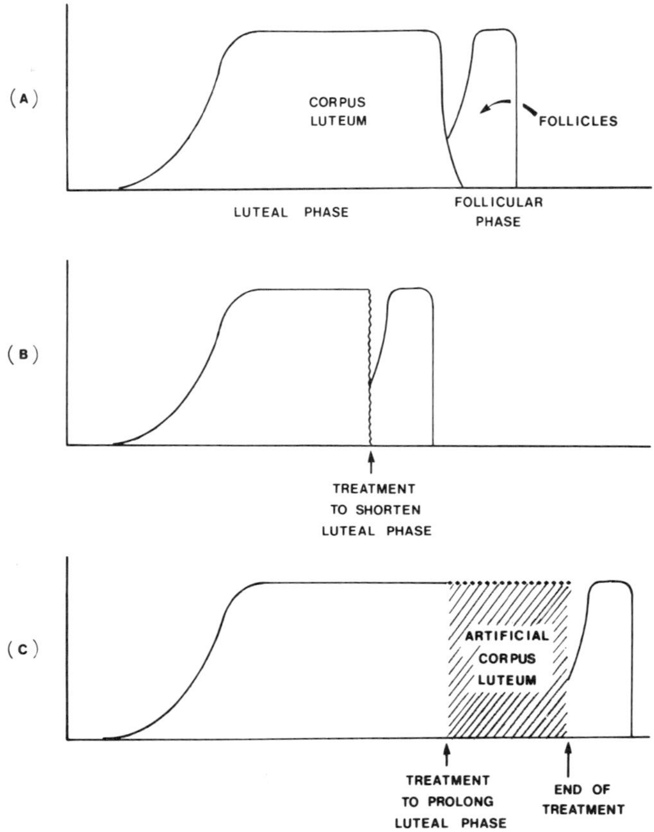

Fig. II.1. Alternative strategies for synchronisation of oestrus. These involve manipulation of the more extensive luteal phase (A), either as an abrupt shortening (B) or more commonly a prolongation (C). The latter involves progesterone or progestagen administration to suppress follicular development by means of negative feedback on pituitary gonadotrophin secretion. A characteristic follicular phase should follow cessation of treatment.

of a treatment to maintain all females in a luteal phase must therefore be in excess of the duration of the luteal phase in that species. On the other hand, if the basis of the treatment is to precipitate the follicular phase prematurely, then the form and/or frequency of treatment must be sufficient to catch animals that initially were distributed at all stages of the cycle.

A concise survey of methods aimed at obtaining synchronisation of oestrus in the major domestic species is pertinent, not least since the efficacy of any new approach should be measured against the shortcomings of techniques that are no longer applied extensively. Taking the species separately, the various methods used will be summarised in chronological order of their investigation. One of the findings to be noted, especially from the numerous studies in cattle, is that synchronisation of oestrus *per se* is not usually a difficult objective to achieve, whereas synchronisation in conjunction with good fertility has for long been an elusive target.

Synchronisation of oestrus in cattle

This section offers the greatest detail, partly since it introduces a review of methods that have been applied to other farm species, and also because the benefits to be derived from successful synchronisation of oestrus in many countries would be most attractive in cattle.

Luteal enucleation

The approach of physically removing the source of progesterone, enucleation of the corpus luteum by digital pressure *per rectum*, suffers from a number of disadvantages, including the risk of haemorrhage and of adhesions developing in the fimbria adjoining the treated ovary. Such a method of terminating the luteal phase of the oestrous cycle (Fig. II.2) can be applied to animals with a mature corpus luteum, but only a proportion (approximately 50–60%) of the animals in a herd is therefore susceptible to the synchronisation procedure. Moreover, enucleation of the corpus luteum requires professional assistance, and even then removal of the luteal tissue may not be complete. When successful, the mean interval between removal of the gland and return to oestrus is about four days, but considerable variation may be found (Jakobsen and Teige, 1956; Dawson, 1961).

Injections of progesterone

Injections of a solution of progesterone (Christian and Casida, 1948) to simulate the luteal phase and blockade the anterior pituitary by negative feedback involve the provision of a subcutaneous reservoir of the hormone in oil, usually maintained by repeated injections for 18–20 days. Although suppression of oestrus is achieved when an appropriate dose of progesterone is administered, such as 50 mg per day in heifers

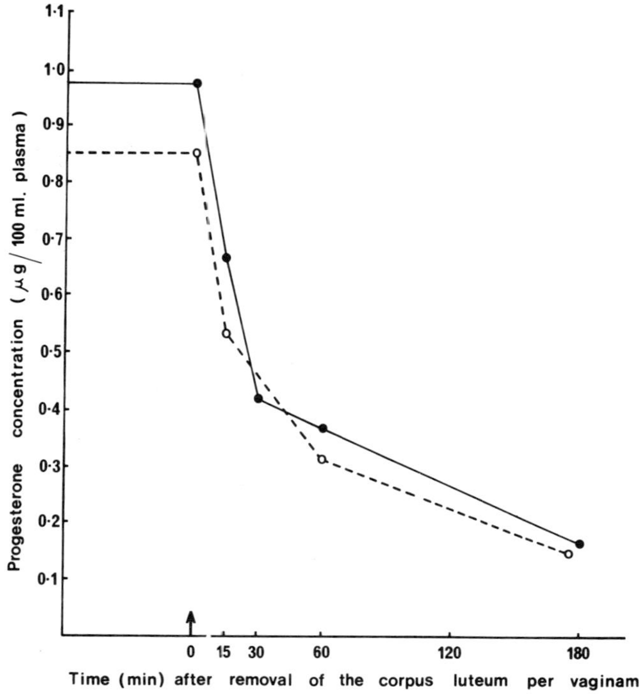

Fig. II.2. The influence of enucleation of the bovine corpus luteum on the concentration of progesterone in peripheral blood plasma. (Redrawn from McCracken, 1963.)

(Ulberg *et al.*, 1951), two significant disadvantages have been noted in addition to the time-consuming schedule of injections. First, that following cessation of treatment, there is no distinct "cut-off" point because of the influence of the subcutaneous depots of hormone. This leads to a considerable spread in the time of return to oestrus, although five days may represent the mean (Ulberg *et al.*, 1951). Second, and of perhaps even greater importance in practice, fertility is depressed at the

Fig. II.3. To illustrate diverse approaches to synchronisation of oestrus in cattle by manipulating or simulating the luteal phase of the oestrous cycle. (a) A mature corpus luteum in the right ovary at an ideal stage for enucleation by application of digital pressure *per rectum* at its base. (Arrows). (b) Solution of progesterone in oil that is injected subcutaneously in order to mimic the activity of a functional corpus luteum. (c) A polyurethane sponge that is soaked in a solution of a progestagen in oil before positioning in the anterior vagina with the aid of a speculum. (d) An intra-vaginal device (PRID) in which silastic rubber containing a progestagen coats a tapering metallic coil. Note the red capsule of oestrogen and progesterone solutions which is essential if the PRID is used for a short-term (9–12 day) treatment.

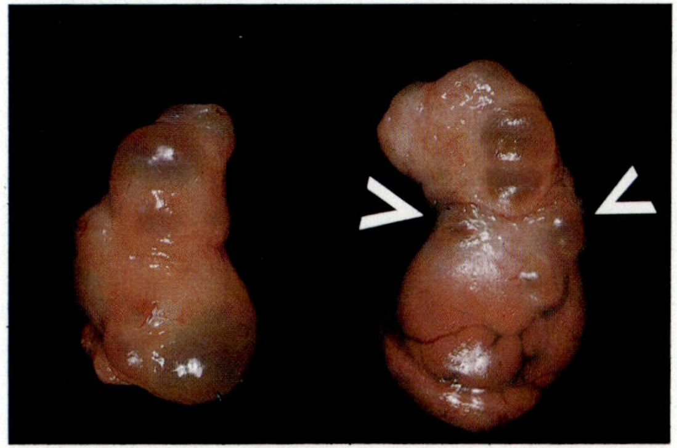

(a)

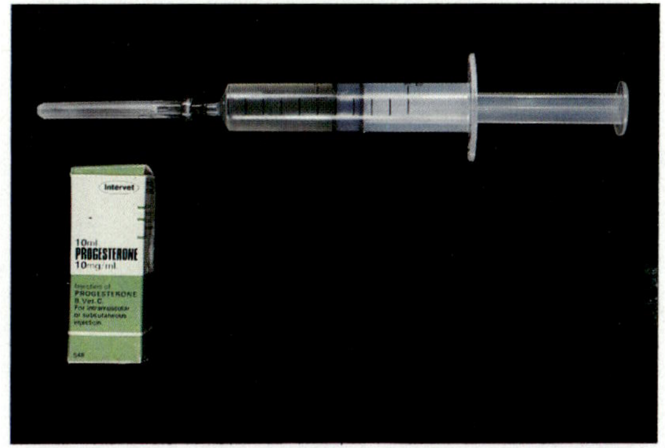

(b)

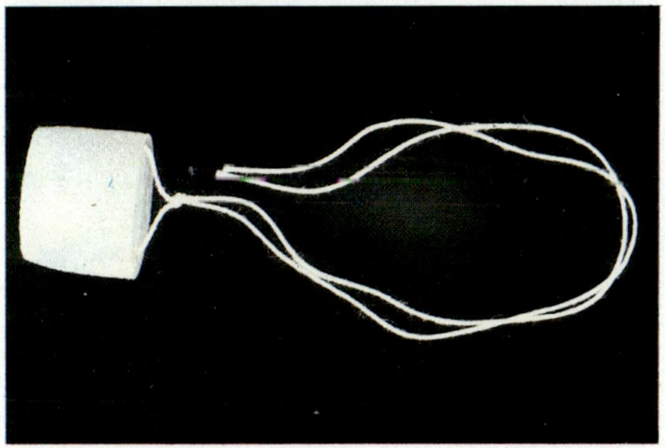

(c)

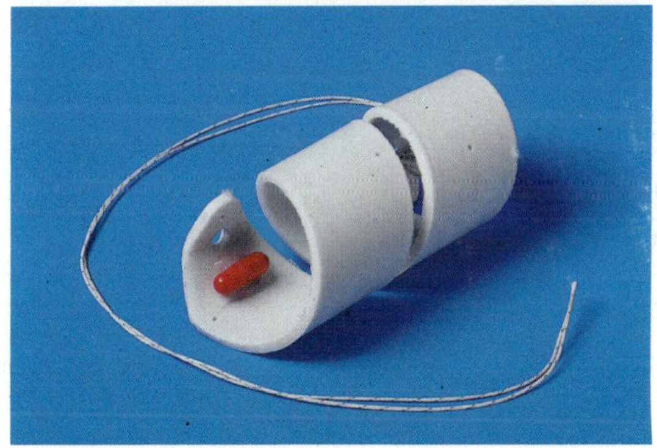

(d)

first oestrus after progesterone therapy (Trimberger and Hansel, 1955; Lamond, 1964), due in part to adverse influences on transport and survival of spermatozoa in the female reproductive tract.

Orally-active progestagens

The advent of orally-active synthetic progestagens (Hansel and Malven, 1960) appeared initially to offset some of the disadvantages of progesterone therapy by repeated injection, and these compounds were also considered more suitable for synchronisation of large batches of animals in feed-lots and other situations of supervised feeding. Melengestrol acetate (MGA) and medroxyprogesterone acetate (MPA) were two such progestagens (Table II.2), the potency of MGA when administered orally being several hundred times that of MPA (Zimbelman and Smith, 1966). Treatment with these synthetic compounds involved feeding for approximately 15 to 18 days, a high proportion of animals showing oestrus 3–5 days after withdrawal of the treatment (Zimbelman and Smith, 1966), but fertility at this synchronised oestrus was invariably poor (Hansel, 1967). For example, in a trial involving 72 animals, Zimbelman and Smith (1966) obtained a conception rate at first service of 42% compared with 82% following service at the second oestrus. In subsequent work, treatment with oestrogens and gonadotrophic hormones was imposed upon the MGA feeding blockade (Smith and Zimbelman, 1968a), but no consistent improvement in conception rate was reported (Smith and Zimbelman, 1968b), nor when insemination was conducted at a predetermined time (Boyd and Tasker, 1971).

Silastic implants

The finding that progestagens would be released into biological tissues from silicone rubber (silastic) impregnated implants (Dziuk and Cook, 1966) presented another possible way of treating cattle with these compounds. Silastic implants containing MGA were inserted subcutaneously in the neck region of cows for 22 to 64 days (Dziuk *et al.*, 1966), and upon removal of the implant, 64% of 45 cows were in oestrus 36 to 72 hours later. Retention of the implants in this particular trial was good, but the problem of diminished fertility at the synchronised oestrus was again encountered (Dziuk *et al.*, 1966; Roche and Crowley, 1972, 1973). However, a trial by Curl *et al.* (1968) with subcutaneous implants containing norethandrolone gave good oestrous synchronisation 48 hours after removal of

Table II.2. The chemical nature, route of administration, and dose of various progestagens used for synchronisation of oestrus in cattle. (After Mauléon, 1974.)

Chemical nature	Name of the compound	Route of administration	Dose per day for one cow (mg)
17α hydroxy-progesterone derivatives	FGA (Fluorogestone acetate, Searle)	Vaginal sponges	200 (total dose)
		Orally	30
		Intramuscular injections	2·4
	MGA (Melengestrol acetate, Upjohn)	Orally	0·5–1
	DHPA (Droxone, Squibb)	Orally	120–150
	MPA (Medroxyprogesterone acetate, Upjohn)	Orally	180
	CAP (Chlormadinone acetate, Syntex)	Orally	5–10
19 Nor progesterone derivatives	SC 21 009 (Searle)	Intramuscular injections	0·14
		Subcutaneous implants	6–12 (total dose)
19 Nor testosterone derivatives	Norethisterone	Orally	240
	Norethynodrel	Intramuscular injections	0·30
	Norethandrolone (Nilevar)	Intramuscular injections	5·4
		Subcutaneous implants	250 (total dose)
		Vaginal sponges	800 (total dose)

the implants in 18 of 22 treated animals, and 68% of these conceived at the first oestrus. The absence of a depot of progesterone or synthetic progestagen upon removal of the implant may offer some advantage in obtaining close synchronisation. In recent trials, the implant has been inserted under the skin of the outer surface of the ear to avoid displacement by rubbing (Wishart and Young, 1974; Wiltbank and Gonzalez-Padilla, 1975).

Intra-vaginal sponges

Progestagens have also been administered intra-vaginally by means of impregnated sponge pessaries which, in theory, permit a more precise treatment of individual animals. Sponges are soaked in oil containing the progestagen, dusted with an antibiotic preparation, positioned deep in the vagina, and left for a period of 18 to 21 days. A good absorption of hormone and suppression of oestrus is obtained with this method (Mauléon and Rey, 1966; Carrick and Shelton, 1967) and, following withdrawal of the sponge, oestrous symptoms are first noted within 24 to 72 hours but fertility at the induced oestrus is generally reduced when compared with control cattle (Hansel and Schechter, 1972). The major disadvantages are (a) that the sponges may be displaced in a proportion of the animals, and (b) that at the time of withdrawal of the sponge, there may be some tearing of the vaginal mucosa. Apart from any discomfort entailed, this may also provide a focal point for subsequent infection. The problem of displacement may be overcome by administering progestagens from a silastic coated coil, the progesterone-releasing intra-vaginal device (PRID, Abbott Laboratories Ltd), that is positioned in the anterior vagina with the aid of a speculum (Mauer *et al.*, 1975; Roche, 1975, 1976a). Other studies have involved the use of an injection of PMSG (750–2000 iu) shortly before or after pessary or coil removal in an attempt to provoke a more concise period of oestrus and to enhance fertility after progestagen treatment (Wishart and Hoskin, 1968); the method appears highly effective for synchronisation of oestrus in heifers.

Short-term progestagens

As an approach to overcoming the problem of diminished fertility following progestagenic treatments of 15–18 days or more in length, a combination of hormonal treatments has been applied during a reduced period of 9–12 days (Wiltbank and Kasson, 1968; Mauléon, 1974; Roche, 1974a; Wishart and Young, 1974; Sreenan, 1975). An example is the use of a synthetic progestagen administered in implants, sponge pessaries or coils for a period of only 9 or 10 days, in conjunction with an injection of 5–7·5 mg oestradiol benzoate and 50–250 mg progesterone given at the onset of treatment. The reason for the oestrogen injection is to cause premature regression of a developing corpus luteum whilst the progesterone injection blocks any imminent ovulation, so that upon completion of the progestagen treatment, all animals should enter a follicular phase irrespective of their ovarian status at the onset of treatment. Reports suggest that this method permits good synchrony and apparently normal fertility (Roche, 1974a; Sreenan and Mulvehill, 1975; Manns and Hafs, 1976), particularly after a 12 day treatment and a single insemination some 56 hours after the end of treatment (Table II.3: Roche, 1976b). As a means of simplifying procedures in the field, Webel (1976) and Roche (1978) have used gelatin capsules attached to the inside of the coil by which means 5 mg oestradiol benzoate and 200 mg progesterone are administered at the start of a 9 or 12 day treatment. The capsules dissolve and release their steroid into the vagina within

Table II.3. The incidence of calving in dairy cows following a 12-day progesterone treatment with intra-vaginal coils for the synchronisation of oestrus. Control cows within the same herds were inseminated at approximately the same time. (Taken from Roche, 1976b.)

	Control Dairy Cows	12-Day Progesterone Treatment	
		Dairy Cows	Heifers
No. of animals treated	384	412	156
No. of coils removed	–	378	153
No. inseminated	365	342	143
% non-return rate	55	53	63
No. of cows present at calving	337	311	139
No. of cows calved to first insemination	169	170	83
No. of cows calved to repeat insemination	71	71	40
% of treated cows that calved	44	41	53
% of cows inseminated that calved	46	45	58
% of cows present at calving that calved	50	55	60

one hour, but a somewhat higher dose may be needed for this route when compared with injection.

Prostaglandin injections

The availability of natural or synthetic prostaglandins has given renewed impetus to attempts to synchronise oestrus in cattle by methods less demanding than those described so far. This is not least since events following treatment with prostaglandins may imitate quite closely part of the natural luteolytic process (Dobson *et al.*, 1975), and may therefore avoid the diminished fertility encountered when progestagens have been employed. The first publications concerned with prostaglandin therapy appeared in the early 1970s (Lauderdale, 1972; Louis *et al.*, 1972; Rowson *et al.*, 1972; Roche, 1974b); the compounds have been more widely employed since the introduction of prostaglandin analogues (Tervit *et al.*, 1973; Cooper and Furr, 1974; and many subsequent references).

The significant finding in the early studies was that prostaglandin $F_{2\alpha}$ ($PGF_{2\alpha}$) introduced non-surgically into the uterine horn adjacent to the corpus luteum caused luteolysis if administered after Day 4 of the oestrous cycle (Hansel and Schechter, 1972; Rowson *et al.*, 1972). Before this stage, the developing bovine corpus luteum is apparently not susceptible to the luteolytic effects of $PGF_{2\alpha}$. To avoid metabolic breakdown of the natural compound during passage through the lungs, $PGF_{2\alpha}$ could not be given systemically in small quantities but had to be instilled into the lumen of the reproductive tract, usually in two doses of 0·5 mg/day 24 hours apart. A high proportion of treated animals was synchronised by this method, oestrus being detected in most animals on the morning of the third day after treatment (Rowson *et al.*, 1972). The availability of potent synthetic analogues of $PGF_{2\alpha}$, such as ICI 79939 (Tervit *et al.*, 1973) and ICI 80996 or 81008 (Cooper and Furr, 1974), which are effective systemically, made synchronisation procedures for large batches of animals a realistic undertaking. A single intramuscular dose of 500 μg of the analogue cloprostenol given at any point during the luteal phase after Days 4 to 5 synchronises oestrus in most animals within 48 to 72 hours after treatment, and ovulation occurs about 24 hours after the onset of this induced oestrus. But because a substantial portion of the natural oestrous cycle is therefore not influenced by such a treatment, two injections of prostaglandin analogue are necessary to catch all animals in a herd when treatment is imposed at unknown stages of the cycle. This requirement is met by injecting prostaglandin twice (Fig. II.4), with an interval of 10 to 12 days between treatments (Cooper and Furr, 1974). Under this régime, 90–95% of the animals at the time of the second injection would be at the sensitive mid-luteal phase of the cycle, when the lifespan of the corpus luteum can be curtailed (Fig. II.5). Animals are inseminated either once at 60–72 hours after treatment or twice at 72 and 96 hours after treatment, the

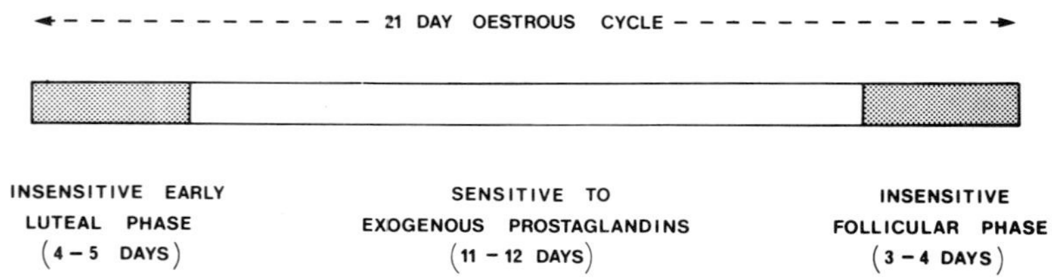

Fig. II.4. The double-spaced injection régime for synchronising oestrus in cattle. A prostaglandin analogue is administered intramuscularly twice some 10–12 days apart so that all animals are in the sensitive portion of the luteal phase at the time of the second injection.

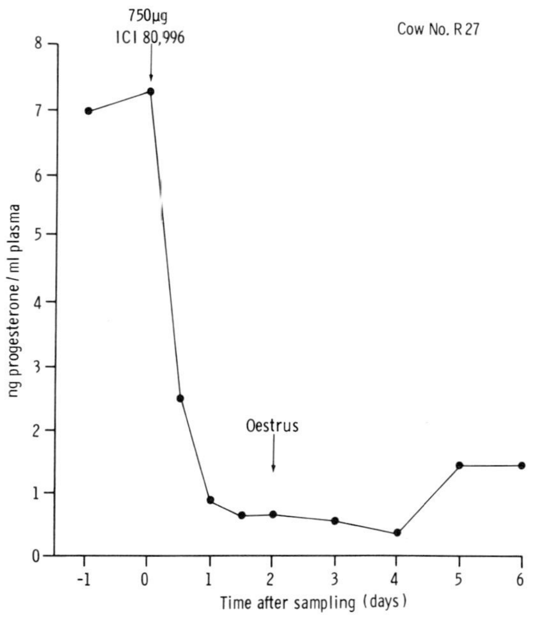

Fig. II.5. To illustrate the rate of luteolysis in the cow after systemic injection of a prostaglandin analogue (ICI 80 996) in the presence of a mature corpus luteum. Note the precipitous decline in the concentration of plasma progesterone. (Courtesy of the late Dr A. L. Walpole.)

double insemination schedule generally leading to better results (Table II.4).

Preliminary data suggested that fertility was not reduced after this treatment régime, and this has since been confirmed in several field trials (Cooper and Rowson, 1975; Lauderdale, 1975; Lamming *et al.*, 1975; Roche, 1976c) but not in all (Hansel and Fortune, 1978). If normal fertility after prostaglandin treatment proves to be the case under a wide range of husbandry conditions, then the intramuscular injection of these compounds to achieve synchronisation of oestrus would seem to be the method of choice in the foreseeable future, assuming that prostaglandins can be marketed at a realistic cost. This form of programming the oestrous cycle would have a number of attractions, particularly if insemination could be undertaken at a predetermined time without reference to the stage of behavioural oestrus.

As an overall comment on the success of these various kinds of therapy, fertility appears to be normal at the synchronised oestrus only after short-term progestagen treatments or those that directly precipitate the onset of a follicular phase (Hunter, 1976). Attempts to simulate or extend the luteal phase with progesterone or synthetic progestagens have invariably been associated with depression of fertility at the first oestrus after the synchronisation treatment (Table II.5). Although normal conception rates can be obtained at subsequent cycles, the compactness of the oestrous distribution is reduced and the cost of maintaining animals through a barren cycle may also be seen as a disadvantage.

Synchronisation of oestrus in sheep

Although repeated injections of a solution of progesterone (Dutt and Casida, 1948; Robinson, 1956; Lamond and Bindon, 1962), feeding of orally-active progestagens (Dziuk *et al.*, 1964; Hinds *et al.*, 1964), and subcutaneous insertion of silastic implants impregnated with progestagens (Dziuk *et al.*, 1968; Dziuk *et al.*, 1972) have all been examined as potential means for controlling the oestrous cycle in sheep, none of these approaches has been sufficiently attractive for large-scale commercial application. Animals return to oestrus about three days after cessation of such treatments, but the accompanying disadvantages correspond closely to those described in detail for cattle, especially that of the lowered conception rate (often only 30–50%) found at the first

Table II.4. Comparison of fertility between control and treated dairy heifers in which oestrus was synchronised with the prostaglandin analogue, Cloprostenol, and insemination performed once or twice. (Taken from Cooper, 1978.)

Natural service			Estrumate[1] plus artificial insemination					
Control			Double insemination			Single insemination		
No. of heifers	No. pregnant	Conception rate (%)	No. of heifers	No. pregnant	Conception rate (%)	No. of heifers	No. pregnant	Conception rate (%)
99	56	57	112	64	57	106	43	41

[1] Estrumate is the trade name under which the prostaglandin analogue Cloprostenol (ICI 80 996) is marketed for use in cattle breeding

Table II.5. Fertility after various forms of treatment for synchronisation of oestrus in cattle.

Nature of treatment[1]	Objective of treatment *vis-à-vis* the oestrous cycle	Fertility[2] at *first* oestrus after treatment
Enucleation of corpus luteum	Induce follicular phase	Usually normal
Subcutaneous injections of progesterone in oil	Simulate *full* luteal phase	Depressed
Oral progestagens (e.g. MPA, MGA)	Simulate *full* luteal phase	Depressed
Silastic implants of progestagens	Simulate *full* luteal phase	Depressed
Vaginal sponges containing progestagens	Simulate *full* luteal phase	Depressed
Short-term progestagen treatments[3]	Simulate *short* luteal phase	Usually normal
Prostaglandin infusion or injection	Induce follicular phase	Normal
Short-term progestagen + prostaglandin	*Short* luteal phase	Normal

[1] More detailed description in text. [2] Fertility termed normal refers to the conception rate that is characteristic of a given herd. [3] In conjunction with treatments for inducing luteolysis.

oestrus following treatment. However, despite some adverse effect on conception, the intra-vaginal administration of progestagens by sponge pessary is used extensively in sheep breeding, both as a procedure for synchronising oestrus and as a component of treatments for advancing the breeding season.

Intra-vaginal sponges

Robinson (1965) first reported the successful use of polyurethane intra-vaginal sponges impregnated with a solution of a potent synthetic progestagen (SC-9880 or Cronolone; G. D. Searle & Co) as a means of obtaining synchronisation of oestrus in cyclic ewes. Sponges containing an appropriate dose of progestagen (30–45 mg Cronolone) in sufficient fluid vehicle need to remain in the vagina for some 14–16 days if this treatment alone is to inhibit oestrus and overcome the random distribution of ovarian stages found in a large flock of animals. Upon withdrawal of the sponges at the end of the treatment period, a high proportion of ewes show oestrus within 24–72 hours, usually on the second day after sponge withdrawal. A single injection of 350–750 iu PMSG given at the time of sponge withdrawal improves the degree of synchronisation in the treated animals and slightly hastens the onset of oestrus (Colas *et al.*, 1973; Gordon, 1975). This 14–16 day treatment with progestagenic compounds is sometimes associated with depression of fertility at the first post-treatment oestrus (Wishart, 1967; Robinson *et al.*, 1968), possibly due in part to a modified pattern of uterine contractions (Hawk and Echternkamp, 1973) and thereby to less effective transport of sperm to the Fallopian tubes (Quinlivan and Robinson, 1969), as well as to a reduced survival of spermatozoa in the progestagen treated ewe. But insemination of adequate numbers of spermatozoa ($2{\cdot}5 \times 10^8$) in 0·1–0·2 ml of undiluted semen or two inseminations at 48 and 60 hours after sponge removal may help to overcome much of the diminished fertility. Pioneer work in this field has been comprehensively summarised in the volume edited by Robinson (1967) and more concisely by Gordon (1975).

There remains a suspicion that the temporal relationship between the onset of oestrus and the preovulatory surge of LH may be disturbed in some treated animals, and that this also contributes to lowered fertility. It is especially in this context that injection of PMSG at the end of treatment may help to improve conception rates (Tables II.6 and II.7).

Table II.6. To illustrate the beneficial influence of an injection of PMSG at the time of intra-vaginal sponge withdrawal on the fertility of ewes treated with progestagens during the breeding season. (Adapted from Colas *et al.*, 1973.)

Nature of treatment	Time of artificial insemination	Conception rate (%)
40 mg FGA alone	First oestrus following treatment	54·5
	Subsequent spontaneous oestrus	72·0[1]
40 mg FGA plus 400 i.u. PMSG	First oestrus following treatment	69·0
	Subsequent spontaneous oestrus	71·8

[1] Significantly greater ($P < 0{\cdot}01$) than after the treated cycle.

Table II.7. The fertility of ewes treated with progestagens and PMSG for the synchronisation of oestrus and breeding at three seasons of the year. The ewes were in farm flocks in Ireland and treated by personnel from University College Dublin. (Taken from Gordon, 1975.)

	Season of treatment		
	Spring	Summer	Autumn
No. of flocks	83	594	37
No. of ewes	2508	21 545	1453
Oestrous response (%)[1]	93	97	97
Conception rate (%)[2]	37	66	78
No. which produced lambs	871	13 795	1094
No. of lambs born	1375	22 396	1885
Pregnancy (%)[3]	34·7	64·0	75·3
Litter size	1·58	1·62	1·72
No. of lambs/100 ewes	55	104	130

[1] Based on number of sheep treated. [2] Based on number of sheep mated. [3] Ewes lambing as percentage of number treated.

Short-term progestagens

Experiments have been aimed at developing short-term progestagenic treatments of 7–8 days' duration, and this approach has been tested in anoestrous ewes with intra-vaginal sponges (Christenson, 1976). As indicated in this section on cattle, short-term treatments with progestagens require more elaborate procedures since the period during which sponges are *in situ* is itself shorter than the luteal phase of the oestrous cycle. One procedure is to inject 10 mg of progesterone at the time of sponge removal followed by 750 iu PMSG the next day, but conception rates may not be increased significantly by this approach when compared with the results of longer treatments (Christenson, 1976).

Prostaglandin injections

As is the case for cattle, the cyclic corpus luteum of the sheep is only sensitive to the luteolytic effects of exogenous prostaglandins during a portion of its lifespan. Regression of the corpora lutea can be induced by a single intramuscular injection of $PGF_{2\alpha}$ (Douglas and Ginther, 1973; Hawk, 1973) or a suitable analogue (Hearnshaw *et al.*, 1974) when these structures are at least three days of age (i.e. Day 4 of the cycle). However, in groups of animals in which oestrous cycles are distributed at random, it is again necessary to use a double-injection régime to achieve synchronisation in all animals, treatments usually being spaced by an interval of 8 or 9 days. Luteolysis is rapid and animals return to oestrus within 72 hours of an intramuscular or subcutaneous injection of prostaglandin analogue (e.g. 100 μg of ICI 80996), a majority of animals showing oestrus between 29 and 48 hours (Trounson *et al.*, 1976) or at a mean of 44 hours (Acritopoulou *et al.*, 1977). Fertility following natural mating at the induced oestrus was not significantly different from controls (Table II.8), conception rates of 66% having been reported (Haresign, 1978).

Table II.8. Fertility of cyclic ewes given two injections of 100 μg of the prostaglandin analogue, ICI 80 996, some 9 days apart to induce synchronisation of oestrus. (Taken from Haresign, 1978.)

		Conception rate to first mating (%)	Prolificacy (%)
Trial 1	Treated	66·0	133
(n = 50)[1]	Control	70·0	120
Trial 2	Treated	65·7	161
(n = 50)[1]	Control	71·4	160

[1] Numbers for both treated and control groups of ewes.

Synchronisation of oestrus in pigs

Although synchronisation of oestrus in pigs, particularly in young stock (gilts) to be bred for the first time could lead to attractive managerial and marketing advantages, a straightforward and acceptable method of oestrous cycle control in gilts is currently unavailable in Britain and North America, due, perhaps surprisingly, to governmental food and drug regulations rather than to biological constraints. Synchronisation of oestrus in sows, on the other hand, can be obtained with reasonable precision simply by weaning the young at an appropriate time.

Injections of progesterone

Much of the early work in this species was conducted in Casida's laboratory at the University of Wisconsin, and is briefly referred to in his valedictory lecture (Casida, 1976). Whilst daily injections of a solution of progesterone in oil (25–100 mg per animal per day) were noted to give excellent suppression of oestrus and ovulation, this form of pituitary and hypothalamic blockade was associated with a high incidence of cystic follicles at the oestrus following treatment (Ulberg *et al.*, 1951) and hence poor or negligible fertility.

Orally-active progestagens

Oral administration of progestagens such as MPA (Provera) incorporated once or twice daily in the food ration, a method of treatment eminently suited to this intensively housed species, was again found to give good suppression of oestrus but to permit development of ovarian cysts if fed at levels of up to 400 mg per animal per day (Nellor, 1960; Dziuk, 1962; First *et al.*, 1963). The appearance of this follicular condition is a sensitive indicator of gonadotrophin imbalance in mature pigs, and may arise during Provera administration due to incomplete blockade of pituitary activity, especially of FSH secretion. At higher dose-levels (500 mg Provera per day), formation of cysts is prevented, but the time between end of treatment and ovulation varies over a period of at least 10 days, and fertility at this oestrus is usually poor (Dziuk and Polge, 1962, 1965).

Non-steroidal pituitary blockade

A satisfactory means of control seemed to be at hand in the early 1960s when it was noted that methallibure (ICI 33828, as it was first termed), a non-steroidal derivative of hydrazine, was an orally-active pituitary inhibitor. When included in the diet at a level of 1 mg per kg liveweight per day (i.e. approximately 100 mg per gilt per day) and administered during a period of 20 days, methallibure should suppress follicular development and oestrus in all animals. The principal effect of the compound is to block synthesis and/or release of FSH (Garbers and First, 1969; Polge and Day, 1969), but this gonadotrophic activity rebounds rapidly and spontaneously following cessation of treatment, leading to oestrus after a normal follicular phase. As first reported by Polge (1964, 1965) and Gerritts and Johnson (1964), the degree of synchronisation is good with 80–90% of animals exhibiting oestrus within 5–7 days (Fig. II.6). Fertility is not impaired, presumably

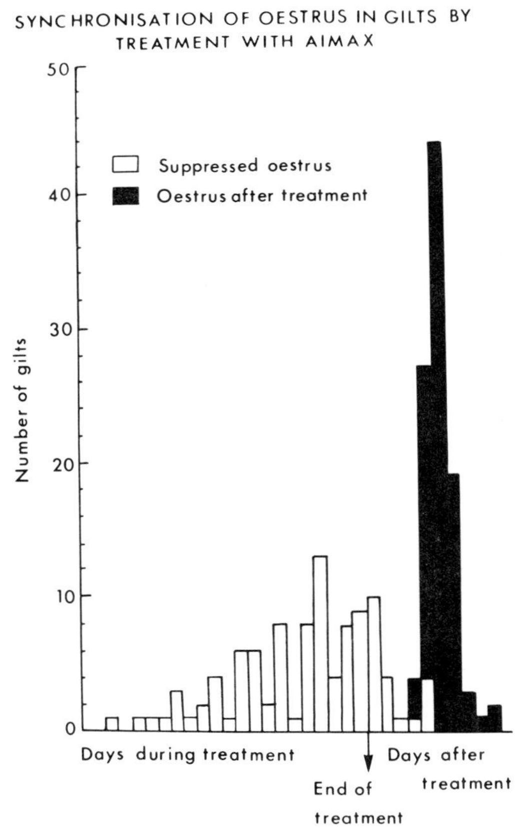

Fig. II.6. The degree of synchronisation of oestrus in pigs obtained after oral treatment of randomly cycling animals for 18–20 days with the non-steroidal pituitary inhibitor, methallibure. White columns represent suppressed heats, black columns the number of gilts showing oestrus. (Courtesy of Dr E. J. C. Polge.)

because this non-steroidal compound does not affect the condition of the reproductive tract. Even greater control over the time of oestrus and ovulation can be obtained if gonadotrophic hormones are administered at the end of methallibure treatment. A subcutaneous injection of 750 iu PMSG at the time of drug withdrawal followed by an ovulating injection of 500 iu HCG given intramuscularly after 96 hours causes ovulation some 40 h later. Artificial insemination is recommended 24 h after the HCG injection (16 hours before ovulation), and this form of programmed breeding has been shown to be applicable and successful under field conditions (Polge *et al.*, 1968). A conception rate of 90% with normal embryonic survival and litter size was reported, this being achieved when insemination was performed without reference to the state of behavioural oestrus.

The optimism arising from this research using methallibure was terminated abruptly when it was found that inadvertent feeding of the compound to pregnant gilts led to teratogenic (skeletal) defects in the developing foetuses (Fig. II.7), particularly when administered for 20 days in the first seven weeks of gestation (King, 1969). Despite the argument that one would hardly be seeking to achieve synchronisation

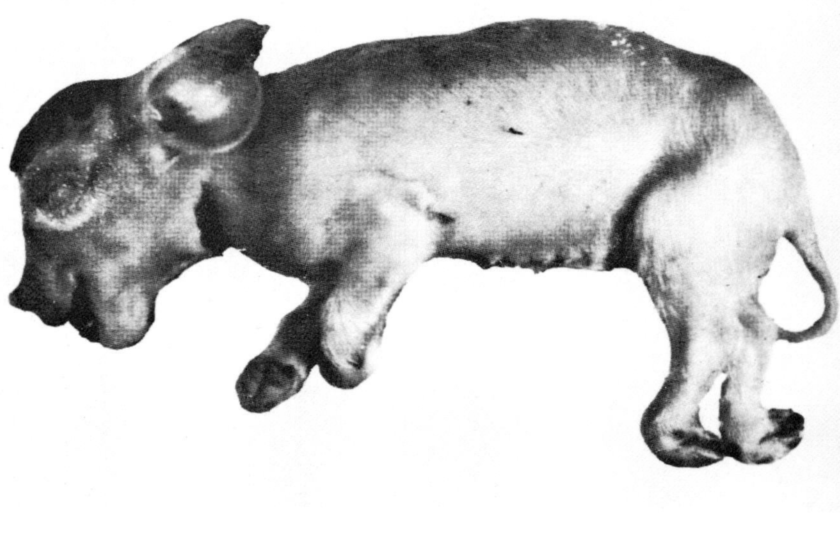

Fig. II.7. A deformed pig foetus showing the teratogenic effects that may arise in the skeletal structures following inadvertent feeding of sows with methallibure during the first seven weeks of gestation. (Courtesy of Dr G. J. King and *Journal of Reproduction and Fertility*.)

of oestrus in pregnant animals (Polge, 1972), the wider implications of making such a drug commercially available caused the British and American authorities to withdraw methallibure from the market for agricultural and veterinary purposes.

Prostaglandin injections

As to a treatment with prostaglandins, the use of the pure compound or prostaglandin analogues to achieve synchronisation of oestrus in pigs is limited by the brief period during which corpora lutea are susceptible to luteolytic effects during the oestrous cycle. Only on Days 11–12 or thereafter can the corpora lutea be induced to regress prematurely by injection of $PGF_{2\alpha}$ (Diehl and Day, 1974; Moeljono *et al.*, 1976) or by the synthetic analogue, ICI 79939 (Guthrie and Polge, 1976a), so there is no simple means of synchronising all the animals in a group in which the cycles are distributed at random. Paradoxically, in view of the above remarks concerning a possible misapplication of methallibure, it has been suggested that a means of prostaglandin use on a field scale could involve first inseminating or mating all the animals and then injecting prostaglandin (cloprostenol) when the corpora lutea were at least 12 days old—thereby inducing synchronous regression and a return to oestrus (Guthrie and Polge, 1978). The pig breeding industry is unlikely to view such a *tour de force* as a method of oestrous synchronisation with commercial attractions, nor indeed other approaches involving induction of accessory corpora lutea (Guthrie and Polge, 1976b). In fact, the only practical application of prostaglandins or their analogues in this species at present is in the use of their luteolytic properties to terminate pregnancy and thereby control the time of farrowing (see Chapter X).

Gonadotrophin injections

Other methods concerned with synchronising oestrus in gilts have been summarised by Webel (1978). Over and above the information presented so far, his concluding remarks suggest that stimulation of gilts at the threshold of puberty by injection of gonadotrophic hormones may be the only currently-reliable method of controlling oestrus and ovulation. Although this approach has been used in an experimental context for many years (Du Mesnil du Buisson, 1954; Dziuk and Gehlbach, 1966; Baker and Coggins, 1968), an extensive application in the pig breeding industry may yet be found if (a) precocious breeding has no deleterious effects on the performance of treated animals in subsequent pregnancies and (b) the cost of a gonadotrophin mixture such as PMSG plus HCG (Schilling and Cerne, 1972) can be kept within reasonable limits.

Synchronisation of oestrus in horses

Synchronisation of oestrus in mares may seem an objective less pressing than in cattle, sheep or pigs, particularly in those countries where artificial insemination has not been permitted to take over the coital rôle of the stallion. However, batch breeding may be required in certain circumstances and synchronisation procedures have been used to facilitate the transplantation of embryos. Follicular activity can be suppressed for 18–20 days with repeated injections of progesterone or administration of oral progestagens, and oestrus occurs 3–7 days after the end of treatment. Other approaches have involved a double injection procedure with prostaglandins in conjunction with two injections of HCG to induce ovulation of mature follicles (review by Allen, 1978).

Persistence of the corpus luteum is a frequent cause of failure to undergo oestrous cycles in this species, and a straightforward means of inducing luteal regression and regulating the time of service would clearly be advantageous. Early reports indicated that a single subcutaneous injection of quite large doses of $PGF_{2\alpha}$ (1·25–10 mg: Douglas and Ginther, 1972) or two intramuscular injections of 300–600 μg of the synthetic prostaglandin analogue ICI 79939 on successive days (Allen and Rowson, 1973) would induce oestrus within 2–4 days when applied in the mid luteal phase. (The developing corpus luteum is refractory to the action of luteolytic agents for the first four or five days). Subsequent studies have shown that mares with persistent luteal function (i.e. prolonged dioestrus) can be treated with prostaglandins or their synthetic analogues by intra-uterine infusion or intramuscular injection to produce a prompt and synchronised return (in 2–4 days) to a fertile oestrus in over 80% of the treated animals (Allen, 1978). Injection of prostaglandins would seem a more suitable means of regulating luteal function than intra-uterine infusion of large volumes of saline (Arthur, 1970), a technique which gives a less predictable result and carries the risk of bacterial infection.

General conclusions

Despite problems of implementation in the field, techniques for the synchronisation of oestrus have an important contribution to make to systems of animal production in many parts of the world, and should feature more widely in both developed and developing countries in the next ten years. However, the choice of method to be applied in a given situation must clearly take into account a number of factors, not least of which are the ease of administration, the effectiveness of treatment, the absence of harmful side effects, and of course the probable cost/benefit ratio. But even after such considerations have been applied quite rigorously, it is difficult to escape the conclusion that synchronisation procedures could lead to improved output, especially where beef and sheep production predominate. Robinson (1975) has stressed that controlled breeding must form part of an overall system of husbandry, and has further suggested that the prerequisites for economic application include:

(a) high value of individual animals and their products;
(b) adequate servicing, training and research facilities to take the procedures on to the farm;
(c) existence of a high level of husbandry and understanding.

Moreover, he cited the practice of the French INRA group at Nouzilly and Professor Ian Gordon's joint project with the Irish Department of Agriculture as good examples of the application of controlled breeding in sheep.

References

Acritopoulou, S., Haresign, W., Foster, J. P. and Lamming, G. E. (1977). Plasma progesterone and LH concentrations in ewes after injection of an analogue of prostaglandin $F_{2\alpha}$. *J. Reprod. Fert.* **49**, 337–340.

Allen, W. R. (1978). Control of oestrus and ovulation in the mare. *In* "Control of Ovulation" (D. B. Crighton, G. R. Foxcroft, N. B. Haynes and G. E. Lamming, eds), pp. 453–468. Butterworths, London and Boston.

Allen W. R. and Rowson, L. E. A. (1973). Control of the mare's oestrous cycle by prostaglandins. *J. Reprod. Fert.* **33**, 539–543.

Arthur, G. H. (1970). The induction of oestrus in mares by uterine infusion of saline. *Vet. Rec.* **86**, 584–586.

Baker, R. D. and Coggins, E. G. (1968). Control of ovulation rate and fertilisation in prepuberal gilts. *J. Anim. Sci.* **27**, 1607–1610.

Boyd, L. J. and Tasker, J. B. (1971). Fertility of oestrus synchronised dairy cattle treated with gonadotropins and inseminated at a predetermined time. *Vet. Rec.* **89**, 632–635.

Carrick, M. J. and Shelton, J. N. (1967). The synchronisation of oestrus in cattle with progestagen-impregnated intravaginal sponges. *J. Reprod. Fert.* **14**, 21–32.

Casida, L. E. (1976). Ovulation studies with particular reference to the pig. *Biol. Reprod.* **14**, 97–107.

Christenson, R. K. (1976). Effect of short-term progestogen treatment on induction of estrus and lambing in anestrous ewes. *J. Anim. Sci.* **43**, 795–801.

Christian, R. E. and Casida, L. E. (1948). The effects of progesterone in altering the estrous cycle of the cow. *J. Anim. Sci.* **7**, 540 (Abstract).

Colas, G., Thimonier, J., Courot, M. and Ortavant, R. (1973). Fertility, prolificacy and fecundity during the breeding season of ewes artificially inseminated after treatment with fluorogestone acetate. *Annls Zootech.* **22**, 441–451.

Cooper, M. J. (1978). Ovulation control in the cow. *In* "Control of Ovulation" (D. B. Crighton, G. R. Foxcroft, N. B. Haynes and G. E. Lamming, eds), pp. 413–420. Butterworths, London and Boston.

Cooper, M. J. and Furr, B. J. A. (1974). The role of prostaglandins in animal breeding. *Vet. Rec.* **94**, 161 (Abstract).

Cooper, M. J. and Rowson, L. E. A. (1975). Control of the oestrous cycle in Friesian heifers with ICI 80 996. *Annls Biol. anim. Biochim. Biophys.* **15**, 427–436.

Curl, S. E., Durfey, W., Patterson, R. and Zinn, D. W. (1968). Synchronisation of estrus in cattle with subcutaneous implants. *J. Anim. Sci.* **27**, 1189 (Abstract).

Dawson, F. L. M. (1961). Corpus luteum enucleation in the cow: therapeutic and traumatic effects. *Vet. Rec.* **73**, 661–669.

Diehl, J. R. and Day, B. N. (1974). Effect of prostaglandin $F_{2\alpha}$ on luteal function in swine. *J. Anim. Sci.* **39**, 392–396.

Dobson, H., Cooper, M. J. and Furr, B. J. A. (1975). Synchronisation of oestrus with ICI 79 939, an analogue of $PGF_{2\alpha}$, and associated changes in plasma progesterone, oestradiol-17β and LH in heifers. *J. Reprod. Fert.* **42**, 141–144.

Douglas, R. H. and Ginther, O. J. (1972). Effect of prostaglandin $F_{2\alpha}$ on length of diestrus in mares. *Prostaglandins* **2**, 265–268.

Douglas, R. H. and Ginther, O. J. (1973). Luteolysis following a single injection of prostaglandin $F_{2\alpha}$ in sheep. *J. Anim. Sci.* **37**, 990–993.

Du Mesnil du Buisson, F. (1954). Possibilité d'ovulation et de fécondation chez la truie avant la puberté. *Annls Endocr. (Paris)* **15**, 333–340.

Dutt, R. H. and Casida, L. E. (1948). Alteration of the estrual cycle in sheep by use of progesterone and its effect upon subsequent ovulation and fertility. *Endocrinology* **43**, 208–217.

Dziuk, P. J. (1962). Control of oestrus and ovulation in farm animals with the aid of orally active progestins. *J. Endocrin.* **24**, xxi (Abstract).

Dziuk, P. J. and Cook, B. (1966). Passage of steroids through silicone rubber. *Endocrinology* **78**, 208–211.

Dziuk, P. J. and Gehlbach, G. D. (1966). Induction of ovulation and fertilisation in the immature gilt. *J. Anim. Sci.* **25**, 410–413.

Dziuk, P. J. and Polge, C. (1962). Fertility in swine after induced ovulation. *J. Reprod. Fert.* **4**, 207–208.

Dziuk, P. J. and Polge, C. (1965). Fertility in gilts following induced ovulation. *Vet. Rec.* **77**, 236–239.

Dziuk, P. J., Hinds, F. C., Mansfield, M. E. and Baker, R. D. (1964). Follicle growth and control of ovulation in the ewe following treatment with 6-methyl-17-acetoxyprogesterone. *J. Anim. Sci.* **23**, 787–790.

Dziuk, P. J., Cmarik, G. and Greathouse, T. (1966). Estrus control in cows by an implanted progestogen. *J. Anim. Sci.* **25**, 1266 (Abstract).

Dziuk, P. J., Cook, B., Niswender, G. D., Kaltenbach, C. C. and Doane, B. B. (1968). Inhibition and control of estrus and ovulation in ewes with a subcutaneous implant of silicone rubber impregnated with a progestogen. *Amer. J. vet. Res.* **29**, 2415–2417.

Dziuk, P. J., Lewis, J. M., Graham, E. F. and Moyer, R. H. (1972). Comparison between natural service and artificial insemination with fresh or frozen sperm at an appointed time in the ewe. *J. Anim. Sci.* **35**, 572–575.

First, N. L., Stratman, F. W., Rigor, E. M. and Casida, L. E. (1963). Factors affecting ovulation and follicular cyst formation in sows and gilts fed 6-methyl-17-acetoxyprogesterone. *J. Anim. Sci.* **22**, 66–71.

Garbers, D. L. and First, N. L. (1969). The effects of injected oestradiol-17β, progesterone and dietary ICI 33828 on ovarian and pituitary functions in the sow and gilt. *J. Reprod. Fert.* **20**, 451–464.

Gerritts, R. J. and Johnson, L. A. (1964). The effect of an orally administered non-steroid on estrus, ovulation and fertility in gilts. *Proc. 5th Int. Congr. Anim. Rep. Artif. Insem., Trento,* **3**, 455–459.

Gordon, I. (1975). Hormonal control of reproduction in sheep. *Proc. Brit. Soc. Anim. Prod.* **4**, 79–93.

Guthrie, H. D. and Polge, C. (1976a). Luteal function and oestrus in gilts treated with a synthetic analogue of prostaglandin $F_{2\alpha}$ (ICI 79 939) at various times during the oestrous cycle. *J. Reprod. Fert.* **48**, 423–425.

Guthrie, H. D. and Polge, C. (1976b). Control of oestrus and fertility in gilts with accessory corpora lutea by prostaglandin analogues, ICI 79 939 and ICI 80 996. *J. Reprod. Fert.* **48**, 427–430.

Guthrie, H. D. and Polge, C. (1978). Treatment of pregnant gilts with a prostaglandin analogue, Cloprostenol, to control oestrus and fertility. *J. Reprod. Fert.* **52**, 271–273.

Hansel, W. (1967). Control of the ovarian cycle in cattle. *In* "Reproduction in the Female Mammal" (G. E. Lamming and E. C. Amoroso, eds), pp. 419–455. Butterworths, London and Boston.

Hansel, W. and Fortune, J. E. (1978). The applications of ovulation control. *In* "Control of Ovulation" (D. B. Crighton, G. R. Foxcroft, N. B. Haynes and G. E. Lamming, eds), pp. 237–263. Butterworths, London and Boston.

Hansel, W. and Malven, P. V. (1960). Estrous cycle regulation in beef cattle by orally active progestational agents. *J. Anim. Sci.* **19**, 1324 (Abstract).

Hansel, W. and Schechter, R. J. (1972). Biotechnical procedures for control of the estrous cycles of domestic animals. *Proc. 7th Int. Congr. Anim. Reprod. and Artif. Insem., Munich,* **1**, 78–96.

Haresign, W. (1978). Ovulation control in the sheep. *In* "Control of Ovulation" (D. B. Crighton, G. R. Foxcroft, N. B. Haynes and G. E. Lamming, eds), pp. 435–451. Butterworths, London and Boston.

Hawk, H. W. (1973). Uterine motility and sperm transport in the estrous ewe after prostaglandin induced regression of corpora lutea. *J. Anim. Sci.* **37**, 1380–1385.

Hawk, H. W. and Echternkamp, S. E. (1973). Uterine contractions in the ewe during progestagen-regulated oestrus. *J. Reprod. Fert.* **34**, 347–349.

Hearnshaw, H., Restall, B. J., Nancarrow, C. D. and Mattner, P. E. (1974). Synchronisation of oestrus in cattle, sheep and goats using a prostaglandin analogue. *Proc. Aust. Soc. Anim. Prod.* **10**, 242–245.

Hinds, F. C., Dziuk, P. J. and Lewis, J. M. (1964). Control of estrus and lambing performance in cycling ewes fed 6-methyl-17-acetoxyprogesterone. *J. Anim. Sci.* **23**, 782–786.

Hunter, R. H. F. (1976). Reproductive physiology in the bovine female: a review. *In* "Beef Cattle Production in Developing Countries" (A. J. Smith, ed.), pp. 79–104. University Press, Edinburgh.

Jakobsen, K. F. and Teige, J. (1956). The effect of enucleation of the corpus luteum in dairy cattle. *Royal Vet. and Agric. Coll., Copenhagen Res. Report* **291**.

King, G. J. (1969). Deformities in piglets following administration of methallibure during specific stages of gestation. *J. Reprod. Fert.* **20**, 551–553.

Lamming, G. E., Hafs, H. D. and Manns, J. G. (1975). Hormonal control of reproduction in cattle. *Proc. Brit. Soc. Anim. Prod.* **4**, 71–78.

Lamond, D. R. (1964). Synchronisation of ovarian cycles in sheep and cattle. *Anim. Breed. Abstr.* **32**, 269–285.

Lamond, D. R. and Bindon, B. M. (1962). Oestrus, ovulation and fertility following suppression of ovarian cycles in Merino ewes by progesterone. *J. Reprod. Fert.* **4**, 57–66.

Lauderdale, J. W. (1972). Effects of $PGF_{2\alpha}$ on pregnancy and estrous cycles of cattle. *J. Anim. Sci.* **35**, 246 (Abstract).

Lauderdale, J. W. (1975). The use of prostaglandins in cattle. *Annls Biol. anim. Biochim. Biophys.* **15**, 419–425.

Louis, T. M., Hafs, H. D. and Morrow, D. A. (1972). Estrus and ovulation after uterine $PGF_{2\alpha}$ in cows. *J. Anim. Sci.* **35**, 247–248.

McCracken, J. A. (1963). Plasma progesterone concentration after removal of the corpus luteum in the cow. *Nature (London)* **198**, 507–508.

Manns, J. G. and Hafs, H. D. (1976). Controlled breeding in cattle: a review. *Can. J. Anim. Sci.* **56**, 121–130.

Mauer, R. E., Webel, S. K. and Brown, M. D. (1975). Ovulation control in cattle with progesterone intravaginal device (PRID) and gonadotrophin releasing hormone (GnRH). *Annls Biol. anim. Biochim. Biophys.* **15**, 291–296.

Mauléon, P. (1974). New trends in the control of reproduction in the bovine. *Livestock Prod. Sci.* **1**, 117–131.

Mauléon, P. and Rey, J. (1966). Effect of fluorogestone acetate absorbed by the vaginal route on oestrus and ovulation in cattle. *Proc. 2nd Int. Congr. Hormonal Steroids, Milan*, p. 348. Excerpta Medica, Amsterdam.

Moeljono, M. P. E., Bazer, F. W. and Thatcher, W. W. (1976). A study of prostaglandin $F_{2\alpha}$ as the luteolysin in swine. I. Effect of prostaglandin $F_{2\alpha}$ in hysterectomised gilts. *Prostaglandins* **11**, 737–743.

Nellor, J. E. (1960). Control of estrus and ovulation in gilts by orally effective progestational compounds. *J. Anim. Sci.* **19**, 412–420.

Polge, C. (1964). Synchronisation of oestrus in pigs by oral administration of compound ICI 33 828. *Proc. 5th Int. Congr. Anim. Rep. Artif. Insem., Trento* **2**, 388–393.

Polge, C. (1965). Effective synchronisation of oestrus in pigs after treatment with ICI compound 33 828. *Vet. Rec.* **77**, 232–236.

Polge, C. (1972). Special problems in the control of the cycle in pigs. *Proc. 7th Int. Congr. Anim. Reprod. Artif. Insem., Munich* **1**, 125–130.

Polge, C. and Day, B. N. (1969). Induction of estrus and ovulation in swine during pituitary suppression with methallibure. *J. Anim. Sci.* **28**, 73–75.

Polge, C., Day, B. N. and Groves, T. W. (1968). Synchronisation of ovulation and artificial insemination in pigs. *Vet. Rec.* **83**, 136–142.

Quinlivan, T. D. and Robinson, T. J. (1969). Number of spermatozoa in the genital tract after artificial insemination of progestagen-treated ewes. *J. Reprod. Fert.* **19**, 73–86.

Robinson, T. J. (1956). The artificial insemination of the Merino sheep following the synchronisation of oestrus and ovulation by progesterone injected alone and with pregnant mare serum gonadotrophin (PMS). *Aust. J. agric. Res.* **7**, 194–210.

Robinson, T. J. (1965). Use of progestagen-impregnated sponges inserted intravaginally or subcutaneously for the control of the oestrous cycle in the sheep. *Nature (London)* **206**, 39–41.

Robinson, T. J. (1967). "The Control of the Ovarian Cycle in Sheep." University Press, Sydney.

Robinson, T. J. (1975). The colloquium in perspective. *Annls Biol. anim. Biochim. Biophys.* **15**, 481–492.

Robinson, T. J., Quinlivan, T. D. and Baxter, C. (1968). The relationship between dose of progestagen and method of preparation of intravaginal sponges on their effectiveness for the control of ovulation in the ewe. *J. Reprod. Fert.* **17**, 471–483.

Roche, J. F. (1974a). Effect of short-term progesterone treatment on oestrous response and fertility in heifers. *J. Reprod. Fert.* **40**, 433–440.

Roche, J. F. (1974b). Synchronisation of oestrus and fertility following artificial insemination in heifers given prostaglandin $F_{2\alpha}$. *J. Reprod. Fert.* **37**, 135–138.

Roche, J. F. (1975). Control of time of ovulation in heifers treated with progesterone and gonadotrophin-releasing hormone. *J. Reprod. Fert.* **43**, 471–477.

Roche, J. F. (1976a). Retention rate in cows and heifers of intravaginal silastic coils impregnated with progesterone. *J. Reprod. Fert.* **46**, 253–255.

Roche, J. F. (1976b). Synchronisation of oestrus in cattle. *World Rev. Anim. Prod.* **12**, 79–88.

Roche, J. F. (1976c). Fertility in cows after treatment with a prostaglandin analogue with or without progesterone. *J. Reprod. Fert.* **46**, 341–345.

Roche, J. F. (1978). Control of oestrus in cattle using progesterone coils. *Anim. Reprod. Sci.* **1**, 145–154.

Roche, J. F. and Crowley, J. P. (1972). Control of oestrus and ovulation in heifers. *Proc. 7th Int. Congr. Anim. Reprod. and Artif. Insem., Munich,* **2**, 859–863.

Roche, J. F. and Crowley, J. P. (1973). The fertility of heifers inseminated at predetermined intervals following treatment with MGA and HCG to control ovulation. *J. Reprod. Fert.* **35**, 211–216.

Rowson, L. E. A., Tervit, R. and Brand, A. (1972). The use of prostaglandins for synchronisation of oestrus in cattle. *J. Reprod. Fert.* **29**, 145 (Abstract).

Schilling, E. and Cerne, F. (1972). Induction and synchronisation of oestrus in prepuberal gilts and anoestrous sows by a PMS/HCG compound. *Vet. Rec.* **91**, 471–474.

Shimizu, H., Toyoda, Y., Takeuchi, S., Kawai, T. and Adachi, S. (1967). Synchronisation of oestrus and subsequent fertility of beef cattle following the intravaginal administration of gestagen. *J. Reprod. Fert.* **13**, 555–558.

Smith, L. W. and Zimbelman, R. G. (1968a). Control of ovulation in cattle with melengestrol acetate. III. Inducing ovulation during MGA treatment. *J. Reprod. Fert.* **16**, 73–79.

Smith, L. W. and Zimbelman, R. G. (1968b). Control of ovulation in cattle with melengestrol acetate. IV. The effects of inducing ovulation with oestradiol cypionate on conception rate. *J. Reprod. Fert.* **16**, 81–89.

Sreenan, J. M. (1975). Effect of long- and short-term intravaginal progestagen treatments on synchronisation of oestrus and fertility in heifers. *J. Reprod. Fert.* **45**, 479–485.

Sreenan, J. M. and Mulvehill, P. (1975). The application of long- and short-term progestagen treatments for oestrous cycle control in heifers. *J. Reprod. Fert.* **45**, 367–369.

Tervit, H. R., Rowson, L. E. A. and Brand, A. (1973). Synchronisation of oestrus in cattle using a prostaglandin $F_{2\alpha}$ analogue (ICI 79 939). *J. Reprod. Fert.* **34**, 179–181.

Thimonier, J., Chupin, D. and Pelot, J. (1975). Synchronisation of oestrus in heifers and cyclic cows with progestagens and prostaglandin analogues alone or in combination. *Annls Biol. anim. Biochim. Biophys.* **15**, 437–449.

Trimberger, G. W. and Hansel, W. (1955). Conception rate and ovarian function following estrus control by progesterone injections in dairy cattle. *J. Anim. Sci.* **14**, 224–232.

Trounson, A. O., Willadsen, S. M. and Moor, R. M. (1976). Effect of prostaglandin analogue Cloprostenol on oestrus, ovulation and embryonic viability in sheep. *J. agric. Sci.* **86**, 609–611.

Ulberg, L. C., Christian, R. E. and Casida, L. E. (1951). Ovarian response in heifers to progesterone injections. *J. Anim. Sci.* **10**, 752–759.

Ulberg, L. C., Grummer, R. H. and Casida, L. E. (1951). The effects of progesterone upon ovarian function in gilts. *J. Anim. Sci.* **10**, 665–671.

Webel, S. K. (1976). Control of the estrous cycle in cattle with a progesterone releasing intravaginal device. *Proc. 8th Int. Congr. Anim. Reprod. Artif. Insem., Krakow*, **3**, 521–523.

Webel, S. K. (1978). Ovulation control in the pig. *In* "Control of Ovulation" (D. B. Crighton, G. R. Foxcroft, N. B. Haynes and G. E. Lamming, eds), pp. 421–434. Butterworths, London and Boston.

Wiltbank, J. N. and Gonzalez-Padilla, E. (1975). Synchronisation and induction of estrus in heifers with a progestagen and estrogen. *Annls Biol. anim. Biochim. Biophys.* **15**, 255–262.

Wiltbank, J. N. and Kasson, C. W. (1968). Synchronisation of estrus in cattle with an oral progestational agent and an injection of estrogen. *J. Anim. Sci.* **27**, 113–116.

Wishart, D. F. (1967). Synchronisation of oestrus in sheep: the use of pessaries. *Vet. Rec.* **81**, 276–287.

Wishart, D. F. and Hoskin, B. D. (1968). Synchronisation of oestrus in heifers using intra-vaginal pessaries impregnated with SC-9880 and PMSG. *J. Reprod. Fert.* **17**, 285–289.

Wishart, D. F. and Young, I. M. (1974). Artificial insemination of progestin (SC 21 009)-treated cattle at predetermined times. *Vet. Rec.* **95**, 503–508.

Zimbelman, R. G. and Smith, L. W. (1966). Control of ovulation in cattle with melengestrol acetate. *J. Reprod. Fert.* **11**, 185–191.

Chapter III

Control of Ovulation and Induction of Superovulation

Introduction

Of the two principal means of modifying the ovarian cycle of mammals—terminating prematurely the lifespan of the corpora lutea or stimulating maturation of Graafian follicles—procedures of ovulation control have undoubtedly excited the greater interest down the years in both the veterinary and animal science communities. A feature of the contemporary research scene however, is, that these two aspects are now tending to be regulated almost simultaneously, since a means of controlled breeding may require shortening of the oestrous cycle and predetermination of the moment of ovulation.

If the time of ovulation could be controlled precisely, this would clearly obviate the need for frequent detection of oestrus as a means of estimating the time of ovulation, and would thereby permit insemination at the correct time in order to avoid the problems of fertility associated with ageing of the gametes (see Chapter V). Moreover, the objective of increasing prolificacy by stimulating ovulation of more than the usual number of follicles (superovulation), especially in cattle and sheep, has attracted much research effort during the last forty years and procedures of superovulation involving injection of pituitary or placental gonadotrophins have been described in detail for the major farm species. It must be noted at the outset, however, that whilst considerable success has been achieved in controlling the time of ovulation, only very moderate success can be reported for attempts to enhance with any precision the number of follicles ovulating in individual animals. The latter objective is still worthwhile, for there exists ample evidence from embryo transplantation experiments that the number of eggs shed (i.e. ovulation rate) rather than uterine capacity may be the primary limitation to fecundity in several breeds of cattle and sheep: twins can be produced in cattle by embryo transfer (Rowson *et al.*, 1971) and an increased litter size in sheep by the same means (Lawson and Rowson, 1972).

Hormone preparations for regulating ovulation

Shortly after the discovery of gonadotrophic hormones was published in 1927 (Aschheim and Zondek; Smith and Engle), much of the work aimed at stimulating follicular growth and/or controlling ovulation in domestic animals used preparations of pituitary gonadotrophins, usually derived from acetone extracts of desiccated pituitary tissue resuspended in a physiological solution at the time of injection. However, most treatments during the last twenty years or so have used preparations of placental gonadotrophins, especially pregnant mare's serum gonadotrophin (PMSG) which is rich in FSH activity and human chorionic gonadotrophin (HCG) which is rich in LH activity. Both gonadotrophins are now known to be of embryonic origin: PMSG reaches its highest titres in the blood between approximately the 60th–90th days of gestation (Fig. III.1; Cole and Hart, 1930; Day and

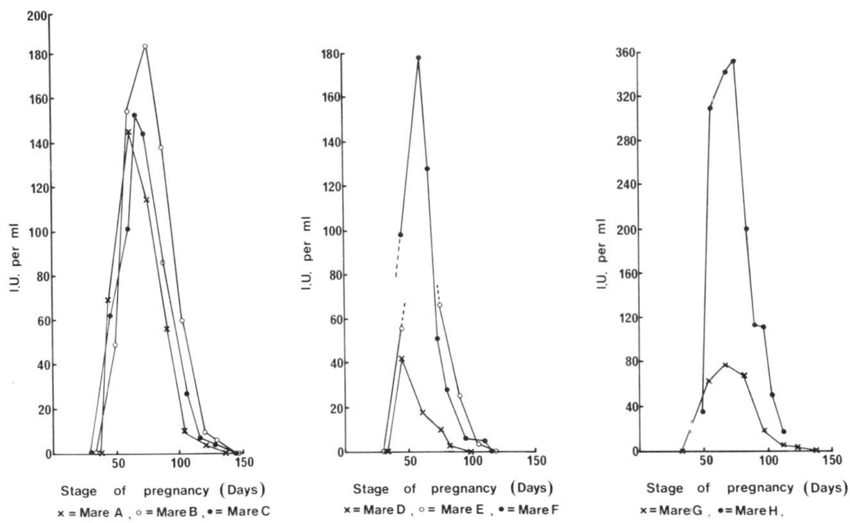

Fig. III.1. Changes in the concentration of pregnant mare serum gonadotrophin (PMSG) in the blood of a series of animals during the first five months of gestation. (Redrawn from Day and Rowlands, 1940.)

Rowlands, 1940; Allen, 1969), and is thought to induce formation of accessory corpora lutea or luteinised follicles necessary for the maintenance of pregnancy. Paradoxically, injection of horses with most preparations of PMSG will not provoke follicular growth or ovulation.

PMSG is more slowly inactivated in peripheral blood than preparations of pituitary gonadotrophins, having a half life of 40–125 hours, but this range conceals two components with different clearance rates (Menzer and Schams, 1979).

The precise function of HCG elaborated by the trophoblast of the very young human embryo and detectable in the urine as soon as eight or nine days after ovulation is uncertain, but undoubtedly it is associated with prolongation of the lifespan of the corpus luteum and therefore with the maintenance of pregnancy. Another fluid source of gonadotrophin, to date used more in experiments with laboratory rodents than farm animals, is human menopausal gonadotrophin (HMG). Prepared from urine, this hormone is rich in FSH activity derived from the pituitary of post-menopausal women due to the absence of significant oestrogenic feedback, the ovary being essentially depleted by this time of mature Graafian follicles.

The hormone-rich fraction in all these materials is more readily available than pituitary tissue, being prepared from blood serum or urine at an appropriate reproductive phase, characterised with respect to biological activity, and stored in lyophilised form, only being reconstituted at the time of use. Although these hormones are of human and equine sources, they continue to be used extensively in cattle, pig and sheep experimentation and in diverse commercial situations. At the time of writing, shortage of source material rather than their strict suitability for inter-species treatments would appear to be the principal reason why their usage may diminish in the future, despite the reservations expressed below when considering the use of PMSG to enhance ovarian responses.

Mechanism of ovulation

Despite a vast body of relevant information (reviewed by Lipner, 1973), the precise mechanism of ovulation in response to endogenous or injected gonadotrophins still remains ill-defined. If the mechanism(s) were more clearly understood, it is likely that attempts to regulate the process of ovulation in farm livestock and in humans could be improved; it therefore seems a topic especially worthy of intensive investigation.

Binding of gonadotrophic hormone to the inner cell layers of the preovulatory Graafian follicle can be demonstrated with radio-labelled tracer preparations (Rajaniemi and Vanha-Perttula, 1972), followed by increased capillary permeability and other local changes in the wall of the follicle—most notably in the structure of the granulosa and basement membrane, but also in the distribution and concentration of proteolytic activity (Espey, 1974; Espey and Coons, 1976). The specific means of breakdown of the wall and release of the oocyte through an aperture in the apical region of the follicle has yet to be determined, although enzymatic dissociation of collagenous tissue has been invoked (Espey, 1976). A morphological sequence of events has been described for a number of mammals, the essential stages in the last hours before ovulation involving:

(1) A final phase of follicular growth, with enhanced accumulation of antral fluid and achievement of the maximum follicular diameter.

(2) A reduction in intra-follicular pressure, the follicle(s) becoming flaccid and pendulous. Ovulation is not explosive in nature, at least not in cows, sheep, pigs and horses, and therefore the term "rupture" is hardly applicable. "Collapse" of the follicle would seem more descriptive of the phenomenon in these species.

(3) Modifications to the vascular network in the follicle wall, with a progressive degree of capillary (petechial) haemorrhage observed over much of the wall (Fig. III.2), whereas a locally defined avascular and translucent region is usually formed towards the apex of the follicle. This change in the vascular bed of the follicle is closely coincident with capillary invasion of the granulosa layer after breakdown of the basement membrane, and is promoted by the preovulatory surge of pituitary gonadotrophins.

(4) A thinning followed by progressive degradation of the apical region of the follicle giving an aperture (Fig. III.3), with or without prior formation of a stigma, followed by rapid release of the non-viscous fluid and then an oozing of the viscous portion of follicular fluid together with the cumulus mass enclosing the egg. Smooth muscle elements in the thecal layer of the follicle wall may contribute towards this phase of expulsion (see Guttmacher and Guttmacher, 1921).

Recent studies have strongly suggested a rôle for autonomic nerves mediating contractility in human and rat Graafian follicles (Owman *et al.*, 1975; Walles *et al.*, 1978) and the contractile proteins, actin and myosin, have also been demonstrated by immunofluorescent techniques in cells of the theca externa (Amsterdam *et al.*, 1977; Walles *et al.*, 1978). Since there is rarely a uniform spatial relationship between the location of the follicular oocyte and the eventual region of its liberation through an aperture in the wall, contractions in the collapsing follicle could function to deliver the oocyte within its cumulus investment to the surface of the ovary and also assist the breaking of cellular connections with the granulosa layer.

At the endocrine level, increased binding of LH to somatic cells within the follicle, especially to the theca interna and mural granulosa (Lindner *et al.*, 1977), occurs after the gonadotrophin surge, such binding being promoted by FSH and probably by a local action of oestrogens (Hunter *et al.*, 1976). The gonadotrophin surge also promotes a gradual change in the preovulatory follicle from predominantly oestrogen secretion by the thecal layer to progesterone secretion by the granulosa cells, even though elements of the theca interna may remain biosynthetically active up to and after the time of ovulation (see Short, 1964; Baird *et al.*, 1975). It therefore appears that steroid, particularly progesterone, production by the follicle itself may be involved in modulating the final enzymatic changes in the follicular wall. Whether the polypeptide hormone, relaxin, contributes to the terminal changes in the follicle has yet to receive systematic examination, but this certainly seems possible, not least since the follicle is already exhibiting some of the characteristics of a corpus luteum. An increase in the concentration of follicular prostaglandins in the fluid compartment has been reported in the preovulatory interval (Le Maire *et al.*, 1973, 1975; Ainsworth *et al.*, 1975; Bauminger and Lindner, 1975), and may facilitate changes both in biochemical pathways and in contractile activity of ovarian tissues. Moreover, treatment with inhibitors of prostaglandins such as indomethacin in late pro-oestrus has been found to block ovulation in rats (Armstrong and Grinwich, 1972; Tsafriri *et al.*, 1972) and rabbits (O'Grady *et al.*, 1972), thereby suggesting a critical rôle for these compounds in the events of follicular breakdown, as has been demonstrated in fish (Jalabert and Szollosi, 1975). The viscosity of the follicular

---

Fig. III.2. The sequence of changes in the wall of mature Graafian follicles during the last hours before ovulation in the pig. Note the extent to which these 9–10 mm diameter follicles protrude from the body of the ovaries, and the progressive breakdown in the thecal vasculature (a, b, c) just before collapse of each follicle at ovulation (d).

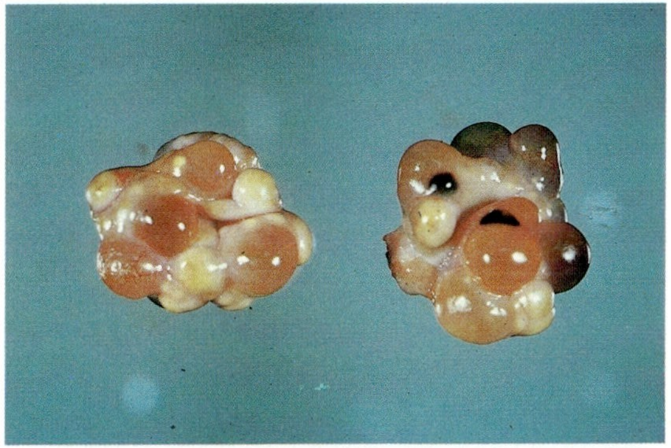

(a)

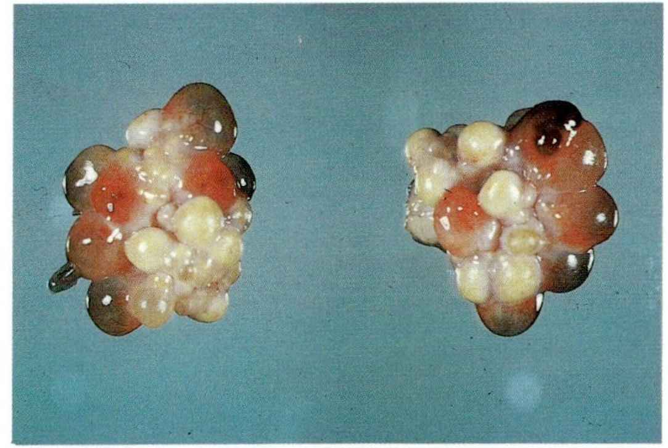

(b)

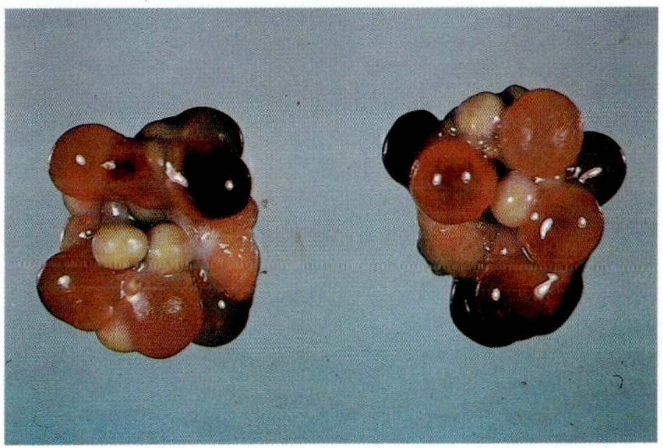

(c)

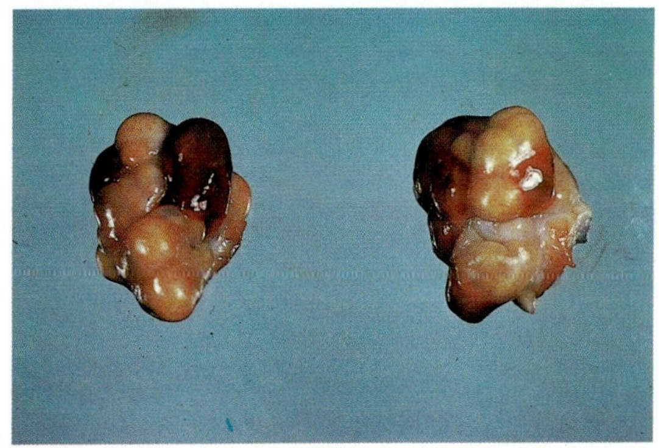

(d)

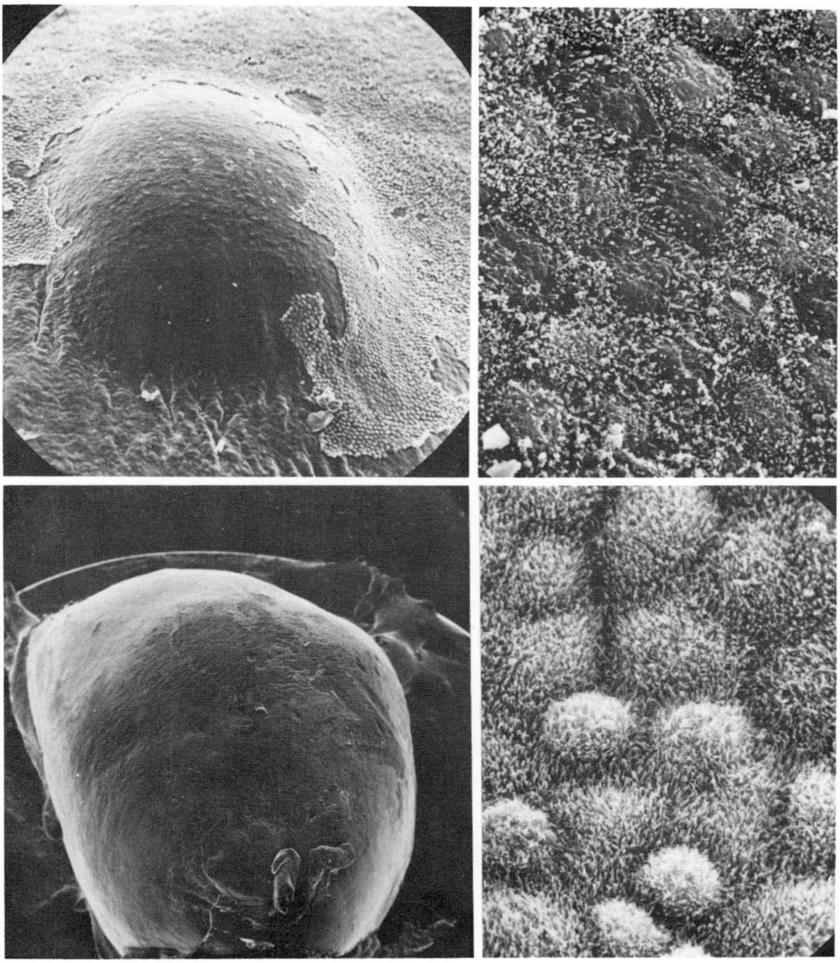

Fig. III.3. Scanning electron micrographs to illustrate some of the preovulatory changes in the wall of Graafian follicles of pigs in which the time of ovulation was controlled by an injection of HCG. Note the desquamation of the follicle wall. (Unpublished studies of J. E. Fléchon and RHFH.)

fluid is likewise changing rapidly in the final hours before ovulation, though whether this is related to dissolution of the granulosa or more to the coagulation potential of the fluid, or indeed to other functional requirements, is not yet clear.

During these alterations in the somatic and fluid tissues of the follicle, the oocyte has resumed its preovulatory maturation. Nuclear changes during this phase of maturation are the most conspicuous and extensively documented (Pincus and Enzmann, 1935; Chang, 1955; Edwards, 1962, 1965; Dziuk, 1965; Hunter and Polge, 1966), but cytoplasmic and membranous components of the maturation process have also been described (Thibault and Gérard, 1970, 1973; Lindner *et al.*, 1974; Hunter *et al.*, 1976; Moor and Trounson, 1977; Thibault, 1977; Moor, 1978). Together, these changes result in the shedding of a viable secondary oocyte with the chromosomes arranged at metaphase and an extruded first polar body (Fig. I.4).

Treatments for controlling the time of ovulation

As noted in Chapter I, the time of ovulation during oestrous and menstrual cycles is regulated by a surge or peak of gonadotrophic hormones, predominantly LH but also FSH, which is provoked by the positive feedback action of oestrogens secreted by the maturing Graafian follicle(s). The occurrence of this peak in cattle, sheep and pigs is closely related to the onset of behavioural oestrus, which is likewise determined by an oestrogen threshold, and the interval then elapsing until ovulation is remarkably constant for a given species (Table I.3). If the moment of onset of behavioural oestrus could be detected with precision, it follows that the time of ovulation could be estimated with accuracy. Unfortunately, however, it is rarely practicable in commercial agricultural systems to be testing for oestrus with intact or vasectomised males sufficiently frequently for this state of affairs to be achieved.

Treatments for controlling the time of ovulation therefore aim to mimic the endogenous surge of gonadotrophin, and usually consist of a single injection of LH in the form of HCG or less frequently as a pituitary fraction rich in LH activity. The injection is administered intravenously or intramuscularly shortly *before* the onset of behavioural oestrus when a mature follicle(s) is present. The important point is that the injection of gonadotrophin must *precede* the endogenous surge if ovulation time is to be controlled by the treatment, and must therefore be given during pro-oestrus. On the other hand, the LH or HCG injection must not be given too soon, for if insufficient follicular development has occurred the animal (a) may not come into behavioural oestrus, (b) may shed primary oocytes rather than mature eggs (see Polge and Dziuk, 1965; Hunter *et al.*, 1976), or (c) may not ovulate at all although luteinisation of the follicle may be initiated.

The stage of pro-oestrus can be judged most readily in pigs on the basis of the degree of vulval swelling together with behavioural symptoms, but in cattle and sheep such a judgement is more difficult. In the latter two species, vulval swelling and more particularly the abundance and viscosity of mucus oozing from the vagina have to be considered in conjunction with behavioural observations. More satisfactory by far, however, in situations in which control of the time of ovulation is required would be first to have regulated the oestrous cycle (Chapter II). The injection of a gonadotrophic preparation to control ovulation is then imposed as a sequel to treatments for regulating follicular development (see Polge *et al.*, 1968). As examples of the doses of gonadotrophin that have proved reliable for inducing ovulation of mature follicles, a single injection of 250–500 iu HCG is effective in sheep, 500 iu HCG in pigs and 2000 iu LH or HCG in cattle (Table III.1). The predicted time of ovulation after such treatment corresponds quite closely with the information in Table I.3; general responses are shown in Table III.2, with some further analysis of the response to HCG in a polyovular species in Table III.3.

Table III.1. Examples of some doses of PMSG and HCG commonly administered to induce superovulation and /or control the time of ovulation.

Species	Dose of PMSG[1] injected subcutaneously (or intramuscularly) i.u.	Dose of HCG injected intramuscularly (or intravenously) i.u.
Cow	2000–3000	500–2000
Sheep	500–800	250–500
Pig	750–1500	500–1000

[1] PMSG is administered as a single dose. Because of its relatively long half-life, there appears to be little advantage in dividing the dose into several injections.

Superovulatory responses to gonadotrophic hormones in mature animals

Numerous reviews on the subject of controlling ovulation are available for cattle (Hammond and Bhattacharya, 1944; Dowling, 1949; Hammond, 1949; Rowson, 1951; Willett, 1953; Gordon *et al.*, 1962; Hafez, 1969; Foote and Onuma, 1970; Ortavant and Thibault, 1970; Dziuk,

Table III.2. Timing of the response to a pro-oestrous injection of human chorionic gonadotrophin (HCG) for the control of ovulation in cattle, sheep and pigs.

Species	Details of treatment	Interval from injection until ovulation Hours	Reference
Cow	500 i.u. HCG after treatment with progestagens	40	Graves and Dziuk, 1968
Sheep	400–2000 i.u. HCG after progesterone and PMSG	20–28	Braden, Lamond and Radford, 1960
	250–500 i.u. HCG after treatment with progestagens	24 ± 2	Dziuk *et al.*, 1964a
	250 or 1000 i.u. HCG during pro-oestrus	26–27	McGovern, Williams and Hancock, 1969
Pig	250–2000 i.u. HCG after oral progestagens	39–40	Dziuk and Baker, 1962
	500 i.u. HCG during pro-oestrus	40	Dziuk *et al.*, 1964b
	500 i.u. HCG to sows after weaning litter	39–41	Buttle and Hancock, 1967
	500 i.u. HCG during pro-oestrus	40–42[1]	Hunter, 1967a
	500 i.u. HCG during pro-oestrus	44–46[2]	Hunter, 1972b

[1] Evidence for breed differences in the timing of response. [2] Evidence of strain or environmental differences in response.

Table III.3. The timing of ovulation in 18 Large White X Essex gilts after an intramuscular injection of 500 i.u. HCG given during pro-oestrus. The duration of ovulation appears to be brief after HCG treatment, but may require several hours under spontaneous conditions. (Adapted from Hunter, 1967a.)

Interval from HCG injection to autopsy (hours)	Number of recent ovulations	Number of large follicles[1] (7–10 mm. diameter)	Percentage ovulation
36·0	0	10	0
36·0	0	7	0
36·0	0	15	0
36·0	0	13	0
40·0	0	16	0
40·5	0	14	0
41·0	0	15	0
41·25	2	14	12·5
41·5	0	15	0
41·5	1	14	6·7
41·5	11	1	91·7
41·75	19	1	95·0
41·75	13	0	100
42·0	6	10	37·5
42·0	7	7	50·0
42·0	11	8	57·9
42·0	15	0	100
42·0	17	0	100

[1] Only follicles with a mean diameter within the range of 7–10 mm were considered destined for ovulation in response to the HCG injection and taken into account in the calculation of percentage ovulation. This was the number of freshly ovulated follicles counted at autopsy expressed as a proportion of the total number of preovulatory and recently-ovulated follicles.

1973; Betteridge, 1977), sheep (Robinson, 1951; Averill, 1958; Gordon, 1958; Dziuk, 1973) and pigs (Phillippo, 1968; Polge, 1969; Dziuk, 1973; Webel, 1978). Even with all this information, reliable methods of obtaining a controlled superovulation using currently available gonadotrophins have been difficult to formulate, due principally to variability in the response to injected hormones. Moreover, increasing the dose of trophic hormone generally increases the variation in ovarian response. In the great majority of experiments in cattle, sheep and pigs in which pituitary or placental hormones have been used, the number of follicles ovulating in response to treatment has been unpredictable for *individual* animals (Table III.4), even though satisfactory curves of dose against *mean ovarian response* can frequently be plotted for groups of animals (Fig. III.4). This stands in contrast to the situation in laboratory species such as hamsters in which an extremely precise response to superovulatory procedures employing PMSG has been reported (Greenwald, 1962). As a partial explanation of this difference, it is probable that the kinetics of follicular maturation are more closely defined during the 4–5 day oestrous cycle of the golden hamster than in the 16–21 day cycles of the larger species.

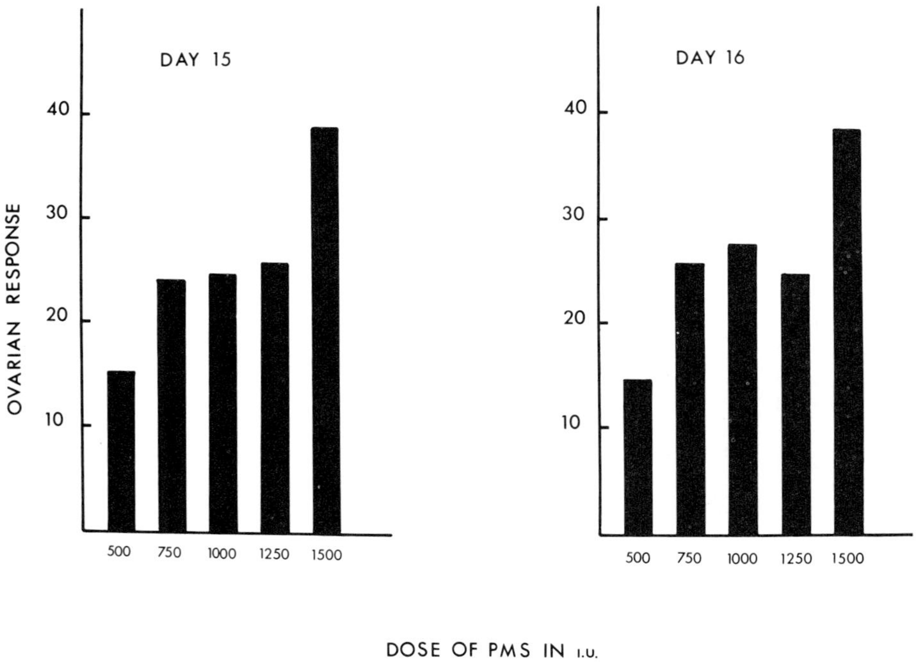

Fig. III.4. Histograms of the mean ovarian response to increasing doses of PMSG injected in the early follicular phase of the oestrous cycle in order to induce superovulation in pigs. A minimum dose of 750 iu PMSG given on Day 15 or 16 seems essential for eliciting a superovulatory response in most animals.

Whilst the timing of gonadotrophin injection in relation to waves of follicular growth is almost certainly critical for initiating an optimal response, a more accurate definition of the biological potency of the hormone preparation would undoubtedly improve the predictability of response. Likewise, attention could be paid to the use of a preparation

Table III.4. To illustrate the marked variation in ovarian response of cows to treatment with a single injection of PMSG given alone or in combination with HCG. Each figure represents the number of ovulations per cow. (Taken from Mauléon *et al.*, 1970a.)

Dose of HCG in i.u.	Dose of PMSG in i.u.																					
	1600						2000						2500					3120				
	1	1	1	1	1		1	6	8	2	6		6	6	10	10	9	6	17	15	1	19
0	13	0	1	1	5		13	9	1	9	16		3	6	7	1	3	8	36	15	1	11
	1	1	1	4	5		1	1	4	1	1	1	1	4	5	10	1	1	9	8	5	1
	5	1	1	1	1		8	1	3	1	1		8	1	18	10	6	9	13	4	16	3
750	1	1	1	1	1		1	1	7	6	2		1	1	1	1	37	13	4	3	1	21
	1	0	1	1	1		12	1	15	2	1		8	1	2	0	1	5	10	7	2	10
	1	1	2	2	2		2	1	1	2	19		1	10	14	13	1	13	24	1	10	15
1500	2	4	7	9	1		2	8	9	8	1		1	5	2	2	1	1	20	6	22	14
	6	1	1	2	4	1	1	1	1	4	4		5	19	1	1	1	18	2	0	1	3
	6	5	1	4	1		19	1	1	1	1		1	6	4	1	7	1	8	6	16	6
3000	1	9	1	1	4		13	1	9	4	18		1	2	11	1	3	1	40	4	3	4
	3	1	2	3	3	1	2	3	5	1	4	1	1	1	9	8	1	7	1	1	7	1

more suitable than PMSG for the stimulation of growth in cattle, sheep or pig ovarian follicles. The disadvantages of using non-specific hormone preparations have been recognised for many years, the essence of the argument being that if species-specific hormones were isolated and in due course synthesised, their employment would reduce the variability of the ovarian response. As of writing, this remains an attractive hypothesis, yet to be put into practice and convincingly tested. There is also the question of the antigenicity of protein hormone preparations, it being occasionally asserted that antibody formation may cause responses to gonadotrophins such as PMSG to diminish after repeated injections (Willett *et al.*, 1953; Jainudeen *et al.*, 1966), although more recent evidence in cattle is to the contrary (Schams *et al.*, 1978; Christie *et al.*, 1979).

The most logical approach to initiating a superovulatory response in mature animals has traditionally been to commence treatment with gonadotrophic hormones in the early follicular phase of the oestrous cycle rather than in the presence of an active corpus luteum (see below). Although, in farm animals, this is very much the shorter of the two phases, there is invariably a population of ripening Graafian follicles of >2–3 mm in diameter in cattle, sheep and pigs that can be brought to maturity under the influence of injected gonadotrophins within a period of four or five days. For, apart from the size and sensitivity of ovarian tissues, the principal factor limiting the number of follicles maturing in normally cyclic animals is the amount of available circulating gonadotrophin, this of course being augmented by the injected material.

In spontaneous oestrous cycles, there is a reduction in the number of "large" follicles ripening during the follicular phase due principally to competition for gonadotrophin, and evidence from studies on attempted superovulation in pigs (Hunter, 1972a) and following unilateral ovariectomy in sheep (Land, 1973) and pigs (Hunter, 1979) suggests that injection of such hormones towards the end of the follicular phase will be too late to increase the number of follicles ovulating; the response to injected PMSG seems to diminish progressively as the follicular phase terminates. However, the population of reserve tertiary follicles available to respond to PMSG administered at the beginning of the follicular phase is considerable; and although under spontaneous conditions a proportion of these may be destined to undergo atresia, there is no reason to suppose that those stimulated by PMSG are going to shed a significant proportion of abnormal oocytes.

Cattle

In cattle with commencement of treatment on Days 16 or 17 of the oestrous cycle, doses of PMSG administered intramuscularly have commonly ranged from 800–2000 iu (Gordon *et al.*, 1962) or 2000–3000 iu (Dowling, 1949; Rowson, 1951; Hafez *et al.*, 1963; Scanlon *et al.*, 1968; see Table III.1). Dividing the total dose of serum gonadotrophin into several injections does not enhance the response (Hafez *et al.*, 1963), which may vary from nil to 55 ovulations (Scanlon *et al.*, 1968). In contrast to the requirement in the presence of an active corpus luteum, injection of an LH preparation is not necessary to ensure ovulation, the follicles ovulating spontaneously in response to a surge of endogenous gonadotrophin. But a single injection of HCG may improve the completeness of the ovulatory response and lead to a more synchronous ovulation in the group of stimulated follicles. Factors influencing the response to PMSG are listed in Table III.5. A

Table III.5. Factors considered to have an important influence on the ovarian (ovulatory) response to systemic injections of PMSG in mature cattle.

Factor	Comment
1. TIMING OF INJECTION IN RELATION TO STAGE OF OESTROUS CYCLE	This will influence the manner of synergism with endogenous gonadotrophic hormones, and the ability of PMSG to interact with waves of follicular growth
2. DOSE OF HORMONE PREPARATION INJECTED	In general, a dose-mean ovarian response relationship can be plotted, but variability in response increases with dose
3. AGE AND BREED OF ANIMAL	Individual animal variation depends in part on the size of the ovaries and sensitivity to exogenous gonadotrophins
4. ENVIRONMENT	The nutritional status of the animal and, to a lesser extent, the season of the year may influence the ovarian response

serious problem may be encountered when the ovulatory response exceeds 9 or 10 ovulations, for the fimbriae may then be unable to encompass the much enlarged ovaries and a proportion of eggs may be lost into the peritoneal cavity. Quite apart from this particular problem, procedures of superovulation can also have a deleterious influence on fertility itself. This situation may be associated with accelerated rates of egg transport in the Fallopian tubes (Rowson, 1951) due to a disturbance in the concentration of circulating oestrogens and progesterone (Booth *et al.*, 1975), and more significantly perhaps due to a protracted period of ovulation with progesterone secretion from the first formed corpus luteum disturbing transport and fertilisation of eggs from later ovulations. Nonetheless, Dowling (1949) reported the recovery of up to 25 apparently normally fertilised eggs from the tubes of treated cattle: with the more stringent criteria of fertilisation and early embryonic development applied these days, these results must be viewed as exceptional.

Apart from variations on the above general theme for stimulating ovulation, such as commencing PMSG injections before or at the end of a progestagen treatment (Mauléon *et al.*, 1970b; Laster, 1973), or before or at the time of manual expression of the corpus luteum, there are only one or two new concepts in this area of work of immediate practical concern. The first involves the use of prostaglandin analogues to induce regression of the corpus luteum at a predetermined stage of the oestrous cycle, so that gonadotrophin treatment can then begin at a precise time with respect to the ensuing follicular phase (Tervit *et al.*, 1973). In this manner, it may be possible to induce superovulation during a wave of active follicular growth or at least before any substantial wave of follicular regression has commenced. Treatment with PMSG between Days 9 and 12 one or two days before induction of luteal regression with $PGF_{2\alpha}$ produces a better response than treatment earlier in the luteal phase (Phillippo and Rowson, 1975; Newcomb and Rowson, 1976; Sreenan and Gosling, 1977; see Fig. III.5). The second approach involves the use of hypothalamic releasing factors, or synthetic gonadotrophin releasing hormones (Schally *et al.*, 1973), to induce secretion of endogenous stores of gonadotrophic hormones from the anterior pituitary. Experiments of this nature involving intracarotid or intramuscular injections of synthetic releasing hormone have been reported by Mauer and Rippel (1972) and Kaltenbach *et al.*, (1974), but a general conclusion based on the evidence to date is that a new form of therapy of practical value in obtaining superovulation is not at hand. The purely experimental approach of local injections of gonadotrophic hormones directly into the ovarian tissue has also failed to yield uniform or useful results (Betteridge, 1974).

As a current perspective on superovulation in cattle, the number of

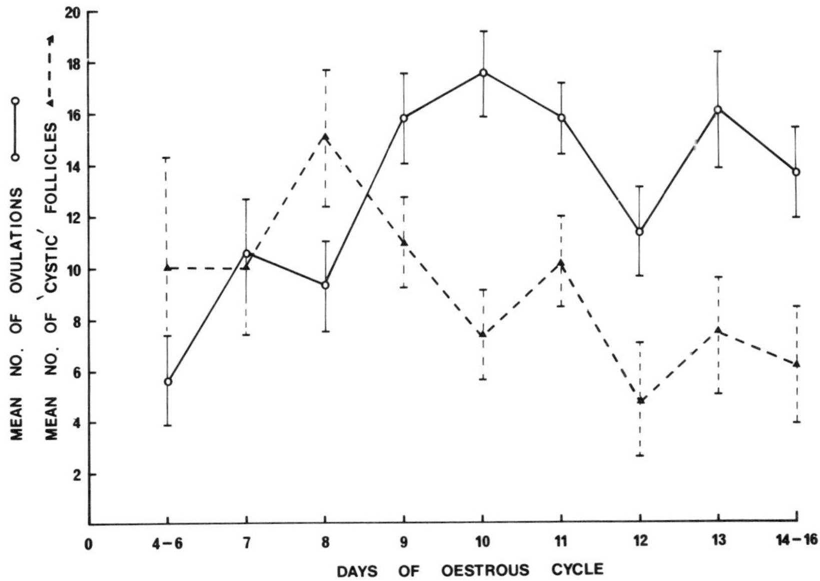

Fig. III.5. Ovarian response of cows to PMSG treatment during the luteal phase of the oestrous cycle one or two days before injection of $PGF_{2\alpha}$ or a synthetic analogue to effect regression of the corpus luteum. (Redrawn from Newcomb and Rowson, 1976.)

calves born to a relatively small number of individual cows, in terms of twins and triplets, has been increased, but the procedure is as yet unsatisfactory as a means of increasing fecundity on a field scale (Gordon *et al.*, 1962; Ortavant and Thibault, 1970; Saumande *et al.*, 1978). Until preparations of hormones or schemes of therapy are available that yield a more precise ovarian response, it is likely that procedures of superovulation in cattle will be used largely to provide eggs and embryos from selected donors for *in vitro* manipulations such as deep-freeze preservation (Chapter VII) or for more immediate transplantation to recipient animals (Chapter VIII). And even with precise control over ovarian response, which might suggest the basis of a scheme for induced twinning in the treated animal, the low incidence of intra-uterine migration of the bovine embryo (Scanlon, 1972) would lead to competition between embryos for the finite number of specialised implantation sites—caruncles—and therefore might diminish the probable outcome of the treatment where more than one ovulation occurred on the same ovary.

Sheep

Procedures of superovulation in sheep have also been widely examined during the last 35 years or so (Hammond *et al.*, 1942; Robinson, 1951; Wallace, 1955; Averill, 1958; Gordon, 1958; Dziuk, 1973), treatment involving injection of horse anterior pituitary (HAP) extracts or PMSG, usually given either in the very late luteal phase of the cycle (Day 12 or 13) or after injections of a solution of progesterone in oil (Hancock and Hovell, 1961; Robinson, 1961) or use of intra-vaginal pessaries containing progestagens. The latter procedures endeavour to render the population of maturing follicles more homogeneous in size before commencing stimulation with exogenous gonadotrophin. PMSG has also been injected during the luteal phase shortly before a luteolytic dose of the prostaglandin analogue Cloprostenol (Trounson *et al.*, 1976). Doses of PMSG have commonly ranged from 500–800 iu, although some treatments have involved as much as 1300 iu (Averill, 1958; Averill and Rowson, 1958; Gordon, 1958; Hancock and Hovell, 1961; Lamond, 1962). Responses have varied from 3 or 4 ovulations to in excess of 50 (Moore and Shelton, 1964), but the mean response may increase from 2·8 ovulations with 700 iu PMSG to 9·1 ovulations with 1300 iu PMSG when given on Day 12 or 13 of the oestrous cycle (Averill, 1958). When a limited degree of superovulation results, it may increase the lambing percentage of less prolific breeds but there seems to be a maternal restriction upon litter size at an average figure of, in general, close to three (Fig. III.6). With very large numbers of ovulations, there is a tendency to total loss after the stage of implantation (Casida *et al.*, 1944). Once again, therefore, the lack of precision of response and lower incidence of fertilisation, together with the overall limitation of uterine capacity, may act to prevent significant increases in flock fecundity after gonadotrophin treatment.

As alternative approaches to increasing the number of eggs ovulated in mature sheep, consideration must be given to nutritional flushing and genetic selection. The former technique involves improving the quantity and composition of feedstuffs available to the ewes in the two or three weeks at the onset of the breeding season that precede mating. The flushing response is usually interpreted as restoring the ovulation rate to a level characteristic of the breed, in situations where ewes have been prevented from becoming overfat during the summer months, rather than elevating it above such a level. Even so, animals subjected to

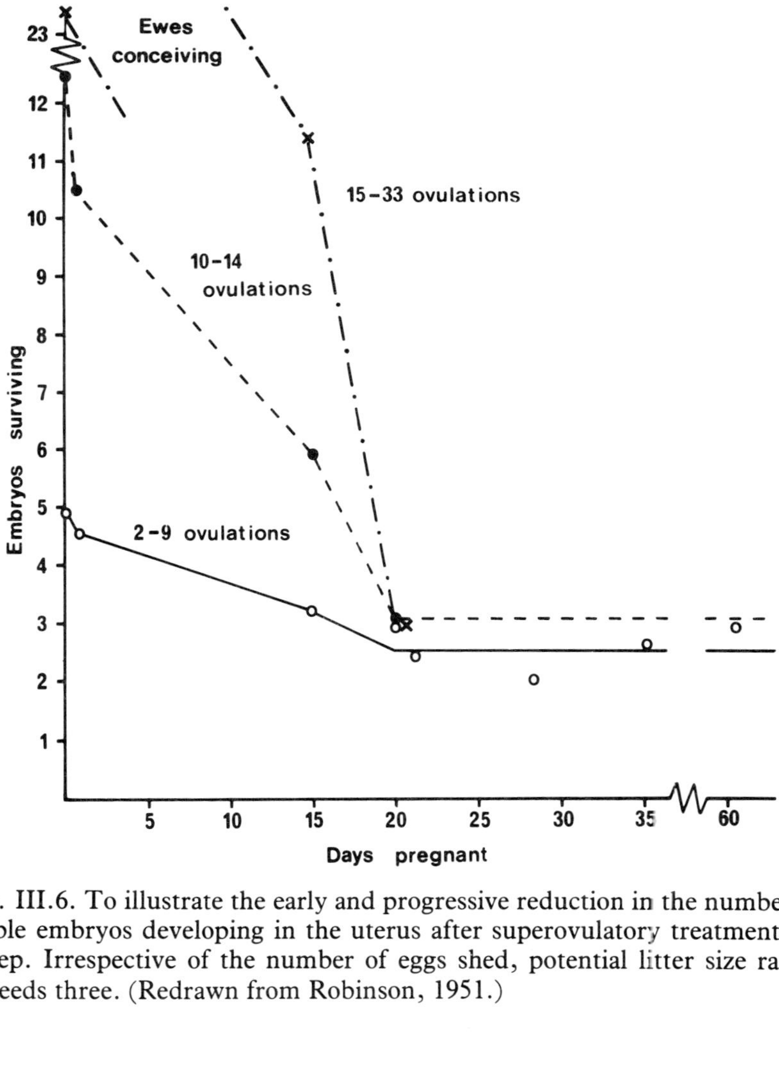

Fig. III.6. To illustrate the early and progressive reduction in the number of viable embryos developing in the uterus after superovulatory treatments in sheep. Irrespective of the number of eggs shed, potential litter size rarely exceeds three. (Redrawn from Robinson, 1951.)

flushing procedures are generally more prolific than untreated control animals. By contrast, the approach of genetic selection for increased reproductive rates is essentially long term in nature, but has been fruitful in Australia when applied to Merino ewes (Turner, 1978).

Pigs

Experiments involving superovulation of mature animals appear to have been somewhat less extensive in pigs, perhaps because of the inherent fecundity of this species, although use of the technique has become fashionable in recent years as a means of analysing components of embryonic mortality (Dziuk, 1968; Pope *et al.*, 1972; Webel and Dziuk, 1974). Nonetheless, dose-response data are now available for a number of breeds, from which it appears that a dose of PMSG in excess of 500 iu injected on Day 15 of the cycle is usually required to induce superovulation: 750–1250 iu PMSG have been the doses most commonly used for this purpose (see Gibson *et al.*, 1963; Hunter, 1964; Longenecker *et al.*, 1965; Phillippo, 1968; Dziuk, 1973, 1977). In animals of the Large White or Landrace breeds, the mean number of ovulations increased from 24 after injection of 750 iu PMSG to 38 after a dose of 1500 iu PMSG (Hunter, 1964; see Table III.6), but the proportion of primary oocytes (i.e. immature or atretic eggs) increased significantly with responses in excess of 25 ovulations (Hunter, 1966). Even with pronounced superovulatory responses, the endogenous surge of gonadotrophins is apparently sufficient to ensure ovulation of mature follicles (Fig. III.7). Treatment with PMSG on Day 15 or 16 shortens the follicular phase by a day or two, oestrus occurring within 3–4 days of the injection. A more controlled ovarian response has been suggested to arise from an initial régime of methallibure feeding (see Chapter II) followed by injection of 750 iu PMSG on the day of withdrawal of the drug and injection of 500 iu HCG 96 hours later (Polge *et al.*, 1968). But predictable responses in *litter size* have not been reported after superovulation treatments, although it is worth recording that the number of viable embryos observed at Day 25 has been increased by 1500 iu PMSG injected on Day 15 of the cycle (Hunter, 1966); and Deneke and Day (1973) have reported an increased litter size at Day 70 of gestation when 1500 iu PMSG were injected after synchronisation of oestrus.

Superovulation during the luteal phase of the cycle

If appropriate preparations of hormones are used, it is possible to induce ovulation at most stages of the oestrous cycle. Ovulation and indeed superovulation can certainly be induced in the presence of active corpora lutea (Fig. III.7), but the follicles stimulated to grow and mature in response to FSH-rich preparations then require exposure to systemically injected LH, (after a period of three or four days), before they will ovulate during the luteal phase of the cycle. Apparently, a

Table III.6. Superovulatory responses of nulliparous pigs (gilts) to a single subcutaneous injection of PMSG administered on Day 15 or 16 of a 21 day oestrous cycle. Half of the animals received, in addition, an intramuscular injection of 500 i.u. HCG at the onset of oestrus, but this was found not to influence the ovulatory response. (Adapted from Hunter, 1964.)

	Day 15				Day 16			
Dose of PMSG in i.u.	No. of corpora lutea	Mean no. of corpora lutea	Day of onset of oestrus	No. of follicles > 7 mm. diam.	No. of corpora lutea	Mean no. of corpora lutea	Day of onset of oestrus	No. of follicles > 7 mm. diam.
500[1]	16		21	–	14		21	–
500[1]	12	15·0	22	–	15	14·5	19	–
500[1]	20		23	2	14		20	–
500	12		19	–	15		28	–
750[1]	23		19	–	14		20	13
750	17	24·0	19	7	18	25·5	20	–
750[1]	42		18	–	23		20	–
750	14		19	–	47		18	–
1000[1]	19		18	–	30		19	–
1000	13	24·3	18	–	21	27·3	20	–
1000[1]	25		19	–	29		19	–
1000	40		19	–	30		19	–
1250[1]	22		20	–	33		19	–
1250	25	25·3	19	–	23	24·5	19	–
1250[1]	33		19	2	22		19	–
1250	21		30	–	20		20	–
1500[1]	74		19	–	49		20	1
1500	24	38·5	19	–	47	38·3	19	–
1500[1]	23		18	–	25		21	4
1500	33		19	2	32		20	–

[1] In addition to PMSG, the animal received an intramuscular injection of 500 i.u. HCG.

sufficient surge of LH release is prevented by the negative feedback effect of ovarian progesterone, in spite of the presence of mature follicles presumably secreting oestrogen (Hunter, 1979). A second snag associated with ovulation induced during the luteal phase is that the animals do not exhibit oestrus, and thirdly the incidence of normal fertilisation is usually extremely low following artificial insemination or forced mating. The underlying reasons are associated with poor sperm transport in the female tract (Austin, 1949), a disturbed transport of eggs in the Fallopian tubes (Rowson, 1951), the frequent development of pyometritis after insemination during the luteal phase (Rowson *et al.*, 1953), and abnormalities of fertilisation itself such as polyspermic penetration of the vitellus (Hunter, 1967b). Nonetheless, eggs may be recovered after luteal-phase ovulations for use in *in vitro* fertilisation or egg transplantation studies, but this approach to manipulation of the ovaries is likely to remain largely an experimental technique.

Induction of ovulation in prepuberal animals

The argument has been presented in many papers and reviews that the interval from birth of a female until attainment of puberty represents a significant proportion of her lifespan during which she is unproductive in terms of conception and subsequent embryonic growth, and yet she may be relatively expensive to maintain during this phase in terms of sustaining her own growth and development. Accordingly, treatments that either hasten the onset of puberty (i.e. precocious breeding) or enable the artificial recovery and transfer of eggs or embryos from selected immature individuals may have a rôle to play in certain systems of animal production. Such treatments, if successful, might contribute to a reduction in the *generation interval* (see Adams, 1954), thereby facilitating the early testing of progeny and so increasing the rate of turnover of selected breeding lines. As a general rule for farm animals, the ovaries of the new born may not respond detectably to injected

---

Fig. III.7. Four pairs of pig ovaries, showing in (a) and (b) the superovulatory response obtained after PMSG injection on Day 15 of the oestrous cycle. Each of the reddish structures is a young corpus luteum whereas the creamy-yellow structures are regressing corpora lutea from spontaneous ovulations in the preceding oestrous cycle. The ovaries in (c) and (d) illustrate ovulations induced in the luteal phase of the cycle in the presence of active corpora lutea.

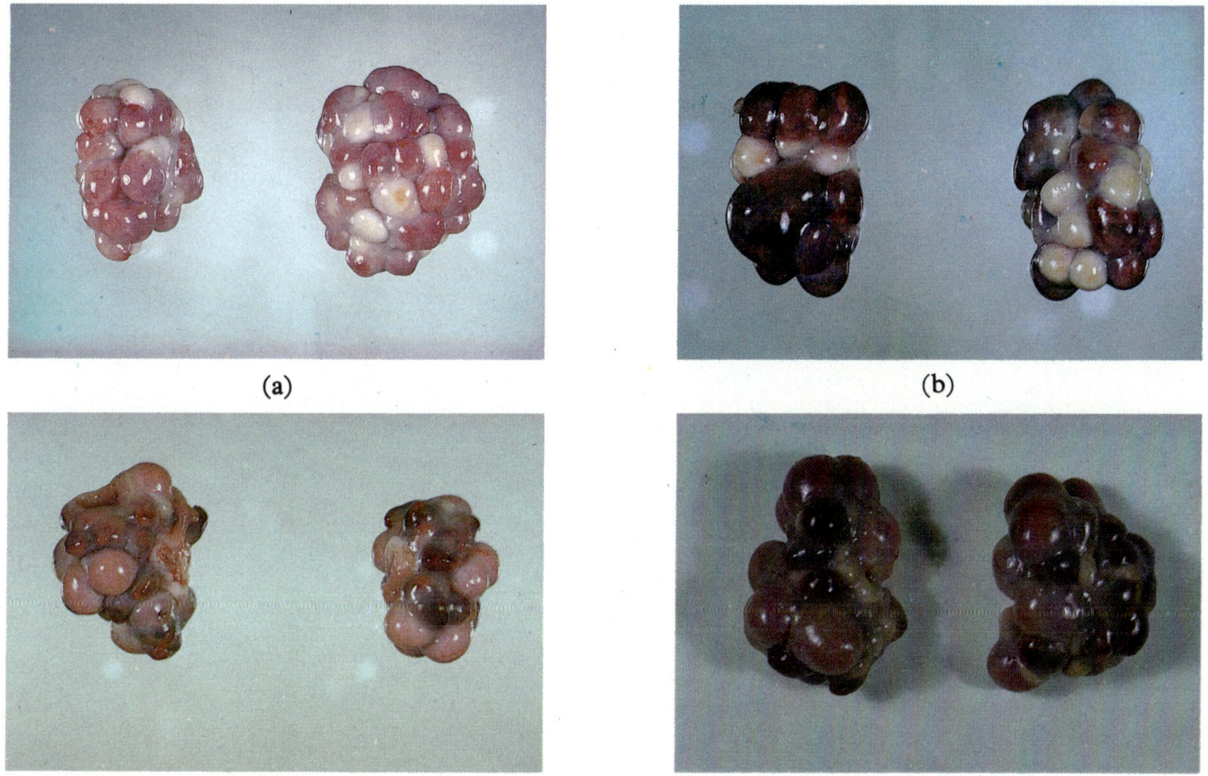

(a) (b) (c) (d)

gonadotrophins, but by eight weeks or so of age there are usually large numbers of growing and vesicular follicles available to be stimulated by appropriate treatments (Table III.7).

Table III.7. Examples of various treatments used for inducing ovulation in prepuberal animals. A treatment that stimulates growth of Graafian follicles (PMSG injection) invariably precedes one for provoking ovulation.

Species	Age of animal (months)	Nature of treatment	Reference
Calf	$1\frac{1}{2}$–3	2000 i.u. PMSG plus 2000–10 000 i.u. HCG 5 days later	Lineweaver and Hafez, 1970
	2–4	1500–2000 i.u. PMSG plus 1500 i.u. HCG 5 days later	Onuma *et al.*, 1970
	3	1800 i.u. PMSG and progestagen blockade for 4 days	Testart *et al.*, 1977
Lamb	1–5	1000 i.u. PMSG plus 500 i.u. HCG 4 days later	Mansour, 1959
	2	350–700 i.u. PMSG plus 500 i.u. HCG 2–3 days later	Land and McGovern, 1968
Pig	3–4	500–1250 i.u. PMSG plus 500 i.u. HCG 4 days later	Dziuk and Gehlbach, 1966
	5	250–2000 i.u. PMSG plus 500 i.u. HCG 2 days later	Baker and Coggins, 1968

Calves

Ovulation and superovulation have been induced in calves aged 1–4 months with injections of placental gonadotrophins or pituitary extracts (Black *et al.*, 1953; Marden, 1953; Jainudeen *et al.*, 1966; Onuma *et al.*, 1969; Lineweaver and Hafez, 1970; Seidel *et al.*, 1971; Testart, 1972; Bedirian and Baker, 1975). As examples of doses of hormone employed, PMSG has been injected in quantities ranging from 1500–3000 iu (Foote and Onuma, 1970; Testart, 1972; Dziuk, 1973) and HCG in quantities ranging from 1500–2000 iu or more (Lineweaver and Hafez, 1970; Testart, 1972). However, numerous experiments during the last twenty years have shown considerable variation in the ovarian response to these treatments, together with a poor incidence of egg recovery and/or fertilisation (Table III.8). As in mature animals, the variation in response arises largely from (a) the nature and biological potency of the hormonal preparations employed and (b) genetically determined variation in ovarian sensitivity between animals. More subtle components of ovarian response that may confound a treatment schedule, such as the progression of waves of follicular development and synergism between exogenous and endogenous gonadotrophins, would also seem to be involved. Indeed, the latter two suggestions probably underlie the fact that the influence of age on ovarian response to gonadotrophin injections in calves is still not clear. Even so, a mean response of 25 ovulations per calf is quite common (Onuma *et al.*, 1970).

More recent approaches to the problem of precocious liberation of oocytes have involved treating 3-month old calves for 4–10 days with intra-vaginal sponges containing progestagens (e.g. fluorogestone acetate; Testart, 1972; Testart *et al.*, 1977), with the aim of obtaining some degree of ovarian uniformity prior to and/or during stimulation. By thus "programming" the ovary and treating with gonadotrophins at a known stage relative to progestagen withdrawal, the range in stages of development among vesicular follicles may be reduced and a more uniform response obtained in a proportion of animals. Nevertheless, two of the disadvantages mentioned above (a and b) remain, so this experimental approach to obtaining increased numbers of eggs from calves is not a long-term solution. The use of preparations of hypothalamic releasing factors to trigger secretion of endogenous gonadotrophins rather than injection of placental gonadotrophins is unlikely to reduce the variation in ovarian response.

As to fertility in situations of induced ovulation, recovery of normally fertilised eggs after the conventional approach of intra-cervical insemination has been difficult and sometimes hazardous in calves aged 3–4 months due to the risk of puncture of the cervical wall; the use of a speculum for insemination may be helpful (Onuma *et al.*, 1970). Moreover, both transport and survival of spermatozoa appear to be less efficient in the genital tract of calves than in mature animals, even though the distance to be traversed by the spermatozoa is considerably reduced compared with that in the tract of a parous cow. Factors contributing to this lower incidence of fertilisation must include the nature and extent of cervical mucus in calves, the pattern and amplitude of myometrial contractions, the glandular proliferation of the endometrium and the leucocytic response to cells in the uterine lumen of animals

Table III.8. Some figures illustrating the variability in response of prepuberal calves to treatment with preparations of gonadotrophic hormones and subsequent insemination. (Adapted from Foote and Onuma, 1970.)

Age of animals (weeks)	No. of animals	Nature of gonadotrophic stimulation	No. of stimulated follicles	No. of ovulations		% of eggs	
				Mean	Range	Recovered	Cleaved
9–24	43	20, 10, 5, 0 and 10 mg FSH on Days 1–5: 100 mg LH on Day 6	–	34	1–88	33	24
4–24	13	Day 1: 2000 i.u. PMSG Day 6: 10 mg NIH-LH	28	16	0–55	27–60	6
8–17	55	Day 1: 1500–2000 i.u. PMSG Day 6: 50 mg PLH	53	43	–	8–75	72

not exhibiting oestrus. In this connection, surgical deposition of semen in the uterus may be beneficial, and normal fertilisation has also been obtained in calves following intra-peritoneal insemination shortly before the estimated time of ovulation (Testart, 1975). Irrespective of the route of semen deposition, only limited evidence for the development of calf embryos has so far been found either *in vitro* (Onuma and Foote, 1969) or after transplantation to mature recipients (Seidel *et al.*, 1971).

Lambs

Ovulation may be induced in lambs as early as eight weeks of age by treatment first with progesterone and then with gonadotrophins (Mansour, 1959). Cleavage of eggs arising from such a form of treatment may follow from intra-uterine insemination (Land and McGovern, 1968). Moreover, the work of Trounson *et al.* (1977) suggested that at least a small proportion of such embryos could develop to term upon transfer to produce viable lambs. The latter treatment involved intramuscular injection of 10–16 week old lambs with 10 mg progesterone daily in an oil solution for a variable number of days (3–18), followed by a subcutaneous injection of 750–1000 iu PMSG at the time of the penultimate progesterone injection. An injection of LH or HCG was not required to induce ovulation, endogenous gonadotrophins proving adequate.

Pigs

The early studies of Casida (1935), and subsequently those of Erickson (1967) and Black and Erickson (1968), characterised the oocyte and/or follicular populations in the prepuberal pig, and Casida (1935) and Du Mesnil du Buisson (1954) were able to obtain ovulation in animals of 15–16 weeks of age by injections of horse, sheep or pig pituitary extracts or serum gonadotrophins. But the ovaries of gilts less than 56 days of age apparently do not develop fully mature follicles or ovulate in response to gonadotrophin treatment.

More recent studies have shown that fertilised eggs can be obtained from prepuberally treated and inseminated gilts of 4–6 months of age (Dziuk and Gehlbach, 1966; Baker and Coggins, 1968; Shaw *et al.*, 1971), but that pregnancy is unlikely to be maintained in most animals due to inadequate pituitary support of the induced corpora lutea; these regress before Day 50 leading to total embryonic loss (Segal and Baker, 1973). Supplementation with exogenous steroids is therefore required to maintain such a prepuberal pregnancy to term (Ellicott *et al.*, 1973). This differs from the situation just described in the lamb where endogenous pituitary gonadotrophins appear adequate to support precociously induced corpora lutea.

Out-of-season breeding

Because the majority of breeds of sheep—as distinct from cattle and pigs—are seasonal breeders under the climatic conditions found in the developed countries of Western Europe, North America, Japan and Australasia, there are attractive arguments in favour of inducing out-of-season breeding as a means of increasing productivity. Such an objective is achieved using procedures that stimulate the relatively quiescent ovaries with gonadotrophic hormones either without pre-treatment (Cole and Miller, 1933), or, more probably, after progesterone therapy involving subcutaneous injections of a solution of progesterone in oil (Dutt, 1953; Robinson, 1954; Gordon, 1958) or after use of intra-vaginal sponge pessaries impregnated with a progestagen (see Chapter II; Roberts and Edgar, 1966; Thimonier *et al.*, 1968). An influence of progesterone, endogenous or exogenous in origin, is essential in order to prime the hypothalamus for the feedback action of oestrogens that must precede overt oestrus and ovulation. The success of progestagen-PMSG treatments in inducing ovarian activity and resumption of fertility is closely correlated with the degree of seasonal anoestrus (Table III.9), the poorest responses to treatment being found in the depth of anoestrus. However, factors that may confound the response in terms of fertility include whether or not the ewe is still suckling offspring from the previous pregnancy (see Chapter XI), and also the fertility of the ram or the quality of the stored semen available at the time of insemination. Nonetheless, the production of three crops of lambs in two years should be a realistic objective when suitable treatments are used, and advancing the breeding season of ewes in late anoestrus should appeal in many production systems (see Gordon, 1975a, b).

As an example of a régime applied to sheep with some success in Ireland, France and the UK, mature ewes are treated with intra-vaginal sponges containing a solution of synthetic progestagen for 14 days. An

Table III.9. To show the influence of season of the year upon conception and number of lambs of a progestagen-PMSG treatment applied to three different groups of ewes. (Taken from Gordon, 1975a.)

Index of productivity	Season of treatment		
	Spring	Summer	Autumn
PREGNANCY %			
First heats	34·7	64·0	75·3
First and second heats	35·0	79·6	90·5
LITTER SIZE			
First heats	1·58	1·62	1·72
Second heats	1·27	1·36	1·62

injection of 500 iu PMSG is given at the time of sponge withdrawal, and oestrus generally follows within 2 days. At least 60% of ewes mated in June or July following this treatment should produce lambs (Table III.10), but an adequate ratio of rams to ewes is desirable (e.g. 1:10) for good fertility at such synchronised heats. In addition, introduction of experienced rams 48 hours after rather than at the time of sponge removal appears beneficial (Gordon, 1975a, b). Ewes failing to conceive to the induced oestrus should return spontaneously approximately one cycle length later.

Since the breeding season of sheep is controlled by the pattern of day length change (Yeates, 1949; Hart, 1950; Hafez, 1952), it is possible to house sheep intensively and use patterns of artificial lighting as an alternative to or in conjunction with hormonal therapy for inducing resumption of cyclic ovarian activity (Robinson *et al.*, 1975). Even though leading to a remarkable output of lambs per ewe per year, this expensive approach to increasing productivity is unlikely to be widely adopted in most contemporary systems of sheep production, but could be suitable where winter housing is practised. On the other hand, there may be a case for turning to breeds such as the Dorset Horn or Finnish Landrace or their crosses that have prolonged breeding seasons or do not show a marked seasonal anoestrus, as a basis for augmenting productivity.

Induction of post-partum *cyclic activity*

Another area of potential improvement in productivity involving treatment with gonadotrophic hormones is that of the early *post-partum*

Table III.10. The influence upon various indices of lambing of a progestagen-PMSG treatment applied to ewes in late anoestrus to induce early lambing. (Adapted from Gordon, 1975b.)

Year of treatment	Number of flocks	Total number of sheep	Lambed to first service			Lambed to first and second services		
			Ewes %	Lambs per conception	Lambs per 100 ewes	Ewes %	Lambs per conception	Lambs per 100 ewes
1971	207	8314	61·6	1·6	100	80·2	1·6	126
1972	307	11 159	65·6	1·6	105	79·3	1·6	124

interval, when the animal is lactating heavily and the uterus may not have completed the morphological and glandular changes associated with involution (see Chapter XI). In the case of cattle, and in particular for beef cows suckling their calves, cyclic ovarian activity and behavioural oestrus may not be observed for two to three months or more, especially if energy intake is low. In such circumstances, it is worth considering the use of gonadotrophin therapy to promote resumption of cyclic activity since the principal reason that gross ovarian activity is in abeyance may be the preferential secretion of prolactin rather than gonadotrophins by the anterior pituitary under the stimulus of intensive suckling activity (Chapter XI).

The objective of gonadotrophin therapy would be to induce follicular maturation, oestrus and ovulation, but the animal would then be left to undergo a spontaneous oestrous cycle before rebreeding at the subsequent oestrus some three weeks later. The specific treatment could be in the form of a short-term progestagen blockade, followed by an injection of 750 iu PMSG at the time of sponge removal (Mulvehill and Sreenan, 1977). The time of imposition of such a *post-partum* therapy will vary according to environmental and managerial factors, but the success in terms of ovarian response is likely to be improved when at least two to three weeks have elapsed from calving. In any event, a calving interval of 11–12 months is probably the objective in a beef breeding herd, so even allowing for the three weeks of the post-treatment cycle, there is still considerable latitude in the choice of time for treatment.

Use of synthetic releasing hormones

The unravelling of the structure and the subsequent synthesis of the hypothalamic releasing factors (see review by Schally *et al.*, 1973), particularly the gonadotrophin releasing factor or hormone (Gn-RH), suggested that a treatment might become available for controlling the number of follicles ovulating more accurately than with hypophyseal hormones themselves. Injecting appropriate quantities of synthetic Gn-RH by the intracarotid or systemic routes has been shown to provoke transient increases in serum LH concentrations in cattle (Kittok *et al.*, 1973; Kaltenbach *et al.*, 1974; Webb *et al.*, 1977), sheep (Reeves *et al.*, 1971; Foster and Crighton, 1974; Shareha *et al.*, 1976) and pigs (Baker *et al.*, 1973; Chakraborty *et al.*, 1973), leading in many instances to ovulation. But a conclusion that would be justified by the findings so far is that whilst the *time* of ovulation may be controlled in this manner, although with little apparent advantage over using preparations of LH or HCG, the actual *number of follicles* ovulating is no more accurately controlled than with gonadotrophin injections. Moreover, it has been reported in several experimental situations that use of Gn-RH to induce ovulation, especially in anoestrous sheep and *post-partum* suckled cows, may lead to the formation of corpora lutea that are not fully functional in terms of progesterone secretion and active lifespan (Haresign *et al.*, 1975; Shareha *et al.*, 1976; Webb *et al.*, 1977). This is possibly because a trophic influence of prolactin is also required at or shortly after the time of ovulation in order to stimulate development of a competent corpus luteum. If, as is now widely believed, the hypothalamic releasing hormone is similar or identical for both FSH and LH, it is not easy to imagine a simple schedule of treatment that would enable control of both the number of follicles maturing and their time of ovulation, even though the serum concentrations of FSH and LH may rise in response to treatment.

However, synthetic releasing factors have been shown to have a practical application in dairy cows with follicular cysts. Animals with this ovarian condition can be treated with a single intramuscular injection of 50, 100 or 250 μg of synthetic Gn-RH, and 60–80% will resume oestrous cycles following treatment; most cows returned to oestrus 18–23 days following treatment (Bierschwal *et al.*, 1975; Cantley *et al.*, 1975; Kesler *et al.*, 1978). In view of the disruption to follicular histology invariably found in the cystic condition, it is of considerable interest that such follicles can apparently respond either by ovulating or by luteinising directly as a sequel to the Gn-RH treatment and the endogenous surge of gonadotrophin. But a more precise identification of the follicles that respond appears necessary before unequivocal statements can be made on the ability of cystic follicles to ovulate, and a description of the process of luteinisation in these ovaries would also be valuable.

References

Adams, C. E. (1954). The experimental shortening of the generation interval. *Proc. Brit. Soc. Anim. Prod.* pp. 97–108.

Ainsworth, L., Baker, R. D. and Armstrong, D. T. (1975). Pre-ovulatory changes in follicular fluid prostaglandin F levels in swine. *Prostaglandins* **9**, 915–925.

Allen, W. R. (1969). The immunological measurement of pregnant mare serum gonadotrophin. *J. Endocrin.* **43**, 593–598.

Amsterdam, A., Lindner, H. R. and Gröschel-Stewart, U. (1977). Localisation of actin and myosin in the rat oocyte and follicular wall by immunofluorescence. *Anat. Rec.* **187**, 311–327.

Armstrong, D. T. and Grinwich, D. L. (1972). Blockade of spontaneous and LH-induced ovulation in rats by indomethacin, an inhibitor of prostaglandin biosynthesis. *Prostaglandins* **1**, 21–28.

Aschheim, S. and Zondek, B. (1927). Hypophysenvorderlappenhormon und Ovarialhormone im Harn von Schwangeren. *Klin. Wschr.* **6**, 1322.

Austin, C. R. (1949). Fertilisation and the transport of gametes in the pseudopregnant rabbit. *J. Endocrin.* **6**, 63–70.

Averill, R. L. W. (1958). The production of living sheep eggs. *J. agric. Sci.* **50**, 17–33.

Averill, R. L. W. and Rowson, L. E. A. (1958). Ovum transfer in the sheep. *J. Endocrin.* **16**, 326–336.

Baird, D. T., Baker, T. G., McNatty, K. P. and Neal, P. (1975). Relationship between the secretion of the corpus luteum and the length of the follicular phase of the ovarian cycle. *J. Reprod. Fert.* **45**, 611–619.

Baker, R. D. and Coggins, E. G. (1968). Control of ovulation rate and fertilisation in prepuberal gilts. *J. Anim. Sci.* **27**, 1607–1610.

Baker, R. D. and Downey, B. R. (1975). Induction of estrus, ovulation and fertility in prepuberal gilts. *Annls Biol. anim. Biochim. Biophys.* **15**, 375–382.

Baker, R. D., Downey, B. R. and Brinkley, H. J. (1973). Induction of ovulation in pigs with gonadotrophin releasing hormone. *J. Anim. Sci.* **37**, 1376–1379.

Bauminger, S. and Lindner, H. R. (1975). Periovulatory changes in ovarian prostaglandin formation and their hormonal control in the rat. *Prostaglandins* **9**, 737–751.

Bedirian, K. N. and Baker, R. D. (1975). Follicular development, oocyte maturation and ovulation in gonadotrophin-treated prepuberal calves. *Can. J. Anim. Sci.* **55**, 193–199.

Betteridge, K. J. (1974). Unilateral stimulation of bovine ovaries by local injection of pregnant mare's serum gonadotrophin. *J. Reprod. Fert.* **37**, 101–104.

Betteridge, K. J. (1977). Techniques and results in cattle. Superovulation. *In* "Embryo Transfer in Farm Animals" (K. J. Betteridge, ed.), pp. 1–9. Canada Department of Agriculture Monograph No. 16.

Bierschwal, C. J., Garverick, H. A., Martin, C. E., Youngquist, R. S., Cantley, T. C. and Brown, M. D. (1975). Clinical response of dairy cows with ovarian cysts to Gn RH. *J. Anim. Sci.* **41**, 1660–1665.

Black, J. L. and Erickson, B. H. (1968). Oogenesis and ovarian development in the prenatal pig. *Anat. Rec.* **161**, 45–56.

Black, W. G., Ulberg, L. C., Christian, R. E. and Casida, L. E. (1953).

Ovulation and fertilisation in the hormone-stimulated calf. *J. Dairy Sci.* **36**, 274–280.

Booth, W. D., Newcomb, R., Strange, H., Rowson, L. E. A. and Sacher, H. B. (1975). Plasma oestrogen and progesterone in relation to super-ovulation and egg recovery in the cow. *Vet. Rec.* **97**, 366–369.

Braden, A. W. H., Lamond, D. R. and Radford, H. M. (1960). The control of the time of ovulation in sheep. *Aust. J. agric. Res.* **11**, 389–401.

Buttle, H. L. and Hancock, J. L. (1967). The control of ovulation in the sow. *J. Reprod. Fert.* **14**, 485–487.

Cantley, T. C., Garverick, H. A., Bierschwal, C. J., Martin, C. E. and Youngquist, R. S. (1975). Hormonal responses of dairy cows with ovarian cysts to Gn RH. *J. Anim. Sci.* **41**, 1666–1673.

Casida, L. E. (1935). Prepuberal development of the pig ovary and its relation to stimulation with gonadotropic hormones. *Anat. Rec.* **61**, 389–396.

Casida, L. E., Warwick, E. J. and Meyer, R. K. (1944). Survival of multiple pregnancies induced in the ewe following treatment with pituitary gonadotropins. *J. Anim. Sci.* **3**, 22–28.

Chakraborty, P. K., Reeves, J. J., Arimura, A. and Schally, A. V. (1973). Serum LH levels in prepubertal female pigs chronically treated with synthetic LH-RH/FSH-RH. *Endocrinology* **92**, 55–61.

Chang, M. C. (1955). The maturation of rabbit oocytes in culture and their maturation, activation, fertilisation and subsequent development in the Fallopian tubes. *J. exp. Zool.* **128**, 379–405.

Christie, W. B., Newcomb, R. and Rowson, L. E. A. (1979). Ovulation rate and egg recovery in cattle treated repeatedly with pregnant mare serum gonadotrophin and prostaglandin. *Vet. Rec.* **104**, 281–283.

Cole, H. H. and Hart, G. H. (1930). The potency of blood serum of mares in progressive stages of pregnancy in effecting the sexual maturity of the immature rat. *Amer. J. Physiol.* **93**, 57–68.

Cole, H. H. and Miller, R. F. (1933). Artificial induction of ovulation and oestrum in the ewe during anoestrum. *Amer. J. Physiol.* **104**, 165–171.

Day, F. T. and Rowlands, I. W. (1940). The time and rate of appearance of gonadotrophin in the serum of pregnant mares. *J. Endocrin.* **2**, 255–261.

Deneke, W. A. and Day. B. N. (1973). Effect of superovulation on litter size of swine at 70 days of gestation. *J. Anim. Sci.* **36**, 1137–1139.

Dowling, D. F. (1949). Problems of the transplantation of fertilized ova. *J. agric. Sci.* **39**, 374–396.

Du Mesnil du Buisson, F. (1954). Possibilité d'ovulation et de fécondation chez la truie avant la puberté. *Annls Endocr. (Paris)* **15**, 333–340.

Dutt, R. H. (1953). Induction of estrus and ovulation in anestrual ewes by the use of progesterone and pregnant mare serum. *J. Anim. Sci.* **12**, 515–523.

Dziuk, P. J. (1965). Time of maturation and fertilisation of the sheep egg. *Anat. Rec.* **153**, 211–224.

Dziuk, P. J. (1968). Effect of number of embryos and uterine space on embryo survival in the pig. *J. Anim. Sci.* **27**, 673–676.

Dziuk, P. J. (1973). Occurrence, control and induction of ovulation in pigs, sheep, and cows. *In* "Handbook of Physiology", Section 7, Endocrinology,

Vol. II, Part 1 (R. O. Greep and E. B. Astwood, eds), pp. 143–152. American Physiological Society, Washington.

Dziuk, P. J. (1977). Reproduction in pigs. *In* "Reproduction in Domestic Animals", 3rd edition (H. H. Cole and P. T. Cupps, eds), pp. 455–474. Academic Press, New York and London.

Dziuk, P. J. and Baker, R. D. (1962). Induction and control of ovulation in swine. *J. Anim. Sci.* **21**, 697–699.

Dziuk, P. J. and Gehlbach, G. D. (1966). Induction of ovulation and fertilisation in the immature gilt. *J. Anim. Sci.* **25**, 410–413.

Dziuk, P. J., Hinds, F. C., Mansfield, M. E. and Baker, R. D. (1964a). Follicle growth and control of ovulation in the ewe following treatment with 6-methyl-17-acetoxyprogesterone. *J. Anim. Sci.* **23**, 787–790.

Dziuk, P. J., Polge, C. and Rowson, L. E. (1964b). Intra-uterine migration and mixing of embryos in swine following egg transfer. *J. Anim. Sci.* **23**, 37–42.

Edwards, R. G. (1962). Meiosis in ovarian oocytes of adult mammals. *Nature, London* **196**, 446–450.

Edwards, R. G. (1965). Maturation *in vitro* of mouse, sheep, cow, pig, Rhesus monkey and human ovarian oocytes. *Nature, London* **208**, 349–351.

Ellicott, A. R., Dziuk, P. J. and Polge, C. (1973). Maintenance of pregnancy in prepuberal gilts. *J. Anim. Sci.* **37**, 971–973.

Erickson, B. H. (1967). Radioresponse of the pre-puberal porcine ovary. *Int. J. Radiat. Biol.* **13**, 57–67.

Espey, L. L. (1974). Ovarian proteolytic enzymes and ovulation. *Biol. Reprod.* **10**, 216–235.

Espey, L. L. (1976). The distribution of collagenous connective tissue in rat ovarian follicles. *Biol. Reprod.* **14**, 502–506.

Espey, L. L. and Coons, P. J. (1976). Factors which influence ovulatory degradation of rabbit ovarian follicles. *Biol. Reprod.* **14**, 233–245.

Foote, R. H. and Onuma, H. (1970). Superovulation, ovum collection, culture and transfer. A review. *J. Dairy Sci.* **53**, 1681–1692.

Foster, J. P. and Crighton, D. B. (1974). Luteinising hormone (LH) release after single injections of synthetic LH-releasing hormone (LH-RH) in the ewe at three different reproductive stages and comparison with natural LH release at oestrus. *Theriogenology* **2**, 87–100.

Gibson, E. W., Jaffe, S. C., Lasley, J. F. and Day, B. N. (1963). Reproductive performance in swine following superovulation. *J. Anim. Sci.* **22**, 858 (Abstract).

Gordon, I. (1958). The use of progesterone and serum gonadotrophin (PMS) in the control of fertility in sheep. I and II. *J. agric. Sci.* **50**, 123–197.

Gordon, I. (1975a). The use of progestagens in sheep bred by natural and artificial insemination. *Annls Biol. anim. Biochim. Biophys.* **15**, 303–315.

Gordon, I. (1975b). Hormonal control of reproduction in sheep. *Proc. Brit. Soc. Anim. Prod.* **4**, 79–93.

Gordon, I., Williams, G. and Edwards, J. (1962). The use of serum gonadotrophin (PMS) in the induction of twin-pregnancy in the cow. *J. agric. Sci.* **59**, 143–198.

Graves, C. N. and Dziuk, P. J. (1968). Control of ovulation in dairy cattle with human chorionic gonadotrophin after treatment with 6α-methyl-17α-acetoxy-progesterone. *J. Reprod. Fert.* **17**, 169–172.

Greenwald, G. S. (1962). Analysis of superovulation in the adult hamster. *Endocrinology* **71**, 378–389.

Guttmacher, M. S. and Guttmacher, A. F. (1921). Morphological and physiological studies on the musculature of the mature Graafian follicle of the sow. *Johns Hopkins Hosp. Bull.* **32**, 394–399.

Hafez, E. S. E. (1952). Studies on the breeding season and reproduction of the ewe. Parts I and II. *J. agric. Sci.* **42**, 189–265.

Hafez, E. S. E. (1969). Superovulation and preservation of mammalian eggs. *Acta endocr.* **62**, Suppl. 140.

Hafez, E. S. E., Sugie, T. and Gordon, I. (1963). Superovulation and related phenomena in the beef cow. I. Superovulatory responses following PMS and HCG injections. *J. Reprod. Fert.* **5**, 359–379.

Hammond, J. (Jr). (1949). Induced twin ovulations and multiple pregnancy in cattle. *J. agric. Sci.* **39**, 222–225.

Hammond, J. (Jr). and Bhattacharya, P. (1944). Control of ovulation in the cow. *J. agric. Sci.* **34**, 1–15.

Hammond, J. (Jr), Hammond, J. and Parkes, A. S. (1942). Hormonal augmentation of fertility in sheep. I. Induction of ovulation, superovulation and heat in sheep. *J. agric. Sci.* **32**, 308–323.

Hancock, J. L. and Hovell, G. J. R. (1961). Transfer of sheep ova. *J. Reprod. Fert.* **2**, 295–306.

Haresign, W., Foster, J. P., Haynes, N. B., Crighton, D. B. and Lamming, G. E. (1975). Progesterone levels following treatment of seasonally anoestrous ewes with synthetic LH-releasing hormone. *J. Reprod. Fert.* **43**, 269–279.

Hart, D. S. (1950). Photoperiodicity in Suffolk sheep. *J. agric. Sci.* **40**, 143–149.

Hunter, R. H. F. (1964). Superovulation and fertility in the pig. *Anim. Prod.* **6**, 189–194.

Hunter, R. H. F. (1966). The effect of superovulation on fertilisation and embryonic survival in the pig. *Anim. Prod.* **8**, 457–465.

Hunter, R. H. F. (1967a). Porcine ovulation after injection of human chorionic gonadotrophin. *Vet. Rec.* **81**, 21–23.

Hunter, R. H. F. (1967b). Polyspermic fertilisation in pigs during the luteal phase of the estrous cycle. *J. exp. Zool.* **165**, 451–460.

Hunter, R. H. F. (1972a). Ovarian response of the pig to gonadotrophins injected at different stages of the oestrous cycle. *Proc. 7th Int. Congr. Anim. Reprod. and Artif. Insem., Munich,* **2**, 1053–1057.

Hunter, R. H. F. (1972b). Ovulation in the pig: timing of the response to injection of human chorionic gonadotrophin. *Res. vet. Sci.* **13**, 356–361.

Hunter, R. H. F. (1979). Ovarian follicular responsiveness and oocyte quality after gonadotrophic stimulation of mature pigs. *Annls Biol. anim. Biochim. Biophys.* **19**, 1511–1520.

Hunter, R. H. F. and Polge, C. (1966). Maturation of follicular oocytes in the pig after injection of human chorionic gonadotrophin. *J. Reprod. Fert.* **12**, 525–531.

Hunter, R. H. F., Cook, B. and Baker, T. G. (1976). Dissociation of response to injected gonadotropin between the Graafian follicle and oocyte in pigs. *Nature, London* **260**, 156–158.

Jainudeen, M. R., Hafez, E. S. E., Gollnick, P. D. and Moustafa, L. A. (1966a). Antigonadotrophins in the serum of cows following repeated therapeutic pregnant mare serum injections. *Amer. J. vet. Res.* **27**, 669–675.

Jainudeen, M. R., Hafez, E. S. E. and Lineweaver, J. A. (1966b). Superovulation in the calf. *J. Reprod. Fert.* **12**, 149–153.

Jalabert, B. and Szollosi, D. (1975). *In vitro* ovulation of trout oocytes: effect of prostaglandins on smooth muscle-like cells of the theca. *Prostaglandins* **9**, 765–778.

Kaltenbach, C. C., Dunn, T. G., Kiser, T. E., Corah, L. R., Akbar, A. M. and Niswender, G. D. (1974). Release of FSH and LH in beef heifers by synthetic gonadotrophin releasing hormone. *J. Anim. Sci.* **38**, 357–362.

Kesler, D. J., Garverick, H. A., Caudle, A. B., Bierschwal, C. J., Elmore, R. G. and Youngquist, R. S. (1978). Clinical and endocrine responses of dairy cows with ovarian cysts to Gn RH and $PGF_{2\alpha}$. *J. Anim. Sci.* **46**, 719–725.

Kittok, R. J., Britt, J. H. and Convey, E. M. (1973). Endocrine response after Gn RH in luteal phase cows and cows with ovarian follicular cysts. *J. Anim. Sci.* **37**, 985–989.

Lamond, D. R. (1962). Oestrus and ovulation following administration of placental gonadotrophins to Merino ewes. *Aust. J. agric. Res.* **13**, 707–717.

Land, R. B. (1973). Ovulation rate of Finn-Dorset sheep following unilateral ovariectomy or chlorpromazine treatment at different stages of the oestrous cycle. *J. Reprod. Fert.* **33**, 99–105.

Land, R. B. and McGovern, P. T. (1968). Ovulation and fertilisation in the lamb. *J. Reprod. Fert.* **15**, 325–327.

Laster, D. B. (1973). Ovulation, fertility and prenatal mortality in heifers treated with PMSG or porcine FSH. *J. Reprod. Fert.* **33**, 275–282.

Lawson, R. A. S. and Rowson, L. E. A. (1972). The influence of breed of ewe and offspring on litter size after egg transfer in sheep. *J. Reprod. Fert.* **28**, 433–439.

LeMaire, W. J., Yang, N. S. T., Behrman, H. R. and Marsh, J. M. (1973). Preovulatory changes in the concentration of prostaglandins in rabbit Graafian follicles. *Prostaglandins* **3**, 367–376.

LeMaire, W. J., Leidner, R. and Marsh, J. M. (1975). Pre- and post-ovulatory changes in the concentration of prostaglandins in rat Graafian follicles. *Prostaglandins* **9**, 221–229.

Lindner, H. R., Amsterdam, A., Salomon, Y., Tsafriri, A., Nimrod, A., Lamprecht, S. A., Zor, U. and Koch, Y. (1977). Intraovarian factors in ovulation: determinants of follicular response to gonadotrophins. *J. Reprod. Fert.* **51**, 215–235.

Lindner, H. R., Tsafriri, A., Lieberman, M. E., Zor, U., Koch, Y., Bauminger, S. and Barnea, A. (1974). Gonadotropin action on cultured Graafian follicles: induction of maturation division of the mammalian oocyte and differentiation of the luteal cell. *Rec. Progr. Hormone Res.* **30**, 79–127.

Lineweaver, J. A. and Hafez, E. S. E. (1970). Ovarian responses in gonadotropin-treated calves. *Amer. J. vet. Res.* **31**, 2157–2166.

Longenecker, D. E., Lasley, J. F. and Day, B. N. (1965). Fecundity in gilts and sows administered PMS. *J. Anim. Sci.* **24**, 924 (Abstract).

Lipner, H. (1973). Mechanism of mammalian ovulation. *In* "Handbook of Physiology", Section 7, Endocrinology, Vol. II, Part 1 (R. O. Greep and E. B. Astwood, eds), pp. 409–437. American Physiological Society, Washington.

McGovern, P. T., Williams, H. L. and Hancock, J. L. (1969). The time of ovulation in the ewe following treatment with human chorionic gonadotrophin. *J. Reprod. Fert.* **20**, 537–540.

Mansour, A. M. (1959). The hormonal control of ovulation in the immature lamb. *J. agric. Sci.* **52**, 87–94.

Marden, W. G. R. (1953). The hormonal control of ovulation in the calf. *J. agric. Sci.* **43**, 381–406.

Mauer, R. E. and Rippel, R. H. (1972). Response of cattle to synthetic gonadotropin releasing hormone. *J. Anim. Sci.* **35**, 249 (Abstract).

Mauléon, P., Mariana, J. C., Benoit, M., Solari, A. and Chupin, D. (1970a). Influence de différentes doses de PMSG et HCG, injectées en phase folliculaire du cycle oestrien sur le nombre et le rendement d'ovulations de vaches. *Annls. Biol. anim. Biochim. Biophys.* **10**, Hors-série, 1, 31–46.

Mauléon, P., Rey, J., Mariana, J. C. and Benoit, M. (1970b). Possibilités de superovulation après contrôle du cycle oestrien par l'acetate de fluorogestone chez les bovins. *Annls. Biol. anim. Biochim. Biophys.* **10**, Hors-série 1, 65–79.

Menzer, C. and Schams, D. (1979). Radioimmunoassay for PMSG and its application to *in-vivo* studies. *J. Reprod. Fert.* **55**, 339–345.

Moor, R. M. (1978). Role of steroids in the maturation of ovine oocytes. *Annls. Biol. anim. Biochim. Biophys.* **18**, 477–482.

Moor, R. M. and Trounson, A. O. (1977). Hormonal and follicular factors affecting maturation of sheep oocytes *in vitro* and their subsequent developmental capacity. *J. Reprod. Fert.* **49**, 101–109.

Moore, N. W. and Shelton, J. N. (1964). Response of the ewe to a horse anterior pituitary extract. *J. Reprod. Fert.* **7**, 79–87.

Mulvehill, P. and Sreenan, J. M. (1977). Improvement of fertility in postpartum beef cows by treatment with PMSG and progestagen. *J. Reprod. Fert.* **50**, 323–325.

Newcomb, R. and Rowson, L. E. A. (1976). Multiple ovulation, egg transplantation—towards twinning. *In* "Principles of Cattle Production" (H. Swan and W. H. Broster, eds), pp. 59-83. Proc. 23rd Easter School in Agric. Science, University of Nottingham, Butterworths, London.

O'Grady, J. P., Caldwell, B. V., Auletta, F. J. and Speroff, L. (1972). The effects of an inhibitor of prostaglandin synthesis (indomethacin) on ovulation, pregnancy and pseudopregnancy in the rabbit. *Prostaglandins* **1**, 97–106.

Onuma, H. and Foote, R. H. (1969). *In vitro* development of ova from prepuberal cattle. *J. Dairy Sci.* **52**, 1085–1087.

Onuma, H., Hahn, J., Maurer, R. R. and Foote, R. H. (1969). Repeated superovulation in calves. *J. Anim. Sci.* **28**, 634–637.

Onuma, H., Hahn, J. and Foote, R. H. (1970). Factors affecting superovulation, fertilisation and recovery of superovulated ova in prepuberal cattle. *J. Reprod. Fert.* **21**, 119–126.

Ortavant, R. and Thibault, C. (1970). Recherches sur l'induction expérimentale de naissances gemellaires chez les bovins. *Annls. Biol. anim. Biochim. Biophys.* **10**, Hors-série 1.

Owman, Ch., Sjöberg, N.-O., Svensson, K.-G. and Walles, B. (1975). Autonomic nerves mediating contractility in the human Graafian follicle. *J. Reprod. Fert.* **45**, 553–556.

Phillippo, M. (1968). Superovulation in the pig. *Adv. Reprod. Physiol.* **3**, 147–166.

Phillippo, M. and Rowson, L. E. A. (1975). Prostaglandins and superovulation in the bovine. *Annls. Biol. anim. Biochim. Biophys.* **15**, 233–240.

Pincus, G. and Enzmann, E. V. (1935). The comparative behaviour of mammalian eggs *in vivo* and *in vitro*. *J. exp. Med.* **62**, 665–675.

Polge, C. (1969). Advances in reproductive physiology in pigs. *J. Aust. Inst. agric. Sci.* **35**, 147–153.

Polge, C. and Dziuk, P. (1965). Recovery of immature eggs penetrated by spermatozoa following induced ovulation in the pig. *J. Reprod. Fert.* **9**, 357–358.

Polge, C., Day, B. N. and Groves, T. W. (1968). Synchronisation of ovulation and artificial insemination in pigs. *Vet. Rec.* **83**, 136–142.

Pope, C. E., Christenson, R. K., Zimmerman-Pope, V. A. and Day, B. N. (1972). Effect of number of embryos on embryonic survival in recipient gilts. *J. Anim. Sci.* **35**, 805–808.

Rajaniemi, H. and Vanha-Perttula, T. (1972). Specific receptor for LH in the ovary: evidence by autoradiography and tissue fractionation. *Endocrinology* **90**, 1–9.

Reeves, J. J., Arimura, A. and Schally, A. V. (1971). Changes in pituitary responsiveness to luteinising hormone-releasing hormone (LH-RH) in anestrous ewes pretreated with estradiol benzoate. *Biol. Reprod.* **4**, 88–92.

Roberts, E. M. and Edgar, D. G. (1966). The stimulation of fertile oestrus in anoestrous Romney ewes. I. and II. *J. Reprod. Fert.* **12**, 561–567.

Robinson, J. J., Fraser, C. and McHattie, I. (1975). The use of progestagens and photoperiodism in improving the reproductive rate of the ewe. *Annls. Biol. anim. Biochim. Biophys.* **15**, 345–352.

Robinson, T. J. (1951). The control of fertility in sheep. II. The augmentation of fertility by gonadotrophin treatment of the ewe in the normal breeding season. *J. agric. Sci.* **41**, 6–63.

Robinson, T. J. (1954). The production of coincident oestrus and ovulation in the anoestrous ewe with progesterone and pregnant mare serum. *J. Endocrin.* **10**, 117–124.

Robinson, T. J. (1961). The time of ovulation and efficiency of fertilisation following progesterone and pregnant mare serum treatment in the cyclic ewe. *J. agric. Sci.* **57**, 129–135.

Rowson, L. E. A. (1951). Methods of inducing multiple ovulation in cattle. *J. Endocrin.* **7**, 260–270.

Rowson, L. E. A., Lamming, G. E. and Fry, R. M. (1953). The relationship between ovarian hormones and uterine infection. *Vet. Rec.* **65**, 335–340.

Rowson, L. E. A., Lawson, R. A. S. and Moor, R. M. (1971). Production of twins in cattle by egg transfer. *J. Reprod. Fert.* **25**, 261–268.

Saumande, J., Chupin, D., Mariana, J. C., Ortavant, R. and Mauléon, P. (1978). Factors affecting the variability of ovulation rates after PMSG stimulation. *In* "Control of Reproduction in the Cow" (J. M. Sreenan, ed.), pp. 195–224. Martinus Nijhoff, The Hague, Boston, London.

Scanlon, P. F. (1972). Frequency of transuterine migration of embryos in ewes and cows. *J. Anim. Sci.* **34**, 791–794.

Scanlon, P., Sreenan, J. and Gordon, I. (1968). Hormonal induction of superovulation in cattle. *J. agric. Sci.* **70**, 179–182.

Schally, A. V., Arimura, A. and Kastin, A. J. (1973). Hypothalamic regulatory hormones. *Science* **179**, 341–350.

Schams, D., Menzer, C., Schallenberger, E., Hoffmann, B., Hahn, J. and Hahn, R. (1978). Some studies on pregnant mare serum gonadotrophin (PMSG) and on endocrine responses after application for superovulation in cattle. *In* "Control of Reproduction in the Cow" (J. M. Sreenan, ed.), pp. 122–143. Martinus Nijhoff, The Hague, Boston, London.

Segal, D. H. and Baker, R. D. (1973). Maintenance of corpora lutea in prepuberal gilts. *J. Anim. Sci.* **37**, 762–767.

Seidel, G. E., Larson, L. L., Spilman, C. H., Hahn, J. and Foote, R. H. (1971). Culture and transfer of calf ova. *J. Dairy Sci.* **54**, 923–926.

Shareha, A. M., Ward, W. R. and Birchall, K. (1976). Effect of continuous infusion of gonadotrophin-releasing hormone in ewes at different times of the year. *J. Reprod. Fert.* **46**, 331–340.

Shaw, G. A., McDonald, B. E. and Baker, R. D. (1971). Fetal mortality in the prepuberal gilt. *Can. J. Anim. Sci.* **51**, 233–236.

Short, R. V. (1964). Ovarian steroid synthesis and secretion *in vivo*. *Rec. Prog. Hormone Res.* **20**, 303–333.

Smith, P. E. and Engle, E. T. (1927). Experimental evidence regarding the rôle of the anterior pituitary in the development and regulation of the genital system. *Amer. J. Anat.* **40**, 159–217.

Sreenan, J. M. and Gosling, J. P. (1977). The effect of cycle stage and plasma progesterone level on the induction of multiple ovulations in heifers. *J. Reprod. Fert.* **50**, 367–369.

Tervit, H. R., Rowson, L. E. A. and Brand, A. (1973). Synchronisation of oestrus in cattle using a prostaglandin $F_{2\alpha}$ analogue (ICI 79 939). *J. Reprod. Fert.* **34**, 179–181.

Testart, J. (1972). Synchronisation des ovulations induites chez le veau femelle impubère. *Proc. 7th Int. Congr. Anim. Reprod. and Artif. Insem., Munich,* **1**, 493–498.

Testart, J. (1975). Obtention et transplantation d'oeufs divisés chez les bovins. Thèse doctorat d'État, Université de Paris VI.

Testart, J., Kann, G., Saumande, J. and Thibier, M. (1977). Oestradiol-17β, progesterone, FSH and LH in prepubertal calves induced to superovulate. *J. Reprod. Fert.* **51**, 329–336.

Thibault, C. (1977). Are follicular maturation and oocyte maturation independent processes? *J. Reprod. Fert.* **51**, 1–15.

Thibault, C. and Gérard, M. (1970). Facteur cytoplasmique nécessaire à la formation du pronucleus mâle dans l'ovocyte de Lapine. *C.R. Acad. Sci., Paris,* **270**, 2025–2026.

Thibault, C. and Gérard, M. (1973). Cytoplasmic and nuclear maturation of rabbit oocytes *in vitro*. *Annls. Biol. anim. Biochim. Biophys.* **13**, 145–156.

Thimonier, J., Mauléon, P., Cognié, Y. and Ortavant, R. (1968). Déclenchement de l'oestrus et obtention précoce de gestations chez des agnelles à l'aide d'éponges vaginales imprégnées d'acetate de fluorogestone. *Annls Zootech.* **17**, 275–288.

Trounson, A. O., Willadsen, S. M. and Moor, R. M. (1976). Effect of prostaglandin analogue Cloprostenol on oestrus, ovulation and embryonic viability in sheep. *J. agric. Sci.* **86**, 609–611.

Trounson, A. O., Willadsen, S. M. and Moor, R. M. (1977). Reproductive function in prepubertal lambs: ovulation, embryo development and ovarian steroidogenesis. *J. Reprod. Fert.* **49**, 69–75.

Tsafriri, A., Lindner, H. R., Zor, U. and Lamprecht, S. A. (1972). Physiological role of prostaglandins in the induction of ovulation. *Prostaglandins* **2**, 1–10.

Turner, H. N. (1978). Selection for reproduction rate in Australian Merino sheep: direct reponses. *Aust. J. agric. Res.* **29**, 327–350.

Wallace, L. R. (1955). Studies in the augmentation of fertility of Romney ewes with pregnant-mare serum. *J. agric. Sci.* **45**, 60–79.

Walles, B., Gröschel-Stewart, U., Owman, Ch., Sjöberg, N.-O. and Unsicker, K. (1978). Fluorescence histochemical demonstration of a relationship between adrenergic nerves and cells containing actin and myosin in the rat ovary, with special reference to the follicle wall. *J. Reprod. Fert.* **52**, 175–178.

Webb, R., Lamming, G. E., Haynes, N. B., Hafs, H. D. and Manns, J. G. (1977). Response of cyclic and post-partum suckled cows to injections of synthetic LH-RH. *J. Reprod. Fert.* **50**, 203–210.

Webel, S. K. (1978). Ovulation control in the pig. *In* "Control of Ovulation" (D. B. Crighton, G. R. Foxcroft, N. B. Haynes and G. E. Lamming, eds), pp. 421–434. Butterworths, London and Boston.

Webel, S. K. and Dziuk, P. J. (1974). Effect of stage of gestation and uterine space on prenatal survival in the pig. *J. Anim. Sci.* **38**, 960–963.

Willett, E. L. (1953). Egg transfer and superovulation in farm animals. *Iowa State College J. Sci.* **28**, 83–100.

Willett, E. L., Buckner, P. J. and McShan, W. H. (1953). Refractoriness of cows repeatedly superovulated with gonadotrophins. *J. Dairy Sci.* **36**, 1083–1088.

Yeates, N. T. M. (1949). The breeding season of the sheep with particular reference to its modification by artificial means using light. *J. agric. Sci.* **39**, 1–43.

Chapter IV

Mating, Sperm Transport in the Female Genital Tract, and Artificial Insemination

Introduction

Although in induced (reflex) ovulators such as the cat, rabbit and ferret, coitus initiates ovulation via a neuro-endocrine link (see Chapter I), the prime function of mating in both induced and spontaneously ovulating animals is to deposit semen in the female genital tract so that viable spermatozoa will ultimately have access to a recently-ovulated egg. The site of semen deposition during mating varies according to species, and this in turn influences the nature and complexity of the subsequent processes of sperm transport to the site of fertilisation in the Fallopian tube. Before dealing with these aspects, which clearly will focus upon the female genital tract, it is important to make a brief statement about the rôle of the male.

Semen

Semen, the fluid produced at ejaculation, is composed of a cellular fraction—the spermatozoa—and a fluid vehicle in which the cells are suspended—the seminal plasma. At the onset of puberty and not before, spermatozoa are formed in the testes by a series of highly specialised mitotic and meiotic divisions originating in the spermatogonia of the germinal epithelium (Fig. IV.1). These divisions result in the production of haploid cells of X or Y genotype with respect to the sex chromosomes, containing very little residual cytoplasm although some of the cytoplasmic organelles such as mitochondria and Golgi derivatives can still be identified. The spermatozoa liberated by smooth muscle contractions of the male tract at ejaculation do not come directly from the testes, but rather from the vas deferens and more particularly from the caudal region of the epididymis (Figs IV.2 and IV.3). After formation in the seminiferous tubules of the testes, sper-

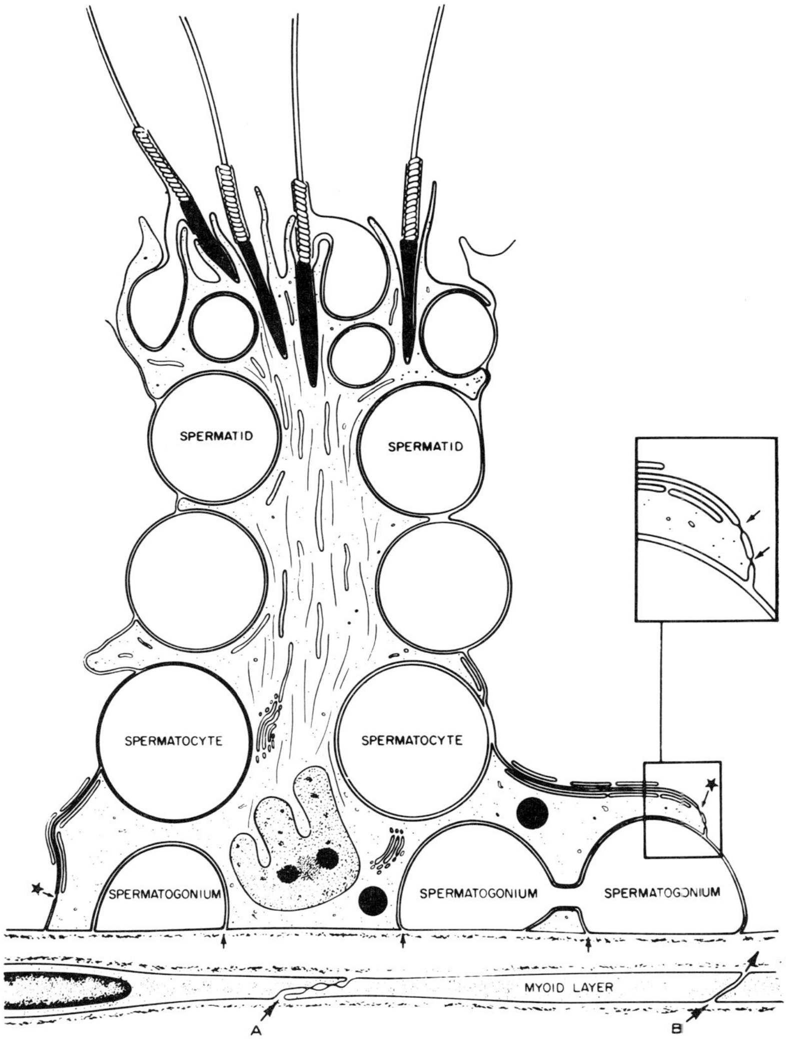

Fig. IV.1. A diagrammatic illustration of the process of spermatogenesis, showing the formation of spermatozoa from spermatids after the reduction divisions associated with the primary and secondary spermatocyte stages. (From Dym and Fawcett, 1970, by courtesy of the Editor, *Biology of Reproduction.*)

matozoa pass as a dilute suspension in tubular fluid via the rete testis and efferent ducts into the highly convoluted but single duct of the epididymis. The sperm suspension is greatly concentrated in the initial section of the epididymis, the caput (Fig. IV.3), and the spermatozoa

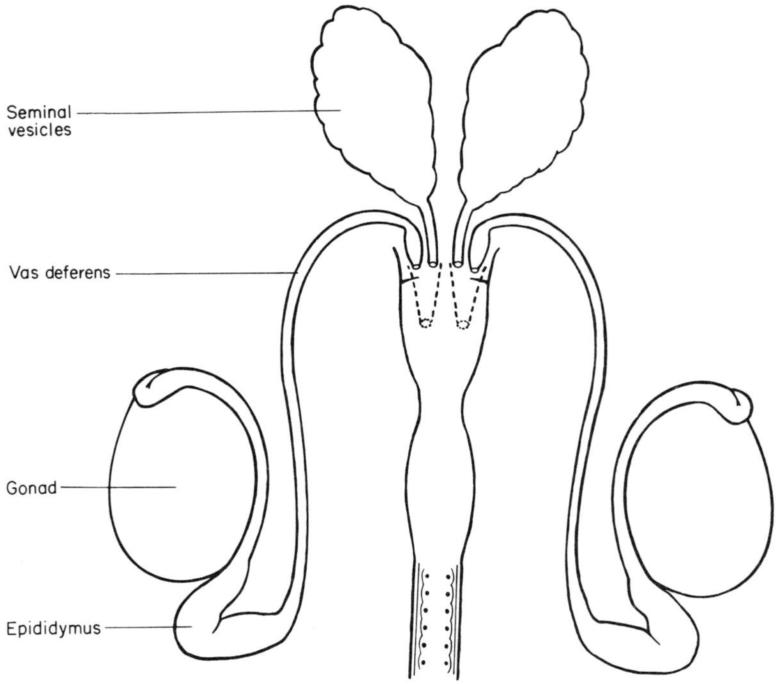

Fig. IV.2. Generalised view of the male reproductive organs, showing the gonads, epididymis and vas deferens, and the accessory glands associated with the pelvic urethra that yield the bulk of the seminal plasma. (Redrawn after Glover, 1974.)

undergo further morphological and physiological maturation in this organ during a period of some 10–21 days or more in the domestic species (Orgebin-Crist, 1969; Glover, 1974; Bedford, 1975). In normal circumstances, it is only after this phase of maturation and when the still immotile spermatozoa have reached the terminal portion of the epididymis, the cauda, that they are competent to be liberated at ejaculation via the vas deferens and urethra. The general structure of a mammalian spermatozoon at ejaculation is shown diagrammatically in Fig. IV.4, but it is worth emphasising that there is an enormous range in the morphology of mammalian spermatozoa (see Roosen-Runge, 1977), and that even within a single ejaculate there is a considerable variation in the shape and functional potential of the sperm cells.

The seminal plasma, on the other hand, derives only a minute fraction of its fluid from the seminiferous tubules and is composed principally of the secretions of the accessory glands. The major acces-

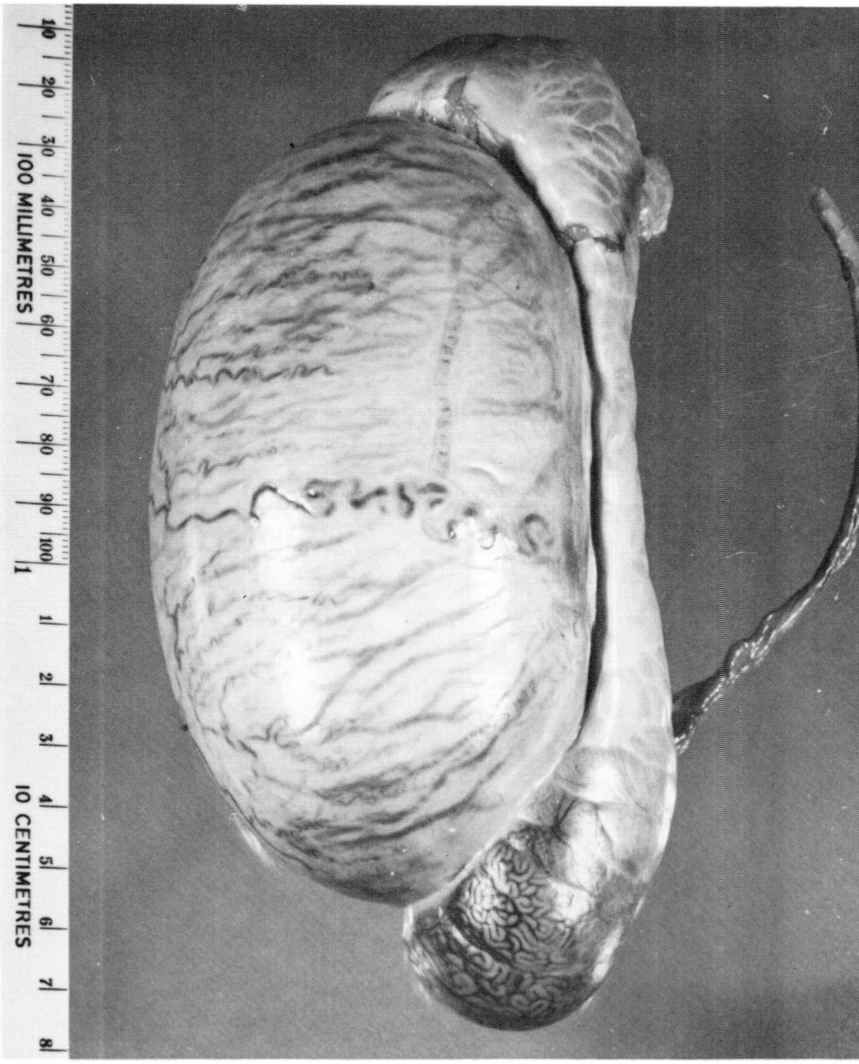

Fig. IV.3. The epididymis of a mature boar showing the close association of this structure with the gonad. The epididymal duct is extensively coiled, as can be noted from its distended caudal region from which spermatozoa pass to the vas deferens.

sory glands are the seminal vesicles, the prostate and the bulbo-urethral (Cowper's) glands (Fig. IV.2), but the ampullae of the vasa deferentia may also secrete a specific seminal fluid. The function of these glands is largely regulated by testicular androgens. There is some contribution

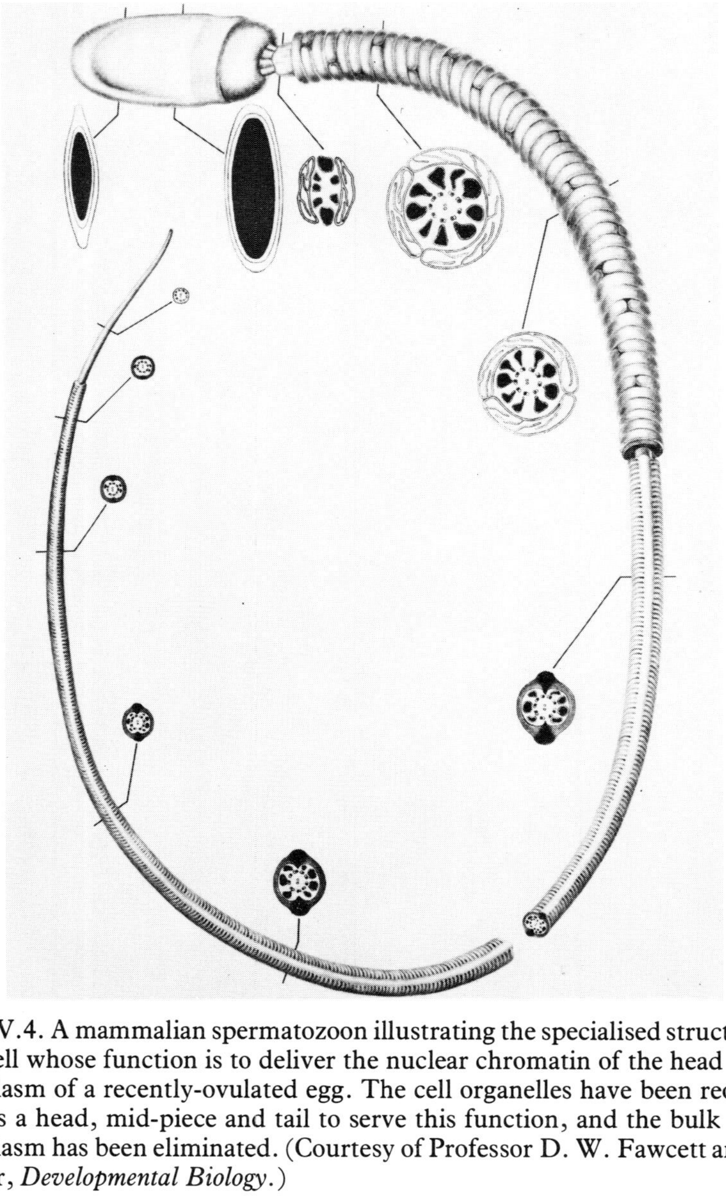

Fig. IV.4. A mammalian spermatozoon illustrating the specialised structure of this cell whose function is to deliver the nuclear chromatin of the head to the cytoplasm of a recently-ovulated egg. The cell organelles have been reorganised as a head, mid-piece and tail to serve this function, and the bulk of the cytoplasm has been eliminated. (Courtesy of Professor D. W. Fawcett and the Editor, *Developmental Biology*.)

also of epididymal secretions to the seminal plasma: in the context of stabilising the sperm plasma membrane and the underlying content of acrosomal enzymes, this secretion together with that of the rete testis may be of considerable significance, but on a volumetric basis it is very minor indeed. The contents of the accessory glands are liberated by powerful contractions at the moment of ejaculation, facilitating mixture of spermatozoa and seminal plasma in the pelvic urethra prior to discharge and providing the initial stimulus to sperm motility. Details of the volume of semen and density of spermatozoa in the ejaculate of some domestic species are presented in Table IV.1. The actual composition of seminal plasma, that is the biochemistry of semen, forms a discipline in its own right (Mann, 1964), and seminal analysis for specific biochemical markers originating in separate accessory glands may be a valuable diagnostic tool in clinical studies of male sub-fertility or infertility (see Mann, 1972).

Semen deposition in female tract

In eutherian mammals, semen is deposited during mating into either (a) the anterior vagina or (b) the uterus (Fig. IV.5). Species representative of the former group include the rabbit, ruminants such as the cow, sheep and goat, and primates; the latter group in which semen enters the uterus directly or is forced through the cervical canal at ejaculation includes the pig, horse, dog and many rodents such as hamsters, rats and mice. The cervix is more complex in the rodents and lagomorphs than is suggested diagrammatically in Fig. IV.5, for the two uterine horns terminate in twin cervices giving a duplex tract. The series of basic forms of the female genital tract from duplex to simplex is illustrated in Fig. I.8. Semen enters the female tract under the propulsive force of the ejaculatory contractions in the male duct system, and further physical aspects that may be important in facilitating successful semen deposition include the pumping action of the penis (although *not* in pigs) and smooth muscle contractions of the female genital system (see below). In addition, semen may undergo a degree of coagulation in the female to aid its initial retention, as in primates and certain rodents, or a gelatinous fraction of the semen may act as a plug as occurs in the pig and guinea pig.

It is worth mentioning at this stage that whilst the practice of artificial insemination is successfully employed in many animal breeding and commercial farming schemes, it is clearly unable to mimic all these aspects occurring at natural mating. Therefore, to obtain levels of fertility comparable to those found in systems where the male has free and continued access to oestrous females, artificial insemination may

Table IV.1. Characteristics of ejaculated semen from normal males and details relevant to mating or insemination. Note particularly the volume of semen and total number of spermatozoa ejaculated by the boar.

Species	Volume of ejaculate (ml)	Expected sperm concentration ($X10^6$/ml)	Total number of sperm per ejaculation ($X10^9$)	Number of collections per week	Site of semen deposition at		Potential number of inseminations per ejaculate	Volume of inseminate after dilution (ml)	Number of motile sperm ($X10^6$)
					mating	A I			
Bull	4–8	1200–1800	4–14	3–4	Anterior vagina	Cervix and/or uterus	400	0·25–1·0	5–15
Ram	0·8–1·2	2000–3000	2–4	12–20	Anterior vagina	External cervical os	40–60	0·05–0·2	50
Boar	150–500[1]	200–300	40–50	2–4	Uterus[2]	Uterus[2]	15–30	50–100	2000
Stallion	30–>150	100–250	3–15	2–6	Uterus[2]	Uterus[2]	5	20–50	1500[3]

[1] Includes gelatinous secretion of the bulbo-urethral glands. [2] The ejaculate makes passing contact with the cervical canal. [3] Total number deposited in three inseminations during the prolonged period of oestrus.

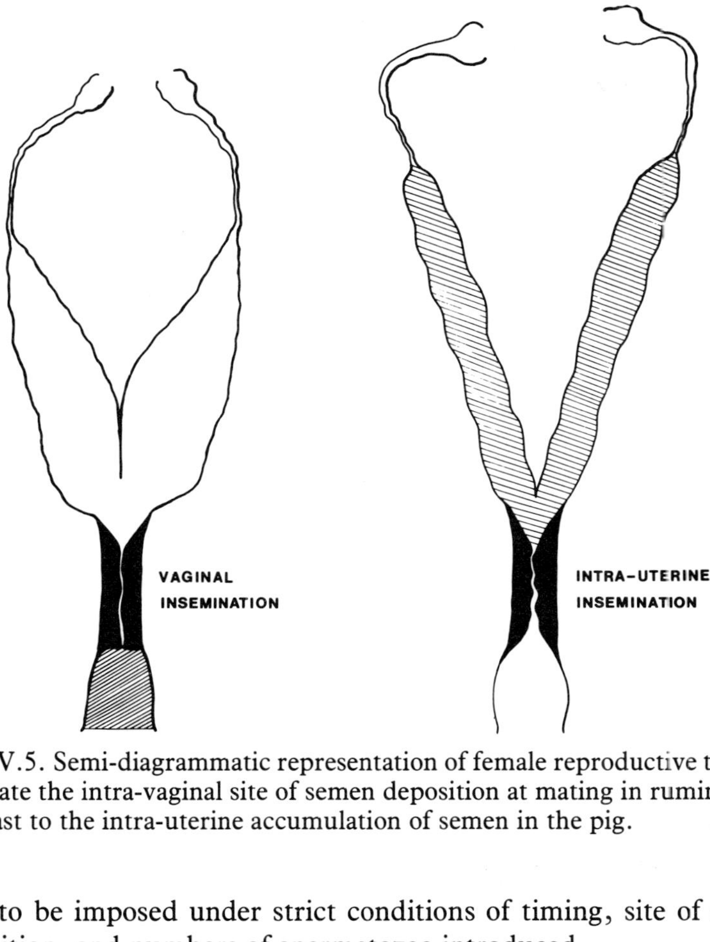

Fig. IV.5. Semi-diagrammatic representation of female reproductive tracts to illustrate the intra-vaginal site of semen deposition at mating in ruminants in contrast to the intra-uterine accumulation of semen in the pig.

need to be imposed under strict conditions of timing, site of semen deposition, and numbers of spermatozoa introduced.

Removal of seminal plasma

Seminal plasma certainly has functional significance at ejaculation and in the earlier phases of sperm transport as a medium of suspension and it may also serve as an important metabolic substrate for spermatozoa with its specialised constituents such as fructose, citric acid and inositol (see Mann, 1964). Nonetheless, seminal plasma appears to be stripped from the surface of the sperm cells before they reach the site of fertilisation, irrespective of the initial site of semen deposition at mating. In fact, in the biological situation close to the time of ovulation, seminal plasma is unlikely to enter the Fallopian tubes in detectable quantities, this statement being supported by both biochemical (Mann *et al.*, 1956b) and physiological evidence (Hunter and Hall, 1974; Hunter *et al.*, 1978). Moreover, instillation of cell-free seminal plasma directly into the tubes delays or inhibits the process of fertilisation (Chang, 1957; Hunter and Hall, 1974), and may also have deleterious effects upon the egg membranes. Consistent with these observations, numerous experiments have demonstrated that suspensions of epididymal spermatozoa, particularly those prepared from the cauda epididymidis, promote perfectly normal fertilisation and embryonic development when inseminated directly into the uterine lumen (Young, 1931; Orgebin-Crist, 1967; Holtz and Smidt, 1976) or Fallopian tubes (Bedford, 1966; Hunter *et al.*, 1976), thereby demonstrating no mandatory rôle for the bulk of the seminal plasma or its specific constituents in the events preceding fusion of the gametes.

Transport of spermatozoa

Before describing similarities and differences in the processes of sperm transport and storage in cows, sheep and pigs, a number of basic facts relevant to the arrival of a population of competent spermatozoa at the site of fertilisation must be presented. The normal location of the recently-ovulated eggs at the time of sperm penetration is at the ampullary-isthmic junction of the Fallopian tube (see Chapter V).

First, irrespective of the precise site of semen deposition, it is necessary to distinguish between *sperm transport*, which implies a passive movement of cells in the genital tract, and *sperm migration* which clearly attributes great importance to the intrinsic motility of the cell. Although early investigators were inclined to view sperm ascent along the female tract as an outcome of the sperm cell's own movements, it is now generally accepted that the rôle of sperm motility becomes critical only in certain regions of the tract (i.e. cervix and utero-tubal junction) and during penetration of the egg membranes. The rapidity with which vanguard spermatozoa reach the upper portions of the Fallopian tubes after mating or artificial insemination has emphasised that sperm motility alone could not have contributed significantly to passage over the relatively large distances involved. The journey is probably effected at all stages by contractile mechanisms in the smooth musculature of the tract, part of the transportation process undoubtedly being assisted by central nervous mechanisms set in motion by coital stimulation. Smooth muscle stimulants such as prostaglandins of the E series are found in the semen of rams and primates (Von Euler, 1936), as in the latter group are certain hydroxy derivatives of prostaglandin E_1 (Kelly *et al.*, 1976), but their function in the process of sperm transport in the female tract is difficult to specify. Nor is it possible among the farm species to draw any simple correlation between the site of semen deposition at mating and the seminal content of prostaglandins (Table IV.2). In any event, when the duration of oestrus is long and mating

Table IV.2. The concentration of prostaglandin E_2 and $F_{2\alpha}$ in the semen of three species in which the ejaculate passes into the uterus at the time of mating. The results were obtained by bioassay on the rat fundus. (Taken from Hunter and Poyser, 1973.)

Species	Method of semen collection	Content of prostaglandin (ng/ml) E_2	Content of prostaglandin (ng/ml) $F_{2\alpha}$
Pig	Manual, directly from penis	$<0{\cdot}27$	$<0{\cdot}34$
Pig	Aspirated from uterine lumen	$<1{\cdot}1$	$<2{\cdot}8$
Dog	Manual, directly from penis	$<1{\cdot}1$	$<2{\cdot}1$
Horse	Artificial vagina	24·1	0·66

takes place early, active migration of spermatozoa from storage regions such as the lower isthmus of the Fallopian tube must be of physiological significance, especially since the enhanced muscular contractions of the uterus and tubes promoted by oxytocin released at coitus would long since have ceased. As a corollary, it may be inferred that rapid post-coital transport of spermatozoa to the ampulla in this latter situation would have had little influence on fertility.

Second, in comparison with the number of spermatozoa deposited in the female tract at mating, there is a massive reduction in sperm numbers as one proceeds towards the ampulla of the Fallopian tube. At a given time, this remarkable sperm gradient may extend from the order of 10^8 cells per ml in fluid bathing the cervical os or utero-tubal junction down to a total of approximately 10^2 cells or less in the vicinity of the eggs (Fig. IV.6). The gradient has been reasoned to reduce the chances of polyspermy by restricting the numbers of spermatozoa that achieve the site of fertilisation. The regions of the female tract that serve to regulate the numbers of spermatozoa, such as the cervix where semen is deposited intra-vaginally (rabbit, ruminants and primates) and the

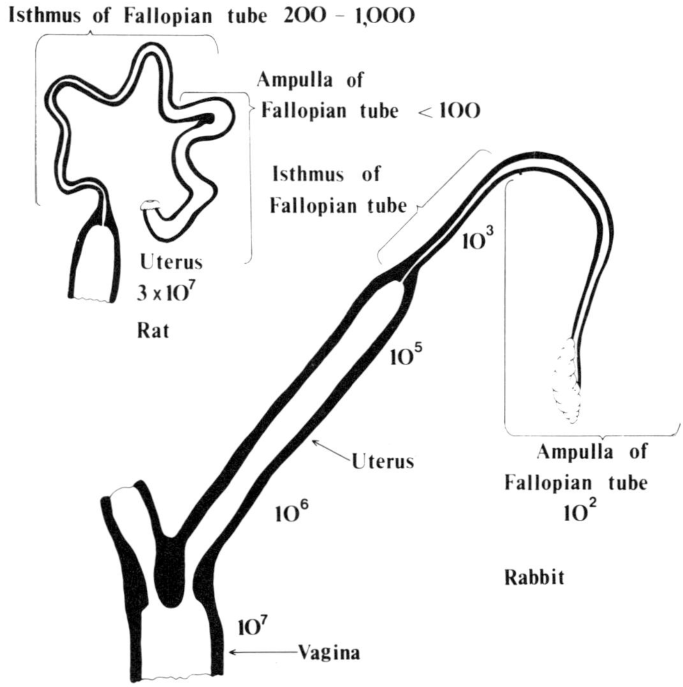

Fig. IV.6. Representation of one side of a female reproductive tract to indicate the steeply-declining gradient in sperm numbers found between the vagina and the site of fertilisation in the Fallopian tube. (Redrawn after Austin, 1965.)

utero-tubal junction and isthmus in the pig, also function as reservoirs for the storage of spermatozoa, at least until the time of ovulation.

Third, during this marked diminution in sperm numbers, the cells are progressively removed from the seminal plasma, and their economy becomes strongly influenced by the luminal fluids of the uterus and Fallopian tubes. This situation appears to have a direct bearing on the metabolic activity of ejaculated spermatozoa, and may also be associated with development of the capacitated state and the concomitant increase in sperm motility (see Bedford, 1970a, Yanagimachi, 1970; Chang and Hunter, 1975).

Sperm selection in female tract

The topic of the viable lifespan of spermatozoa in the female genital tract is clearly of fundamental importance in any consideration of fertility. But because the ejaculate contains a widely heterogeneous collection of cells varying in morphology, motility, degree of maturation (ripeness), agglutination potential and other factors, an analysis of viability needs to consider whether any form of sperm selection is operating in the female tract such that the more vigorous spermatozoa have a greater chance of reaching the site of fertilisation. Studies using mixtures of differentially labelled live and dead spermatozoa (Baker and Degen, 1972), frozen-thawed spermatozoa (Polge *et al.*, 1970) or diploid marker spermatozoa (Mortimer, 1977) have all strongly suggested that there is a progressive selection of cells between the site of deposition and the site of fertilisation, which is perhaps best viewed as an elimination of less viable spermatozoa. The interpretation here is that the spermatozoa are selecting themselves (i.e. auto-selection) due to fitness or rather to lack of fitness (i.e. poor motility), especially at the various "barriers" in the reproductive tract, rather than there being some specific selection on the basis of, for example, subtle membrane or antigenic characteristics.

As far as sperm transport in mammals is concerned, only the work of Cohen's group argues that there is some form of active selection operating in the female tract against spermatozoa with an imperfect genome, especially in the region of the utero-tubal junction. Cohen argues for selection of a small population of viable spermatozoa on the basis of calculations of chiasma frequency and sperm redundancy values in a range of mammals (Cohen, 1967), and also from the results of studies with labelled spermatozoa in mice and rabbits (Cohen and McNaughton, 1974; Cohen and Werrett, 1975). The mechanism whereby (a) the virgin female is programmed to select spermatozoa and (b) the specific detection and elimination of defective spermatozoa take place remains to be enunciated. An alternative explanation of Cohen's data might simply be that the populations of spermatozoa recovered from successively higher regions of the female tract were indeed "fitter" in that they contained fewer non-viable, poorly motile or moribund spermatozoa than those from the lower regions of the tract or ejaculate. As such they would of course be selected populations, but not ones produced by any active selection against genetically abnormal gametes by the female tract (Mortimer, 1978).

In some contrast to views on selection which attribute importance to sperm progression through the female tract, it is worth pointing out that normal fertilisation has been obtained after intraperitoneal insemination in several species of mammal including rabbits (Dauzier and

Thibault, 1956), guinea pig (Rowlands, 1957), cattle (Skjerven, 1955) and pigs (Hunter, 1978). Following insemination of spermatozoa into the body cavity, a progressive and selective influence of the female tract upon the sperm population would not have been exerted, for only the fimbriated extremity and ampullary region of the tube intervene between this site of experimental insemination and the site of fertilisation. Embryonic development has been shown to follow a normal course after intraperitoneal insemination.

Quantitative aspects of sperm transport

The number of spermatozoa reaching the site of fertilisation at a given time after preovulatory mating or insemination has a direct bearing on the probability of conception. Accordingly, the paragraphs that follow record some details on the timing and efficiency of sperm transport derived from experimental observations in the large farm species.

Cow

The rate at which spermatozoa may be transported from the anterior vagina to the Fallopian tubes of cattle has been the subject of controversy for many years (Thibault *et al.*, 1973). On the one hand, evidence has been presented suggesting rapid transport of live or dead spermatozoa to the upper Fallopian tube in a matter of minutes (e.g. Van Demark and Moeller, 1951; Van Demark and Hays, 1954), inferring the importance of smooth muscle contractions in this situation. Conversely, and from other laboratories, only a slow and regulated progression of sperm cells over a period of hours has been noted (Thibault, 1973). In the biological situation in which spermatozoa would be deposited during oestrus and hence at least 10 to 12 hours before ovulation, the rate of sperm transport is unlikely to become a critical factor determining fertility. But under conditions of artificial insemination, and particularly when this forms part of an oestrous synchronisation and/or gonadotrophin treatment programme (see Chapter II), the rate and efficiency of sperm transport to the site of fertilisation are of fundamental importance.

At mating, in contrast to the conditions found under artificial insemination, the bull deposits approximately 4–8 ml of semen in the anterior vagina (Table IV.1) and the male secretions usually bathe the external cervical os immediately upon ejaculation. Penetration of gametes into the cervical canal may depend upon muscular activity in the female tract, which will be enhanced by a surge of oxytocin released at coitus, but entry into the canal is also dependent on vigorous sperm motility; in both situations, the condition of the cervical mucus is critical for successful colonisation of the cervix (Mattner, 1966, 1968; Gibbons, 1969; Bedford, 1970b), the latter region providing the principal post-coital reservoir of spermatozoa in ruminants (Mattner, 1968), at least during the first 24 hours (Thibault, 1973). Presumptive sperm storage in the uterine glands of cattle has been documented, but a further important region of storage for 48 hours or more also occurs in the folds of the lower isthmus and region of the utero-tubal junction (Thibault, 1973; Thibault *et al.*, 1973), where spermatozoa may be protected from phagocytosis. The utero-tubal junction does not present a significant barrier to spermatozoa or fluids in oestrous or periovulatory cattle, for Rowson (1955) has demonstrated patency of this junction to radio-opaque fluid.

On the basis of meticulous histological work, it is thought unlikely that significant numbers of spermatozoa enter the Fallopian tubes of oestrous cows within 2 hours of mating, although they are found in the ampulla by 8 hours (Thibault *et al.*, 1973). The cautions should be added, however, that in this study the reproductive tracts of only two cows were examined prior to 8 hours *post-coitum*, and as animals were mated in the first half of oestrus, the possibility of an accelerated transport of spermatozoa nearer the time of ovulation should not be neglected. Nonetheless, the earlier reports of Van Demark and Moeller (1951) and Van Demark and Hays (1954) that spermatozoa can achieve the oviducal ampulla within 2·5–3·3 minutes of mating or insemination may require re-interpretation. As suggested above, there would be little biological advantage in a mechanism having evolved for such rapid transport, for the male gametes would still be awaiting release of the egg(s) at ovulation.

The distribution of spermatozoa in the isthmus of the bovine Fallopian tube is very regular at 8 hours after mating, but muscular activity has clustered the sperm by 18 hours (Thibault, 1973). Sperm counts on sectioned material reveal that only a small fraction of the ejaculate reaches the upper tube (Thibault *et al.*, 1973), and from the point of view of avoiding the pathological condition of polyspermic penetration of the egg cytoplasm, it is important that this should be so. Microscopical examination of the eggs and their surrounding membranes suggests that very few spermatozoa are present in the ampulla at the time of fertilisation, and counts of spermatozoa at a series of levels in the isthmus reveal an abrupt diminution in cell numbers a few centimetres below the ampulla (Thibault, 1973). But it is still possible that the contents of the ovulating follicle, the so-called products of ovulation (Harper, 1973), may stimulate a release of spermatozoa from the lower regions of the Fallopian tube as apparently occurs in rabbits (Harper, 1973; Overstreet and Cooper, 1975) and guinea pigs (Yanagimachi and Mahi, 1976).

Sheep

Sperm transport in the ewe has several parallels with the situation found in cattle in terms of the rapidity of coitus, the site of semen deposition in the anterior vagina, the regulatory and storage rôles of the cervix, and to a lesser extent the temporal relationship between oestrus and ovulation. In addition, however, the adverse effects of stress on sperm transport to the Fallopian tubes and on the incidence of fertilisation have been examined more comprehensively in the ewe than in any of the other farm species. The same comment is also true of analyses of sperm transport and depressed fertility following procedures for synchronisation of oestrus (Chapter II).

Colonisation of the cervix by spermatozoa from the vaginal pool of semen and especially from semen actually bathing the cervical os depends very largely on sperm motility (see Lightfoot and Restall, 1971). This does not preclude the possibility of a fraction of the ejaculate entering directly into the cervical canal due to major contractions of the genital tract at mating, a situation which could give a gross displacement of the contents of the canal. Such a phenomenon might afford a partial explanation of discrepancies in estimates of the timing of sperm transport to the Fallopian tubes. But there is no evidence for significant amounts of seminal plasma entering the uterus following mating of ruminants or rabbits (see Walton, 1930; Asch *et al.*, 1977). The physical properties of the cervical mucus are under endocrine control and are critical for successful colonisation by spermatozoa, for the cells swim from the inseminate into the intermicellar fluid phase of the mucus at the external cervical os. Once within the flow lines of the mucus (Fig. IV.7), spermatozoa are constrained to swim along them and are thus guided to the mucosa of the cervical crypts where a sperm reservoir is formed (Tampion and Gibbons, 1962; Mattner, 1966).

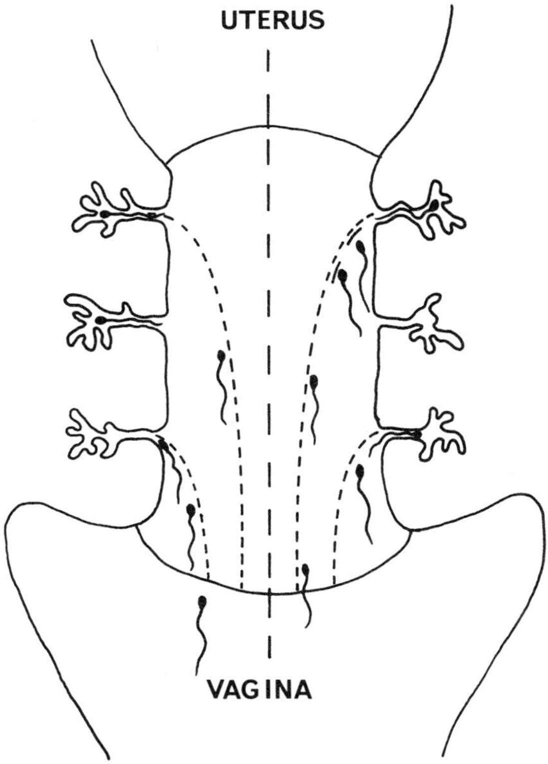

Fig. IV.7. Representation of the lines of strain in the mucus originating from the cervical crypts at oestrus, to suggest the manner in which motile spermatozoa entering the cervical canal are orientated to swim towards the crypts. A primary reservoir of spermatozoa is thought to be formed in this manner in the cervix of ruminants. (Reproduced from Bedford, 1970b.)

Whilst such a pattern of migration towards epithelial surfaces may account for the majority of spermatozoa entering the cervix, some sperm are almost certainly not constrained to enter the crypts, but rather pass through the mucus in the central portion of the canal, thereby gaining the uterus with some rapidity, possibly almost immediately after mating.

It should be noted in passing that one of the explanations of a diminished conception rate in sheep and cattle treated with progestagens invokes a change in the condition of the cervical mucus such that the efficiency of establishment of a cervical population of spermatozoa and their subsequent migration are impaired at the first oestrus after cessation of treatment. Nonetheless, the extent to which the cervix acts as a long-term reservoir for those spermatozoa that actually effect fertilisation needs to be questioned. It could be argued from evidence on the optimum time of mating that a prolonged interval, possibly extending up to 24 hours, is required for accumulation of sufficient spermatozoa in the Fallopian tubes to ensure the maximum chance of fertilisation; and certainly the time relationship between the onset of oestrus and ovulation in sheep (see Chapter III) would endorse this viewpoint. In such a situation, it is questionable whether there would be a prolonged migration of spermatozoa from a cervical reservoir to the site of fertilisation. More probable would be a mobilisation of spermatozoa from reserves in the uterus (Thibault and Levasseur, 1973) or from the region of the utero-tubal junction and isthmus.

Compared with the number of cells deposited at ejaculation, the number of spermatozoa entering the uterus by migration from the cervix is massively reduced and, as inferred above, some selection of spermatozoa on the basis of motility would have occurred. The reduction in sperm numbers across the cervix has been shown to be one hundred fold or greater in several mammals (e.g. rabbit and sheep: Braden, 1953; Braden and Austin, 1954). Counts made on the tracts of artificially inseminated sheep indicate an even more abrupt gradient between vaginal and uterine populations, numbers being reduced by a factor of 10^3 or 10^4 (Quinlivan and Robinson, 1969; Robinson, 1973; Table IV.3). In the cervix itself, there is an initial rapid build-up of about 1·5 million spermatozoa within an hour of a single insemination, and a population of about 0·5 million is maintained for up to 36 hours (Quinlivan and Robinson, 1969).

Table IV.3. Mean numbers of spermatozoa recovered from the reproductive tract of ewes at intervals after insemination with 500 X 10^6 spermatozoa. (Data from Quinlivan and Robinson, 1969.) As ewes were inseminated soon after the onset of oestrus, the numbers of spermatozoa in the isthmus 24 hours later suggest that this region of the Fallopian tube forms the functional sperm reservoir that is drawn upon at ovulation.

Time after insemination (h)	Vagina X10^6	Cervix X10^6	Uterus X10^3	Isthmus X10^3	Ampulla X10^3
1	12·3	1·5	10	0·2	0·05
12	0·6	0·4	17	0·7	0·3
24	0·6	0·7	62	3·8	0·7
36	0·5	0·5	28	0·5	0·2
48	0·1	0·1	12	0·2	0·2

Transport of spermatozoa through the uterus depends principally on contractile activity of the myometrium producing segmental waves of propulsion in the uterine wall. The contractions may be enhanced by the action of seminal prostaglandins, the ejaculate of the ram being rich in such smooth muscle stimulants (Von Euler, 1936). However, any significant influence of coitally-released oxytocin on uterine contractions is unlikely to be detectable by the time spermatozoa liberated from the cervix have entered the uterine lumen, but background myometrial activity is always present during oestrus. Disturbances to the endocrine state of ewes following, for example, progestagen treatment (Quinlivan and Robinson, 1969; Raynaud, 1973) or during feeding on clover pastures rich in phyto-oestrogens (Lightfoot *et al.*, 1967; Kaltenbach and Davies, 1970) can lead to an impaired pattern of uterine contractions and sperm transport, and hence to diminished fertility.

As to the timing of sperm transport into the tubes and thus to the site of fertilisation, Dauzier and Wintenberger (1952) and Dauzier (1958) have concluded that spermatozoa do not usually proceed beyond the utero-tubal junction within 2 hours of mating; and Thibault and Wintenberger-Torres (1967) have noted that the duration of sperm transport from the vagina to the tubal isthmus is greater than 1 hour in free mating ewes. On the other hand, Quinlivan and Robinson (1969) have found spermatozoa in both the isthmus and the ampulla of ewes within an hour of insemination; and Mattner (1963) and Mattner and Braden (1963) have reported a phase of rapid transport in animals conditioned to experimental procedures, with spermatozoa present in the Fallopian tubes of a series of ewes some 8–15 minutes after mating. This phase of rapid transport was followed by a prolonged phase of slow transport with increasing numbers of spermatozoa accumulating in the upper reaches of the tract. Mattner and Braden have reasoned that the phase of rapid transport is especially susceptible to the effects of environmental stress (i.e. fright), thus reducing the rapidity of sperm transport in such situations. The deleterious effect of environmental stress on sperm transport in the ewe has been confirmed by Thibault and Wintenberger-Torres (1967), and may be an important factor influencing fertility in artificial insemination programmes in this species.

As there is a continuing controversy over the rate of sperm transport to the site of fertilisation in ruminants, two comments seem appropriate here for the sake of perspective. First, any division of the rate of sperm transport into rapid and slow phases is clearly arbitrary, and not particularly helpful when attempting to describe a dynamic situation (i.e. a continuing prolonged migration). Second, since ovulation does not occur until the end of oestrus or for some hours thereafter, the rapidity of sperm transport to the Fallopian tubes in the early part of oestrus may not be biologically significant. Indeed, the first spermatozoa reaching the tubes as a result of precocious transport may not be involved in the process of fertilisation, but rather undergo early death as has been noted in rabbits (Overstreet and Cooper, 1978). Rapid transport of spermatozoa to the tubes would be valuable in animals mated late in oestrus, and indeed there is circumstantial evidence that the rate of sperm transport in mammals may increase at or just after the time of ovulation (Braden and Austin, 1954; Yanagimachi and Chang, 1963; Turnbull, 1966). In any event, the optimum time for mating has been cited as about 12 hours before ovulation (Dziuk, 1970), this figure being based on competitive mating experiments involving two males serving ewes at different times during oestrus.

Pig

Because a comprehensive review of the physiological aspects of sperm transport, migration and storage in the domestic pig has been presented elsewhere (Hunter, 1975a, b), the paragraphs that follow are relatively brief and many references of historical interest have been omitted.

Coitus in gilts and sows is a protracted affair, on occasions lasting for 8–10 minutes or more whilst the spiral tip of the penis is "locked" in the muscular folds of the cervix. During this period, semen is propelled by muscular contractions of the male tract almost directly into the pars indivisa of the bicornuate uterus, although passing contact is made with the proximal region of the cervical canal (Fig. IV.8). The exceptional

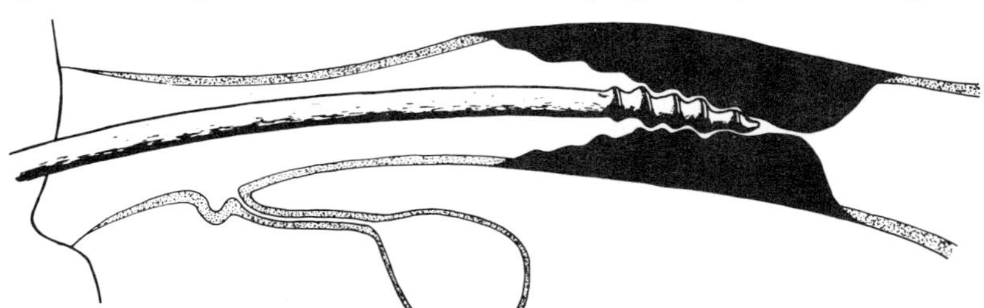

Fig. IV.8. Diagrammatic representation of the manner in which the spiral glans of the boar's penis engages the oedematous and muscular folds of the cervix during mating. As a consequence of this specialisation, the bulk of the voluminous ejaculate passes almost directly into the uterus.

volume of the ejaculate, frequently amounting to nearly half a litre, is emitted in fairly distinct fractions commencing with a watery pre-sperm secretion, followed by a sperm-rich fraction, and terminating with a post-sperm fraction (Glover, 1955; Polge, 1956). Together with the gelatinous portion of the ejaculate which serves to occlude much of the cervical canal, this sequence of seminal fractions may then be repeated in a second wave, and the uterine cornua may be considerably distended by the time of completion of mating. The fluctuation in sperm density during the prolonged ejaculation has been considered of significance for efficient transport of spermatozoa to the Fallopian tubes (Du Mesnil du Buisson and Dauzier, 1955): whether or not this proves to be so, the utero-tubal junction may be bathed by semen before the end of mating. The total number of spermatozoa produced by the boar in a single ejaculate may be as high as $7{\cdot}83 \times 10^{10}$ cells (Rodolfo, 1934), with a concentration varying between $0{\cdot}1$ and 10×10^{8} cells per ml (Polge, 1956).

Because of the oedematous barrier formed by the processes of the utero-tubal junction during oestrus (Fig. IV.9), and hence the severe limitation on access to the lumen of the tubes, a high concentration of cells must blockade the uterine side of this region for a time sufficient to enable formation of a sperm reservoir in the caudal region of the isthmus (Hunter, 1973a): motility is almost certainly important for negotiation of the utero-tubal junction by sufficient numbers of spermatozoa to ensure subsequent fertility. Some form of selection on passage of cells into the Fallopian tubes may also be imposed by the utero-tubal junction early in oestrus, possibly in the form of discrimination against spermatozoa with a damaged plasma membrane (see Polge *et al.*, 1970).

The rapid passage of semen to the utero-tubal junction helps to explain the presence of spermatozoa in the upper reaches of the Fallopian tubes within an hour or so of mating. The shortest transit to be documented for boar spermatozoa to the upper half of the tubes is a period of 15 minutes after mating (Burger, 1952), and sperm penetration of the zona pellucida has been noted within 2 hours of intra-cervical insemination (Hunter and Dziuk, 1968). But as mating under natural conditions would precede ovulation by many hours, the biological importance of such rapid transport of spermatozoa to the site of fertilisation is again questionable. The optimum time of mating or insemination has been suggested as 12–16 hours before ovulation (Boender, 1966; Polge *et al.*, 1968; Dziuk, 1970).

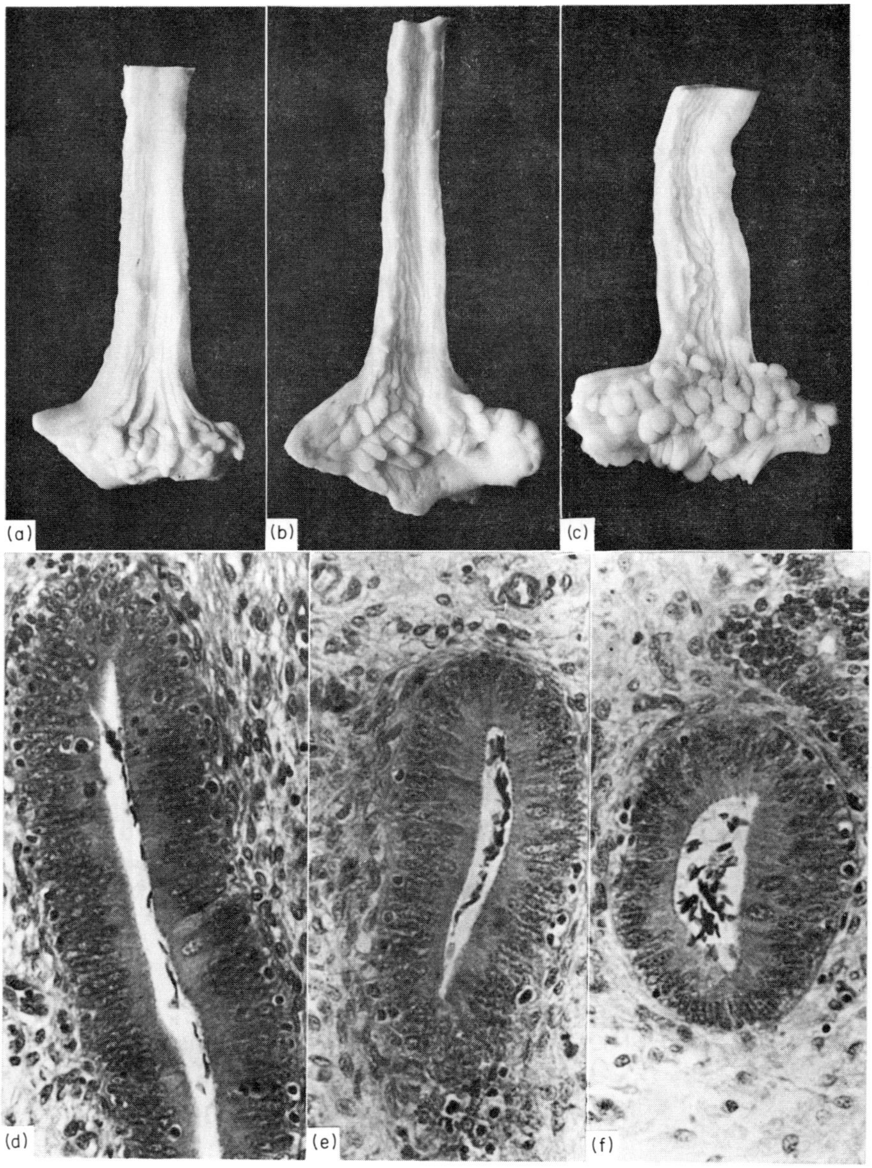

Fig. IV.9. The region of the utero-tubal junction in the pig to show the swollen and oedematous condition of the terminal folds of the isthmus (the polypoid processes) found during oestrus (c); (a) and (b) are sections taken during the early and mid follicular phase. Spermatozoa tend to accumulate in the diverticulae of the utero-tubal junction (d, e, f).

Numerous studies in gilts and sows indicate that only a very small proportion of the massive number of spermatozoa deposited in the female tract at mating or artificial insemination can be recovered from the Fallopian tubes. However, if a sufficient volume of semen is inseminated so that fluid reaches the utero-tubal junction, the number of spermatozoa entering the tubes is directly related to the concentration of spermatozoa in the inseminate (Rigby, 1964; Baker *et al.*, 1968). Moreover, the concentration of sperm at the site of fertilisation is probably a function of the number of spermatozoa persisting at the utero-tubal junction. Although this junction may impede the rapid transport of large numbers of spermatozoa from the uterus to the site of fertilisation, it has also been reported to act as a "reservoir" from which spermatozoa gradually proceed to the isthmus (Du Mesnil du Buisson and Dauzier, 1955; Rigby, 1964). There is good evidence to suggest, however, that the isthmus of the Fallopian tube functions as the principal physiological sperm reservoir in the pig (Hunter and Léglise, 1971), and counts of spermatozoa accumulating in the zona pellucida of fertilised eggs as they descend the isthmus can be invoked to endorse this argument (Hunter, 1972).

Since the site of fertilisation is at the ampullary-isthmic junction, the isthmus must exert its extremely powerful muscular function to constrict the lumen and produce a sperm gradient decreasing from 10^8 cells per ml of semen on the uterine side of the utero-tubal junction down to approximately 10^2 cells or less at the site of fertilisation (Hunter and Léglise, 1971; and Fig. IV.10). Thus, in situations of natural mating in this species, *regulation* rather than *facilitation* of sperm transport in the female tract should be seen as a major physiological requirement, for excessive numbers of spermatozoa at the site of fertilisation lead to polyspermic penetration of the eggs and early embryonic loss (Hunter, 1973b). The actual means of sperm ascent within the isthmus involves waves of contraction in the myosalpinx, as well as active migration by motile spermatozoa. Studies by Gaddum-Rosse and Blandau (1973) have also suggested the importance of "pro-ovarian ciliary currents" in the isthmus in propelling spermatozoa towards the isthmo-ampullary junction. Thus, within the compartment of the oviduct shortly before and after the time of ovulation, waves of both muscular and ciliary activity in the ampulla and isthmus appear to differ in their direction of progression, the net effect being to transport male and female gametes in opposing directions to the site of fertilisation.

Boar spermatozoa do not remain alive in the vagina or cervix for more

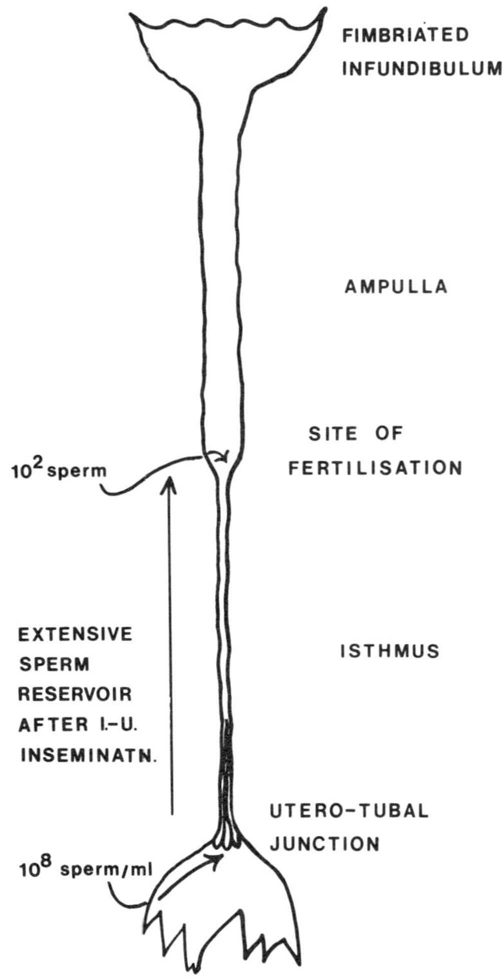

Fig. IV.10. Semi-diagrammatic representation of a Fallopian tube in the pig to illustrate the extent of the gradient in sperm numbers that is obtained in the region of the isthmus in the hours after mating.

than 5 hours, but they can survive elsewhere in the female tract for 24–48 hours, or as long as 70–72 hours in the region of the utero-tubal junction and terminal portions of the lower isthmus (Fig. IV.11). In the uterine horns themselves, not only does the large volume of semen disappear within about 2 hours, but sperm motility is drastically reduced and those spermatozoa that evade phagocytosis are not viable for more than 30 hours (Du Mesnil du Buisson and Dauzier, 1955).

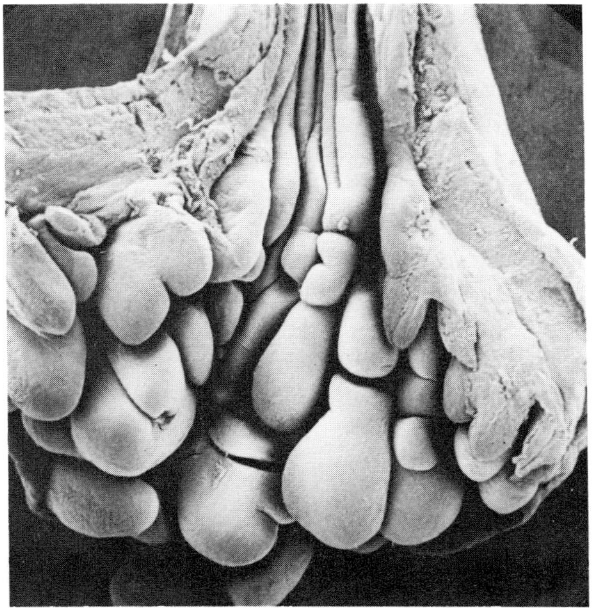

Fig. IV.11. Scanning electron micrograph of the utero-tubal junction and lower isthmus of an oestrous pig to show the complex oedematous arrangement of this region and the prominent polypoid processes. (Courtesy of Dr J. E. Fléchon.)

Horse

Throughout the period of oestrus in the mare, the reproductive tract presents characteristic oedematous features, although the cervix is usually relaxed and flaccid (Day, 1942). During copulation, the glans of the penis becomes greatly engorged and abuts against the cervical os, presumably increasing the degree of dilatation in the canal. The muscular contractions at ejaculation together with the pumping action of the penis thereby transmit much of the semen through the canal into the body of the uterus. In this situation, it is difficult to appreciate the biological necessity for the development of a negative pressure in the uterus during coitus, but such a feature has been claimed, the inference being that passage of semen into the uterus is facilitated. Of greater physiological importance might be the concentration of prostaglandin E_2 in stallion semen (Table IV.2), which would promote contractile activity in the myometrium and other smooth-muscle compartments of the female tract.

The semen of the stallion is voluminous, measurements commonly ranging from 18–170 ml. The volume is influenced by the breed of stallion, and the composition of the semen is also subject to marked seasonal fluctuations. Various fractions can usually be distinguished in the semen: initially, a watery, pre-sperm secretion, followed by a milky, sperm-rich fraction, and terminating with a highly viscous, gelatinous material (Mann *et al.*, 1956a). The considerable volume of fluid deposited in the female tract reduces one of the problems of sperm transport for, shortly after mating, the uterine horns of the mare including the regions adjoining the Fallopian tubes are filled with the products of ejaculation (Mann *et al.*, 1956b). Variation is also found in the density of spermatozoa in the semen, values ranging from $0{\cdot}3$ to 8×10^8 cells per ml. A single ejaculate may contain from 4 to 13×10^9 spermatozoa.

Among the larger species of domestic livestock, there is evidence that spermatozoa of the stallion exhibit one of the longest periods of survival in the female tract. The viability of spermatozoa in the mare has been studied by several authors, the approach being dependent upon the extended period of oestrus, with ovulation occurring towards the end of this time (Day, 1942). Although few observations exist in the literature on the motility of spermatozoa after periods of storage in the female tract, the evidence for prolonged maintenance of fertilising capacity is overwhelming. Day recorded pregnancies in mares inseminated 5 days before ovulation and, in one instance, conception followed a sperm survival interval of 6 days. These figures received support from the subsequent study of Burkhardt (1949), but although the latter author makes the statement that ". . . in services given early in oestrus, the sperms are quite capable of reaching the fimbriae of the Fallopian tube and maintaining fertilising capacity for 90 hours or more . . .", a specific site of sperm storage was not indicated. Once again, the isthmus of the tube seems a likely candidate, not least since the uterus has been considered a hostile environment for spermatozoa (Parker *et al.*, 1975).

Concerning the fate of the seminal plasma and spermatozoa, only relatively small volumes of fluid (3–5 ml) can be recovered from the uterine horns when 30–50 minutes have elapsed since mating, and the concentration of spermatozoa suspended in this fluid may have been reduced significantly (Mann *et al.*, 1956b). Moreover, although there is chemical evidence suggesting that very small amounts of seminal plasma may negotiate the utero-tubal junction and enter the Fallopian tubes within 50 minutes of mating (Mann *et al.*, 1956b), the fate of most ejaculated fluid remains uncertain; retention in the endometrial crypts and/or leakage through the cervical canal are reasonable possibilities. Whether the prolonged periods of sperm storage are associated with specific endometrial glands, or with persistence in the lower portions of the Fallopian tube, also remains to be studied systematically.

Artificial insemination

The practice of artificial insemination is widely accepted to have made a major impact on cattle breeding schemes, especially in the more developed countries with a well organised insemination service, even though paradoxically the scope for artificial insemination would appear greater in the less affluent countries. As a technique for improving performance in farm livestock, it gained acceptance quite rapidly after commercial introduction and has been used to considerable economic advantage, particularly in the breeding of dairy cattle. In Britain today, for example, approximately 70% of the national herd is bred by artificial insemination. In fact, this method of propagating cattle, and to a much lesser extent sheep and pigs, represents by far the most notable success to date of applying reproductive technology to the animal production industry. The critical factor underlying the potential of the technique is that the ejaculate of a mature animal contains many times more spermatozoa than are required for successful fertilisation in a single female.

The range of benefits that can accrue from the successful introduction of artificial insemination is briefly summarised in Table IV.4. Although the attractiveness of such benefits will vary according to circumstances, perhaps the most compelling is the availability of semen from the males proven to be best for desirable production traits, and thus the opportunity for increasing the number of offspring from the best sires. However, as far as the present chapter is concerned, the intention is not to repeat such information at length (see Perry, 1968; Foote, 1974; Gomes, 1977; Salisbury *et al.*, 1978), but rather to note to what extent the technique simulates or differs from natural mating in terms of the stages between semen collection and fertilisation.

Collection of semen

Semen is collected from bulls and rams using an artificial vagina or, less commonly, by means of electro-ejaculation. In the boar, by contrast,

Table IV.4. Some benefits associated with the use of artificial insemination; these come under the headings of livestock improvement, disease control and economic aspects.

1. Enables the widespread use of outstanding sires, and dissemination of valuable genetic material even to small farms.
2. Facilitates progeny testing under a range of environmental and managerial conditions, thereby further improving the rate and efficiency of genetic selection.
3. Leads to improved performance and potential of the national herd, and permits coordination of breeding policy on a national basis.
4. Permits cross-breeding to change the production emphasis, such as from milk to beef.
5. Accelerates introduction of new genetic material by export of semen, and reduces international transport costs.
6. Enables the use of deep-frozen semen after the donor is dead, thus aiding preservation of selected lines.
7. Permits use of semen from incapacitated or oligospermic males.
8. Reduces the risk of spreading sexually-transmitted diseases.
9. Provides an essential research tool for investigating many aspects of male and female reproductive physiology.

semen is usually collected simply by applying manual (digital) pressure to the spiral tip of the penis. The male may be aroused by the presence of a "teaser" female in oestrus, either spontaneously so or after oestrogen therapy, but sexual excitement can also be brought about by conditioning the male to appropriate stimuli, enabling the use of a "dummy" female which will be mounted. Teaser females are commonly used with bulls and rams, but boars respond well to a dummy, particularly if it exudes male odours derived from saliva, urine and preputial secretions.

Shortly after mounting, the animal commences pelvic thrusts and in the case of bulls and rams, the penis is deflected into the artificial vagina. This is a cylindrical device of appropriate dimensions with a smooth rubber lining, the space between the lining and the wall being filled with warm water (37°C) until the required pressure is obtained (Fig. IV.12). The artificial vagina may be lubricated shortly before use, but it is well known to personnel at artificial insemination centres that individual bulls have quite marked preferences for the circumstances of collection, including the condition of the artificial vagina. The device is angled downwards at or immediately after ejaculation, and the male secretions are collected in a calibrated glass tube attached to the end of the water jacket. Insulation is most important to prevent rapid cooling of the semen, for mammalian spermatozoa are notoriously susceptible to temperature shock. In the case of boars, and after electro-ejaculation of bulls and rams, the semen is collected directly from the tip of the urethra into the insulated collection vessel. It is also important during this procedure, especially in boars, to avoid bacterial contamination from the fluid contents of the preputial sac running along the outside of the penis and dripping into the semen. In any event, the pre-sperm fraction of the boar ejaculate is usually discarded since it is generally heavily contaminated with bacteria which may depress fertility. The gelatinous fraction is also routinely removed from the boar ejaculate immediately after collection by filtering through cotton gauze or muslin. Otherwise, the gelatinous material absorbs fluid, swells, and rapidly reduces the available volume of semen.

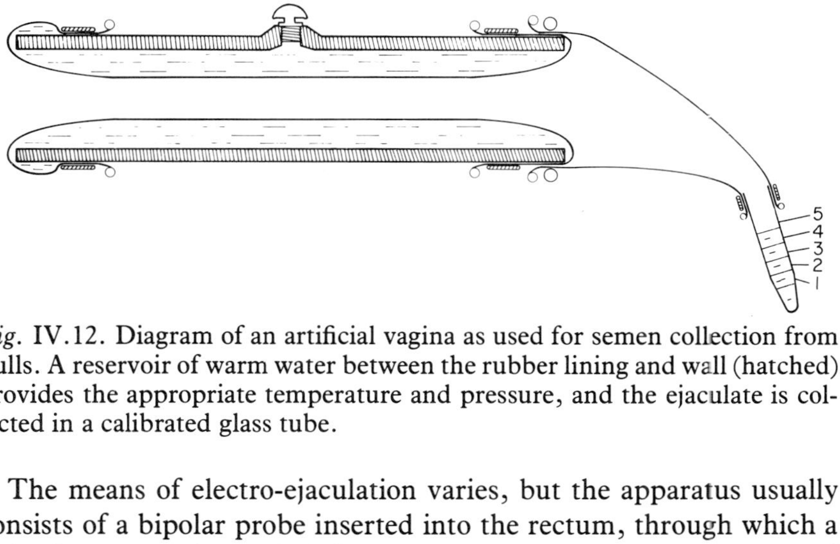

Fig. IV.12. Diagram of an artificial vagina as used for semen collection from bulls. A reservoir of warm water between the rubber lining and wall (hatched) provides the appropriate temperature and pressure, and the ejaculate is collected in a calibrated glass tube.

The means of electro-ejaculation varies, but the apparatus usually consists of a bipolar probe inserted into the rectum, through which a pulsating current is applied; for this procedure, the male does not need to be aroused by the presence of an oestrous female. Whilst the technique is satisfactory for semen collection in bulls and rams, it tends to be favoured when the male is in some way incapacitated (i.e. damaged feet) and cannot mount with confidence or comfort. It may thus be possible to extend the working lifespan of an ageing but genetically superior male. The suggestion has also been made that by electro-ejaculating young males before they would routinely commence to mount, one may be able marginally to shorten the evaluation time of such animals, but the practice is little used in this context. It is worth noting that semen collected by electro-ejaculation contains more accessory secretions and is therefore more dilute than after spontaneous ejaculation, even though fertilising capacity is apparently quite normal (Polge, 1972).

Evaluation and dilution of semen

Semen samples collected with the intention of widespread propagation by means of artificial insemination must clearly undergo some form of laboratory evaluation, and routine checks are performed on the ejaculates of all animals standing at insemination centres. Straightforward visual checks ensure that the semen sample is not contaminated with blood or urine, or more grossly with particulate matter (dust, hair) whereas microscopic checks include estimations of sperm concentration, motility, morphology, agglutination, and the proportion of live relative to dead cells. The volume of the ejaculate may vary with the degree of maturity of the male and, more particularly, with the extent of sexual preparation and frequency of collection. The latter factor may also have a significant influence on the concentration of spermatozoa, this being estimated most accurately under the microscope with a haemocytometer slide or more routinely with a calibrated optical instrument. The concentration of spermatozoa will certainly influence the potential for dilution of the ejaculate, since there are generally accepted figures for the minimum number of spermatozoa to be inseminated to achieve conception. Oligospermic males can be used quite successfully by concentrating the sperm suspension before insemination.

Sperm motility and morphology are best examined under a phase-contrast microscope, but judgement of the former criterion requires the use of a microscope with heated stage so that estimates of swimming ability can be made at 35°–37°C. Strong, progressive motility is an important index of viability in the sperm population, and is usually seen as a swirling, wave-like motion in the highly concentrated ejaculates of the bull and ram. Whilst motility can be measured by various devices (see Katz and Dott, 1975) or by a photographic method (Revell and Wood, 1978), it is usually assessed subjectively by an experienced observer. Poor or zero motility may indicate the lethal effects of temperature shock or contamination in the collecting vessel, whereas moderate but not good motility places the observer in a dilemma since the correlation between motility and fertilising ability is not good. Turning to morphological assessments, grossly abnormal forms of spermatozoa such as those with double heads or tails, separated heads and tails, or a high incidence of cytoplasmic (kinoplasmic) droplets will be clearly identifiable (Fig. IV.13), and their proportion in the ejaculate can be estimated. Cytoplasmic droplets are usually indicative of an incomplete phase of sperm maturation in the epididymis, and may suggest that semen collections are being made too frequently. More complex forms

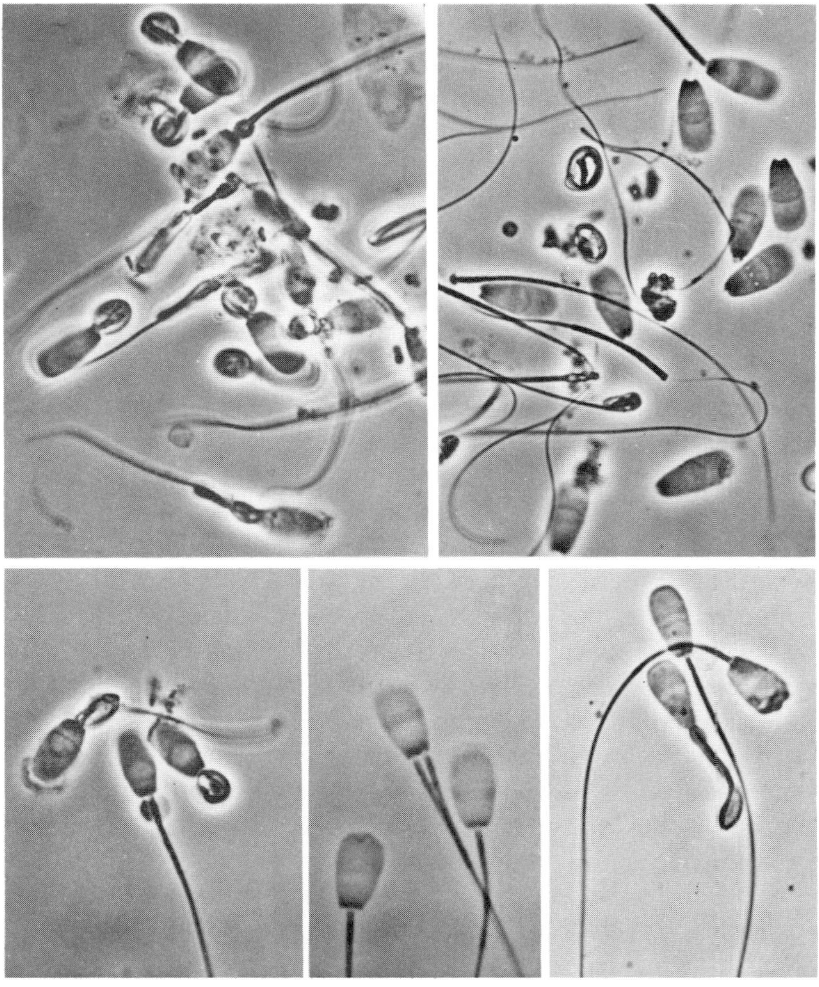

Fig. IV.13. Diverse morphological abnormalities observed in ejaculated spermatozoa of the large farm species. (Courtesy of Dr H. M. Dott.)

of morphological abnormality such as knobbed spermatozoa (Donald and Hancock, 1953; Buttle and Hancock, 1965) which reflects a deformed acrosome (see Cran and Dott, 1976) can also be detected. Although males yielding such abnormal spermatozoa may frequently be sterile, this is not always the case.

As already pointed out, the ejaculate represents a heterogeneous population of cells, a proportion being under-ripe and others already degenerating. The proportion of dead cells in the ejaculate is normally monitored by a staining procedure such as the nigrosin-eosin technique (Campbell *et al.*, 1953, 1956), the underlying principle being that the eosin stain is prevented from entering live cells due to the integrity of the plasma membrane. Dead or dying spermatozoa, on the other hand, with a degenerating cell membrane are permeable to the stain, and by a representative sampling and counting procedure, the proportion of dead cells in the ejaculate can be estimated. The relevance of this estimate lies once again in the dilution potential of the ejaculate, for it is the number of live or motile spermatozoa which governs the extent of possible dilution. Estimates of the ratio of live to dead cells may also be of considerable value in seasonal breeders such as rams, in which ejaculates collected at the beginning and end of the breeding season may be of poor quality.

Having outlined various microscopical procedures for evaluating some of the qualities of the ejaculate, it nevertheless requires to be stressed that there is no absolute relationship between these indices and the potential fertilising ability of the sample. Indeed, a study on samples of freshly-collected and frozen-thawed bull semen indicated that many of the standard laboratory tests may frequently be of little use as a means of *predicting* non-return values after insemination (Linford *et al.*, 1976). Thus, the conception rate is the only functional evaluation of the fertilising ability of a sample of semen. The dilemma is to choose between the proportion of samples of moderate quality to be discarded, and the potential wastage of fertilising ability so incurred. But before leaving the reader with doubts as to the wisdom of routine checking of ejaculates, it should be stressed that grossly abnormal samples would be detected straight away and discarded, whereas without such checks their extensive use after dilution might have compromised fertility in a population of several hundred or more cows.

Whilst the technology of deep-freezing spermatozoa for a period of quarantine or long term preservation is discussed in Chapter VII, dilution of semen samples is a routine procedure before insemination in many countries and is therefore referred to at this stage. The volume of the bull ejaculate may commonly be some 4–8 ml (Table IV.1), and since insemination usually uses a volume of 0·25–1·0 ml, it is obvious that dilution (extension) of the highly concentrated fresh semen is a necessary step before a single ejaculate can be used on a large number of females. If the semen is to be used soon after collection, dilution with a simple isotonic salt solution may suffice, but when a period of short-term storage is planned, a more complex medium is required. The essential components of such a medium can be summarised as follows:

(1) A metabolic substrate—usually a sugar.
(2) Electrolytes in concentrations appropriate for the protection of sperm cells against changes in pH (i.e. buffering capacity) and osmotic pressure.
(3) A high molecular weight component, such as those available in egg yolk or milk, to protect the sperm cells against the damaging effects of chilling.
(4) Antibiotics.

A medium used successfully for many years for the storage of semen for 1–3 days at 15–20°C was the Illini Variable Temperature (IVT) medium, named after developmental work at the University of Illinois. The ingredients here are glucose, citrate and bicarbonate, egg yolk and antibiotics, and the medium is saturated with gaseous carbon dioxide, thereby immobilising the spermatozoa. After dilution with media containing such essential components, bull semen may then be cooled to temperatures close to but above 0°C and boar semen to 5°C. The decrease in temperature reduces the motility and associated metabolism of the sperm cells, enabling chilled bull or boar semen to be used satisfactorily in artificial insemination for at least three days. However, some decline in fertilising capacity occurs during this period, in part due to loss of substances such as phospholipid material from the cell surface and also because of irreparable decay of enzyme systems resulting from the effects of dilution.

Techniques of insemination

Whereas the actual site of artificial insemination in the genital tract of ewes and sows does not differ significantly from that of semen deposition at natural mating, this is not true of the approach used in cows. In the latter case, the glass or disposable plastic insemination pipette or straw is guided to the cervical os with manual assistance through the wall of the rectum for raising the extremity of the cervix (Fig. IV.14), and the pipette is then passed through the cervical canal in most animals, although this may prove difficult with heifers. The inseminate

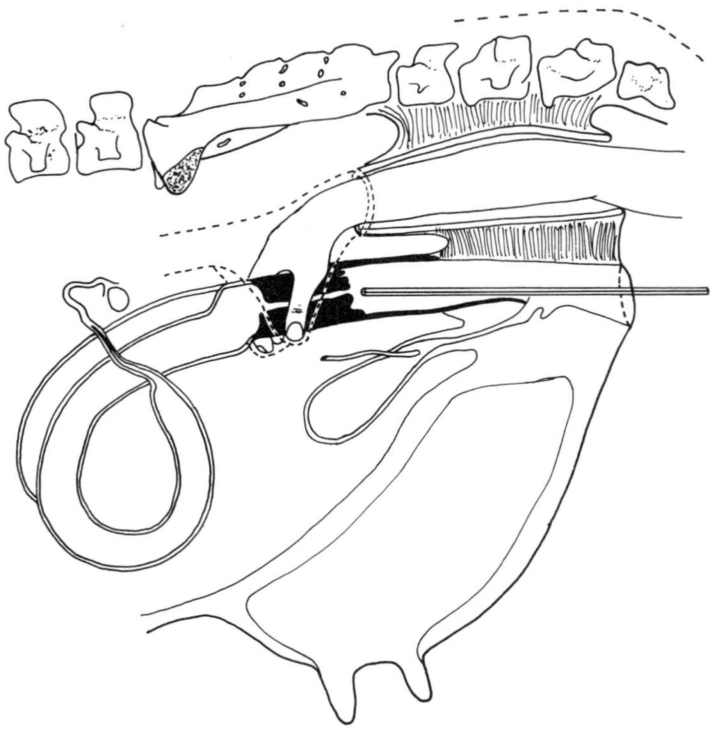

Fig. IV.14. Diagram of the procedure for artificial insemination in cattle, in which a hand inserted *per rectum* is used to raise and orientate the cervix, thereby facilitating introduction of a pipette or straw containing the inseminate into the anterior cervix or body of the uterus.

can thus be expelled directly into the uterus by gentle pressure on the syringe or gun attached to the pipette, but it is a common practice gradually to withdraw the pipette during this procedure so that at least a proportion of the inseminate is deposited within the cervical folds. This may aid the formation of a primary reservoir of spermatozoa within the mucus secreting glands. Nonetheless, by entering the cervix and/or the uterus, a deeper site of insemination is achieved than that obtained by the bull, thereby enabling a very substantial reduction in the numbers of spermatozoa to be made.

The cervix in sheep is composed of more closely and tortuously arranged muscular folds than in cattle, and it is rarely possible to introduce the insemination pipette beyond the external cervical os—perhaps as far as the first or second fold. But, with the aid of a vaginal speculum, one can usually ensure that at least a proportion of the inseminate enters the cervical canal by means of the syringe or insemination gun. Using a supporting cradle, ewes may be orientated head downwards during these procedures.

Although several devices for insemination of pigs were examined during the 1950s, the rubber catheter with spiral tip which mimics very closely the morphology of the boar's penis has now found widespread adoption (Melrose and O'Hagan, 1961). It is introduced into the muscular folds of the cervix (Fig. IV.8), and advanced by anticlockwise rotation until it is fully locked in the folds. A collapsible polythene bottle containing the inseminate is then attached to the catheter, and semen expelled by gentle manual pressure. Great care is required at this stage to avoid expelling the semen too rapidly or forcibly, particularly if the uterine cornua or indeed innermost portion of the cervical canal are being occluded by contractile activity; this would result in much of the semen being forced back into the vagina where it is wasted. The muscular activity of the lower tract can be monitored quite sensitively when the polythene bottle is in the hands of an experienced inseminator, and by following the waves of uterine activity, semen can frequently be introduced into the uterus with little more than the influence of gravity.

The insemination procedure should never be rushed in any of the species, and careful deliberate movements on the part of the inseminator will avoid producing conditions of stress which may inhibit the subsequent phases of sperm transport. It is doubtless a fact that appropriate manipulation of the female tract by an experienced inseminator can release a surge of oxytocin, and this coupled with an intra-uterine site of insemination should aid the transport of spermatozoa to the Fallopian tubes.

Conception rates

As the incidence of fertilisation itself cannot be monitored in practical situations, some criterion is needed for deciding whether or not conception has occurred. The most frequent index used is the failure of the animal to return to oestrus a cycle length or so after mating or insemination, this being taken as presumptive evidence of successful fertilisation. However, since early embryonic loss attains quite significant proportions in domestic animals (Chapter VI), and because complete embryonic loss may lead to a *delayed* return to oestrus, it is important to judge successful conception at an interval greater than one cycle length after insemination. Such *non-return* rates in cattle are usually expressed during the 30–60 day interval, or more critically at 60–90 days. In a sense, this is a discussion of negative evidence and the more meaningful results of pregnancy diagnosis (Chapter IX) will doubtless be applied more widely in the future. However, as examples of conception rates after artificial insemination of cattle, sheep and pigs, the figures presented in Table IV.5 give some idea of the success of the techniques. In

Table IV.5. Examples of conception rates to insemination with stored semen obtained in Britain. The figures represent the proportion of animals becoming pregnant.

Species	Range %	Mean %
Cow	60–73	71
Sheep	50–70[1]	62
Pig	50–88[2]	77

[1] The figures in sheep are influenced markedly by the time of year and by treatments for synchronising oestrus. [2] These figures were provided by the Meat and Livestock Commission for herds with the most accurate records, and the mean is probably not typical of the average pig enterprise.

specific situations, the conception rate will usually be seen as a function of the herd (i.e. expression of management and environmental factors as well as the intrinsic potential fertility), and also as a function of the semen sample and skills of a particular inseminator.

It is important to note that in cattle subjected to superovulation procedures and in pigs inseminated with frozen-thawed boar semen, an increase in the total number of spermatozoa deposited appears essential to avoid diminished fertilisation or conception rates. On the other hand, accurate knowledge of the time of ovulation, and deposition of semen at the optimum interval before ovulation, should permit some reduction in numbers of spermatozoa inseminated.

References

Asch, R. H., Balmaceda, J. and Pauerstein, C. J. (1977). Failure of seminal plasma to enter the uterus and oviducts of the rabbit following artificial insemination. *Fert. Steril.* **28**, 671–673.

Austin, C. R. (1965). "Fertilization." Prentice-Hall Inc., Englewood Cliffs, New Jersey.

Baker, R. D. and Degen, A. A. (1972). Transport of live and dead boar spermatozoa within the reproductive tract of gilts. *J. Reprod. Fert.* **28**, 369–377.

Baker, R. D., Dziuk, P. J. and Norton, H. W. (1968). Effect of volume of semen, number of sperm and drugs on transport of sperm in artificially inseminated gilts. *J. Anim. Sci.* **27**, 88–93.

Bedford, J. M. (1966). Development of the fertilising ability of spermatozoa in the epididymis of the rabbit. *J. exp. Zool.* **163**, 319–330.

Bedford, J. M. (1970a). Sperm capacitation and fertilization in mammals. *Biol. Reprod., Suppl.* **2**, 128–158.

Bedford, J. M. (1970b). The saga of mammalian sperm from ejaculation to syngamy. *In* "Mammalian Reproduction" (H. Gibian and E. J. Plotz, eds), pp. 124–182. Springer-Verlag, New York.

Bedford, J. M. (1975). Maturation, transport, and fate of spermatozoa in the epididymis. *In* "Handbook of Physiology" Section 7, Endocrinology 5, Male Reproductive System (D. W. Hamilton and R. O. Greep, eds), pp. 303–317. American Physiological Society, Washington.

Boender, J. (1966). The development of A.I. in pigs in the Netherlands and the storage of boar semen. *Wld. Rev. Anim. Prod., Special Issue*, **2**, 29–44.

Braden, A. W. H. (1953). Distribution of sperms in the genital tract of the female rabbit after coitus. *Aust. J. biol. Sci.* **6**, 693–705.

Braden, A. W. H. and Austin, C. R. (1954). The number of sperms about the eggs in mammals and its significance for normal fertilisation. *Aust. J. biol. Sci.* **7**, 543–551.

Burger, J. F. (1952). Sex physiology of pigs. *Onderstepoort J. vet. Res., Suppl. No.* **2**, 218 pp.

Burkhardt, J. (1949). Sperm survival in the genital tract of the mare. *J. agric. Sci.* **39**, 201–203.

Buttle, H. R. L. and Hancock, J. L. (1965). Sterile boars with "knobbed" spermatozoa. *J. agric. Sci.* **65**, 255–260.

Campbell, R. C., Hancock, J. L. and Rothschild, Lord (1953). Counting live and dead bull spermatozoa. *J. exp. Biol.* **30**, 44–49.

Campbell, R. C., Dott, H. M. and Glover, T. D. (1956). Nigrosin eosin as a stain for differentiating live and dead spermatozoa. *J. agric. Sci.* **48**, 1–8.

Chang, M. C. (1957). A detrimental effect of seminal plasma on the fertilizing capacity of sperm. *Nature, London* **179**, 258–259.

Chang, M. C. and Hunter, R. H. F. (1975). Capacitation of mammalian sperm: biological and experimental aspects. *In* "Handbook of Physiology",

Endocrinology V, (D. W. Hamilton and R. O. Greep, eds), Chap. 16, pp. 339–351. American Physiological Society, Washington.
Cohen, J. (1967). Correlation between sperm "redundancy" and chiasma frequency. *Nature, London* **215**, 862–863.
Cohen, J. and McNaughton, D. C. (1974). Spermatozoa: the probable selection of a small population by the genital tract of the female rabbit. *J. Reprod. Fert.* **39**, 297–310.
Cohen, J. and Werrett, D. J. (1975). Antibodies and sperm survival in the female tract of the mouse and rabbit. *J. Reprod. Fert.* **42**, 301–310.
Cran, D. G. and Dott, H. M. (1976). The ultrastructure of knobbed bull spermatozoa. *J. Reprod. Fert.* **47**, 407–408.
Dauzier, L. (1958). Physiologie du déplacement des spermatozoides dans les voies génitales femelles chez la brebis et la vache. *Annls Zootech.* **7**, 281–306.
Dauzier, L. and Thibault, C. (1956). Recherche expérimentale sur la maturation des gamètes mâles chez les mammifères, par l'étude de la fécondation "*in vitro*" de l'oeuf de lapine. *Proc. 3rd Int. Congr. Anim. Reprod. and Artif. Insem., Cambridge*, **1**, 58–61.
Dauzier, L. and Wintenberger, S. (1952). Recherches sur la fécondation chez les mammifères: la remontée des spermatozoides dans le tractus génital de la Brebis. *C. R. Soc. Biol.* **146**, 67–70.
Day, F. T. (1942). Survival of spermatozoa in the genital tract of the mare. *J. agric. Sci.* **32**, 108–111.
Donald, H. P. and Hancock, J. L. (1953). Evidence of gene-controlled sterility in bulls. *J. agric. Sci.* **43**, 178–181.
Du Mesnil du Buisson, F. and Dauzier, L. (1955). Distribution et résorption du sperme dans le tractus génital de la truie: survie des spermatozoides. *Annls Endocr.* **16**, 413–422.
Dym, M. and Fawcett, D. W. (1970). The blood-testis barrier in the rat and the physiological compartmentation of the seminiferous epithelium. *Biol. Reprod.* **3**, 308–326.
Dziuk, P. (1970). Estimation of optimum time for insemination of gilts and ewes by double-mating at certain times relative to ovulation. *J. Reprod. Fert.* **22**, 277–282.
Fawcett, D. W. (1975). The mammalian spermatozoon. *Devl Biol.* **44**, 394–436.
Foote, R. H. (1974). Artificial insemination. *In* "Reproduction in Farm Animals", 3rd edition (E. S. E. Hafez, ed.), pp. 409–431. Lea and Febiger, Philadelphia.
Gaddum-Rosse, P. and Blandau, R. J. (1973). In vitro studies on ciliary activity within the oviducts of the rabbit and pig. *Amer. J. Anat.* **136**, 91–104.
Gibbons, R. A. (1969). The function of the cervix. *J. Reprod. Med.* **3**, 126–128.
Glover, T. D. (1955). The semen of the pig. *Vet. Rec.* **67**, 36–40.
Glover, T. D. (1974). Recent progress in the study of male reproductive physiology. *In* "Reproductive Physiology", MTP International Review of

Science, Physiology, Series 1, Vol. 8 (R. O. Greep, ed.), pp. 221–277. Butterworths, London.

Gomes, W. R. (1977). Artificial insemination. *In* "Reproduction in Domestic Animals", 3rd edition (H. H. Cole and P. T. Cupps, eds), pp. 257–284. Academic Press, New York and London.

Hamilton, W. J. (1957). "Textbook of Human Anatomy". Macmillan and Co., London.

Harper, M. J. K. (1973). Stimulation of sperm movement from the isthmus to the site of fertilisation in the rabbit oviduct. *Biol. Reprod.* **8**, 369–377.

Holtz, W. and Smidt, D. (1976). The fertilising capacity of epididymal spermatozoa in the pig. *J. Reprod. Fert.* **46**, 227–229.

Hunter, R. H. F. (1972). Fertilisation in the pig: sequence of nuclear and cytoplasmic events. *J. Reprod. Fert.* **29**, 395–406.

Hunter, R. H. F. (1973a). Transport, migration and survival of spermatozoa in the female genital tract: species with intra-uterine deposition of semen. *In* "Sperm Transport, Survival and Fertilising Ability" (E. S. E. Hafez and C. Thibault, eds), pp. 309–342. INSERM, Paris.

Hunter, R. H. F. (1973b). Polyspermic fertilisation in pigs after tubal deposition of excessive numbers of spermatozoa. *J. exp. Zool.* **183**, 57–64.

Hunter, R. H. F. (1975a). Physiological aspects of sperm transport in the domestic pig, *Sus scrofa*. I. Semen deposition and cell transport. *Brit. vet. J.* **131**, 565–573.

Hunter, R. H. F. (1975b). Physiological aspects of sperm transport in the domestic pig, *Sus scrofa*. II. Regulation, survival and fate of cells. *Brit. vet J.* **131**, 681–690.

Hunter, R. H. F. (1978). Intraperitoneal insemination, sperm transport and capacitation in the pig. *Anim. Reprod. Sci.* **1**, 167–179.

Hunter, R. H. F. and Dziuk, P. J. (1968). Sperm penetration of pig eggs in relation to the timing of ovulation and insemination. *J. Reprod. Fert.* **15**, 199–208.

Hunter, R. H. F. and Hall, J. P. (1974). Capacitation of boar spermatozoa: synergism between uterine and tubal environments. *J. exp. Zool.* **188**, 203–214.

Hunter, R. H. F. and Léglise, P. C. (1971). Polyspermic fertilisation following tubal surgery in pigs, with particular reference to the rôle of the isthmus. *J. Reprod. Fert.* **24**, 233–246.

Hunter, R. H. F. and Poyser, N. L. (1973). Cited in Hunter, R. H. F. (1973). Transport, migration and survival of spermatozoa in the female genital tract: species with intra-uterine deposition of semen. *In* "Sperm Transport, Survival and Fertilizing Ability" (E. S. E. Hafez and C. Thibault, eds), pp. 309–342. INSERM, Paris.

Hunter, R. H. F., Holtz, W. and Henfrey, P. J. (1976). Epididymal function in the boar in relation to the fertilising ability of spermatozoa. *J. Reprod. Fert.* **46**, 463–466.

Hunter, R. H. F., Holtz, W. and Herrmann, H. (1978). Stabilising rôle of epididymal plasma in relation to the capacitation time of boar spermatozoa. *Anim. Reprod. Sci.* **1**, 161–166.

Kaltenbach, C. C. and Davies, H. L. (1970). Fertilisation, sperm transport, and early embryonic loss in ewes grazed on cultivars of subterranean clover. *Aust. J. agric. Res.* **21**, 107–114.

Katz, D. F. and Dott, H. M. (1975). Methods of measuring swimming speed of spermatozoa. *J. Reprod. Fert.* **45**, 263–272.

Kelly, R. W., Taylor, P. L., Hearn, J. P., Short, R. V., Martin, D. E. and Marston, J. H. (1976). 19-Hydroxyprostaglandin E_1 as a major component of the semen of primates. *Nature, London* **260**, 544–545.

Lightfoot, R. J. and Restall, B. J. (1971). Effects of site of insemination, sperm motility and genital tract contractions on transport of spermatozoa in the ewe. *J. Reprod. Fert.* **26**, 1–13.

Lightfoot, R. J., Croker, K. P. and Neil, H. G. (1967). Failure of sperm transport in relation to ewe infertility following prolonged grazing on oestrogenic pastures. *Aust. J. agric. Res.* **18**, 755–765.

Linford, E., Glover, F. A., Bishop, C. and Stewart, D. L. (1976). The relationship between semen evaluation methods and fertility in the bull. *J. Reprod. Fert.* **47**, 283–291.

Mann, T. (1964). "The Biochemistry of Semen and of the Male Reproductive Tract". Methuen and Co. Ltd., London.

Mann, T. (1972). Advances in male reproductive physiology. *Fert. Steril.* **23**, 699–707.

Mann, T., Leone, E. and Polge, C. (1956a). The composition of the stallion's semen. *J. Endocr.* **13**, 279–290.

Mann, T., Polge, C. and Rowson, L. E. A. (1956b). Participation of seminal plasma during the passage of spermatozoa in the female reproductive tract of the pig and horse. *J. Endocr.* **13**, 133–140.

Mattner, P. E. (1963). Spermatozoa in the genital tract of the ewe. II. Distribution after coitus. *Aust. J. biol. Sci.* **16**, 688–694.

Mattner, P. E. (1966). Formation and retention of the spermatozoan reservoir in the cervix of the ruminant. *Nature, London* **212**, 1479–1480.

Mattner, P. E. (1968). The distribution of spermatozoa and leucocytes in the female genital tract in goats and cattle. *J. Reprod. Fert.* **17**, 253–261.

Mattner, P. E. and Braden, A. W. H. (1963). Spermatozoa in the genital tract of the ewe. I. Rapidity of transport. *Aust. J. biol. Sci.* **16**, 473–481.

Melrose, D. R. and O'Hagan, C. (1961). Investigations into the techniques of insemination in the pig. *Proc. 4th Int. Congr. Anim Reprod., The Hague* **4**, 855–859.

Mortimer, D. (1977). The survival and transport to the site of fertilization of diploid rabbit spermatozoa. *J. Reprod. Fert.* **51**, 99–104.

Mortimer, D. (1978). Selectivity of sperm transport in the female genital tract. *In* "Spermatozoa, Antibodies and Infertility" (J. Cohen and W. F. Hendry, eds), pp. 37–53. Blackwells, Oxford and London.

Orgebin-Crist, M. C. (1967). Sperm maturation in rabbit epididymis. *Nature, London* **216**, 816–818.

Orgebin-Crist, M. C. (1969). Studies on the function of the epididymis. *Biol. Reprod.* **1**, 155–175.

Overstreet, J. W. and Cooper, G. W. (1975). Reduced sperm motility in the isthmus of the rabbit oviduct. *Nature, London* **258**, 718–719.

Overstreet, J. W. and Cooper, G. W. (1978). Sperm transport in the reproductive tract of the female rabbit. I. The rapid phase of transport. *Biol. Reprod.* **19**, 101–114.

Parker, W. G., Sullivan, J. J. and First, N. L. (1975). Sperm transport and distribution in the mare. *J. Reprod. Fert., Suppl.* **23**, 63–66.

Perry, E. J. (1968). "The Artificial Insemination of Farm Animals", 4th edition. Rutgers University Press, New Jersey.

Polge, C. (1956). Artificial insemination in pigs. *Vet. Rec.* **68**, 62–76.

Polge, C. (1972). Increasing reproductive potential in farm animals. *In* "Reproduction in Mammals" (C. R. Austin and R. V. Short, eds), Book 5, pp. 1–31. University Press, Cambridge.

Polge, C., Day, B. N. and Groves, T. W. (1968). Synchronisation of ovulation and artificial insemination in pigs. *Vet. Rec.* **83**, 136–142.

Polge, C., Salamon, S. and Wilmut, I. (1970). Fertilizing capacity of frozen boar semen following surgical insemination. *Vet. Rec.* **87**, 424–428.

Quinlivan, T. D. and Robinson, T. J. (1969). Numbers of spermatozoa in the genital tract after artificial insemination of progestagen-treated ewes. *J. Reprod. Fert.* **19**, 73–86.

Raynaud, F. (1973). Physiologie du col de l'utérus de la brebis. Effet d'un progestagène de synthèse: l'acétate de fluorogestone. *Annls Biol. anim. Biochim. Biophys.* **13**, 335–346.

Revell, S. G. and Wood, P. D. P. (1978). A photographic method for the measurement of motility of bull spermatozoa. *J. Reprod. Fert.* **54**, 123–126.

Rigby, J. P. (1964). The fate of spermatozoa in the genital tract of the sow following artificial insemination. *Proc. 5th Int. Congr. Anim. Reprod. and Artif. Insem., Trento* **4**, 421–425.

Robinson, T. J. (1973). Factors involved in the failure of sperm transport and survival in the female reproductive tract. *J. Reprod. Fert., Suppl.* **18**, 103–109.

Rodolfo, A. (1934). The physiology of reproduction in swine. II. Some observations on mating. *Philipp. J. Sci.* **55**, 13–18.

Roosen-Runge, E. C. (1977). "Process of Spermatogenesis in Animals". University Press, Cambridge.

Rowlands, I. W. (1957). Insemination of the guinea pig by intraperitoneal injection. *J. Endocr.* **16**, 98–106.

Rowson, L. E. (1955). The movement of radio-opaque material in the bovine uterine tract. *Brit. vet. J.* **111**, 334–342.

Salisbury, G. W., Van Demark, N. L. and Lodge, J. R. (1978). "Physiology of Reproduction and Artificial Insemination of Cattle", 2nd edition. W. H. Freeman and Company, San Francisco.

Skjerven, O. (1955). Conception in a heifer after deposition of semen in the abdominal cavity. *Fert. Steril.* **6**, 66–67.

Tampion, D. and Gibbons, R. A. (1962). Orientation of spermatozoa in mucus of the cervix uteri. *Nature, London* **194**, 381.

Thibault, C. (1973). Sperm transport and storage in vertebrates. *J. Reprod. Fert., Suppl.* **18**, 39–53.

Thibault, C. and Levasseur, M. C. (1973). Conservation et survie prolongée des spermatozoides dans les voies génitales femelles des vertébrés. *Annls Biol. anim. Biochim. Biophys.* **13**, 267–284.

Thibault, C. and Wintenberger-Torres, S. (1967). Oxytocin and sperm transport in the ewe. *Int. J. Fertil.* **12**, 410–415.

Thibault, C., Gérard, M. and Heyman, Y. (1973). Transport et survie des spermatozoides chez la vache. *In* "Transport, survie et pouvoir fécondant des spermatozoides chez les vertébrés", **26**, 343–356. INSERM, Paris.

Turnbull, K. E. (1966). The transport of spermatozoa in the rabbit doe before and after ovulation. *Aust. J. biol. Sci.* **19**, 1095–1099.

Van Demark, N. L. and Hays, R. L. (1954). Rapid sperm transport in the cow. *Fert. Steril.* **5**, 131–137.

Van Demark, N. L. and Moeller, A. N. (1951). Speed of spermatozoan transport in reproductive tract of estrous cow. *Amer. J. Physiol.* **165**, 674–679.

von Euler, U. S. (1936). On the specific vaso-dilating and plain muscle stimulating substances from accessory genital glands in man and certain animals (prostaglandin and vesiglandin). *J. Physiol., London* **88**, 213–234.

Walton, A. (1930). On the function of the rabbit cervix during coitus. *J. Obstet. Gynaec. Brit. Emp.* **37**, 92–95.

Yanagimachi, R. (1970). The movement of golden hamster spermatozoa before and after capacitation. *J. Reprod. Fert.* **23**, 193–196.

Yanagimachi, R. and Chang, M. C. (1963). Sperm ascent through the oviduct of the hamster and rabbit in relation to the time of ovulation. *J. Reprod. Fert.* **6**, 413–420.

Yanagimachi, R. and Mahi, C. A. (1976). The sperm acrosome reaction and fertilization in the guinea-pig: a study in vivo. *J. Reprod. Fert.* **46**, 49–54.

Young, W. C. (1931). A study of the function of the epididymis. III. Functional changes undergone by spermatozoa during their passage through the epididymis and vas deferens in the guinea-pig. *J. exp. Biol.* **8**, 151–162.

Chapter V

Function of the Fallopian Tubes in Relation to Gametes and Embryos

Introduction

The physiological functions of the Fallopian tubes include facilitating the processes of gamete transport and maturation, providing the environment for fertilisation, sustaining metabolic aspects of early embryonic development, and regulating the rate and mode of progression of embryos to the uterus. Only when semen deposition and ovulation have occurred in a suitable temporal relationship, and after appropriate changes have taken place in the gametes, is the opportunity for normal fertilisation presented. The site of significant cellular changes in spermatozoa and eggs culminating in sperm penetration of the egg investments is within the luminal fluid of the Fallopian tubes, where the free-living (unattached) embryo is then maintained for two to three days or more after fertilisation. This chapter considers the economy of the gametes and embryos in some detail, not least since disturbances in tubal function and/or a degree of ageing in the gametes within the tubal environment underlie many instances of diminished fertility or indeed infertility.

Egg transport from ovary to site of fertilisation

The functional anatomy of the Fallopian tubes has been reviewed extensively in several publications (Hafez and Blandau, 1969; Greep, 1973; Johnson and Foley, 1974) and more recently has been summarised (Hunter, 1977a). With the notable exception of at least one primate (i.e. man), the fimbriated extremity of the mammalian tube appears to be extremely efficient in securing safe passage of the freshly-ovulated egg(s) from the surface of the collapsing Graafian follicle(s) into the tubal ostium (Fig. V.1). This is in some contrast to the situation in birds where instances of peritonitis have been noted

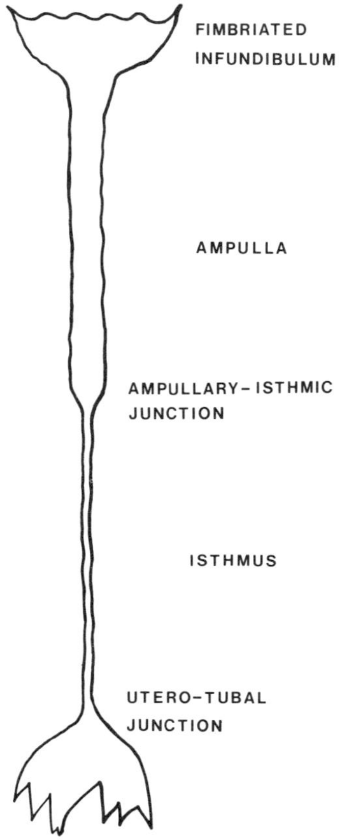

Fig. V.1. Semi-diagrammatic representation of the Fallopian tube to indicate the regions designated infundibulum, ampulla and isthmus, and the relative positions of the ampullary-isthmic and utero-tubal junctions.

when the egg has entered the body cavity rather than the oviduct; however, apart from differences in the structure of the duct systems, the relative sizes of mammalian and avian eggs are of a different order of magnitude. Moreover, the fimbriae of the mammalian oviduct are intimately applied to the ovaries at the time of ovulation (Fig. V.2), and in a number of species extend around the ovary in the form of a bursa. Ovarian bursae are not found in cattle, sheep or pigs, but there is such a specialisation in the horse and in many small rodents.

Essential to the mechanism of egg entry into the Fallopian tube in mammals is the physical interaction between the egg within its investment of follicular cells and the cilia that are densely arrayed on the inner surface of the fimbriated infundibulum (Fig. V.2). The activity of the cilia is regulated by ovarian steroid hormones, and by the time of ovulation there is a closely synchronised and direction-orientated beat of these structures towards the tubal ostium. As shown by the elegant studies of Blandau (1973), the egg is literally pulled or stripped from the surface of the ovulating follicle by the action of the cilia and rapidly propelled into the ampulla. But the cilia can only obtain a purchase on the eggs when they are still encompassed by cumulus (i.e. follicular) cells, as is invariably the situation at the time of ovulation. Eggs that have been artificially denuded of their follicular cells (or microspheres that closely simulate eggs in terms of size and specific gravity) that are placed on the surface of the fimbria are seen simply to rotate and not to be displaced towards the tubal ostium (Blandau, 1969). Egg passage into the tube is probably also aided by the constant to-and-fro movement of the ovaries relative to the fimbriated infundibulum, this being caused by the heightened muscular activity of the supporting mesenteries.

Although the ampullary cilia beat towards the ampullary-isthmic junction shortly after ovulation (Blandau and Verdugo, 1976) and clearly exert a propulsive influence on the cumulus mass, transport of eggs from the tubal ostium to the site of fertilisation at the ampullary-isthmic junction in domestic animals, or to the ampullary site of fertilisation in primates (see Hunter, 1977a), is brought about in part by waves of contractile activity in the smooth muscle compartments of the myosalpinx. These usually consist of distinct layers of circular and longitudinal muscle fibres (Fig. V.3) enabling progressive segmental activity in the wall of the tube to displace the eggs, still within their cellular investments, in an abovarian direction. Progression of eggs is not direct, but rather consists of a series of intermittent rushes in step with waves of contraction in the ampulla (Harper, 1961a; Wintenberger-Torres, 1961). Such smooth muscle activity is also influenced by the ratio of ovarian steroids, and as noted many years ago (Corner, 1923; Seckinger, 1923) is normally accentuated close to the time of ovulation. This enables egg transport to the site of fertilisation to be accomplished in a matter of minutes (i.e. 6–15 minutes in cats and rabbits, Harper, 1961a, b; Blandau, 1969; <45 minutes in pigs, Hunter, 1974), and before any significant degree of postovulatory ageing of the egg would have commenced. Nonetheless, some dispersal of cumulus cells would be expected, and a continuation of the morphological rearrangement of the corona radiata cells (Fig. V.4) that was initiated in the preovulatory follicle would be found. Such depolymerisation and loosening of the egg investments may have functional significance in the events of sperm penetration.

After this initial phase of transport, temporary arrest of the egg(s) occurs at the ampullary-isthmic junction and appears to be brought about by a physiological, not anatomical, sphincter in the wall of the tube. The dense adrenergic innervation in this region, especially in the circular muscle coat, enables a powerful constriction to be activated shortly after ovulation (see Brundin, 1964, 1965). But over and above this fact, the periovulatory waves of abovarian contractile activity tend to fade at the ampullary-isthmic junction, as does the directional beat of the cilia, both of which are opposed by corresponding functions in the isthmus at this stage of the oestrous cycle (see Hunter, 1977a).

Final phase of sperm transport

As noted in Chapter IV, mating would normally occur before ovulation in cattle, sheep, pigs and horses, and indeed in most mammals, if mature males had free access to an appropriately limited number of females. Sperm transport at least as far as the isthmus would have proceeded following such a preovulatory mating but, in the light of recent studies, it is uncertain whether sufficient viable spermatozoa

---

Fig. V.2. Portions of the reproductive tract and ovaries of pigs photographed during mid-ventral laparotomy. (a) Ovaries and Fallopian tubes exposed approximately 24 hours before the time of ovulation. Note the intense vascular engorgement of the fimbriae which have been displaced from the ovaries to display the mature Graafian follicles (9–10 mm diameter) and the still-prominent corpora albicantia. (b) To contrast the extreme vascularity of the ampullary portion of the tube with that of the isthmus that is characteristic of the periovulatory period, and is closely associated with the enhanced secretory function of the ampullary epithelium. (c) Ovary, fimbria and ampulla examined approximately 30 hours before ovulation to illustrate the influence of follicular oestrogens on the capillary bed of the upper regions of the tube. Once again the fimbria has been displaced to reveal ovarian follicular development. (d) To illustrate the influence of double silk ligatures positioned at the ovarian end of the ampulla on the accumulation of fluid in the lumen of this portion of the tube. Note that ligatures have not been positioned in the region of the utero-tubal junction. The fluid distension of the ampulla has occurred within a period of 12 hours.

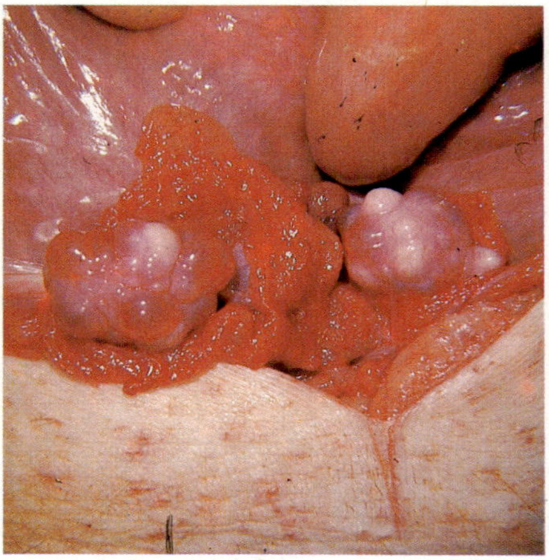

(a)

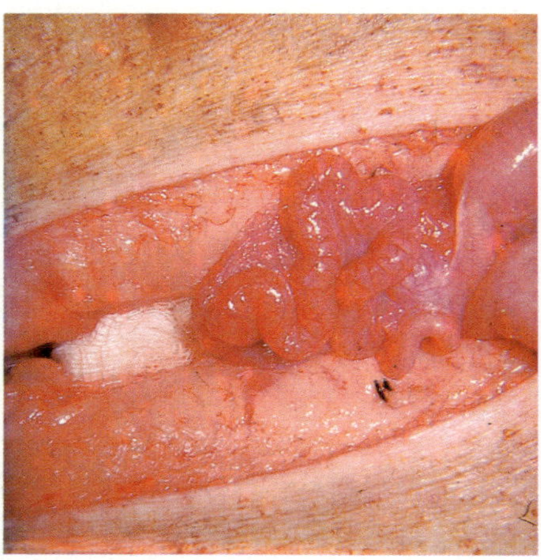

(b)

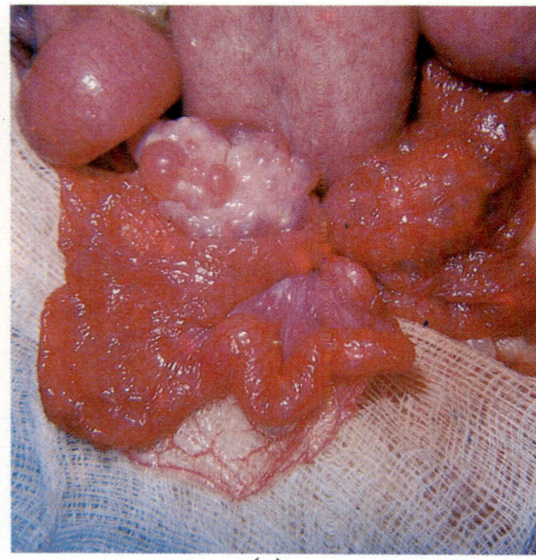

(c)

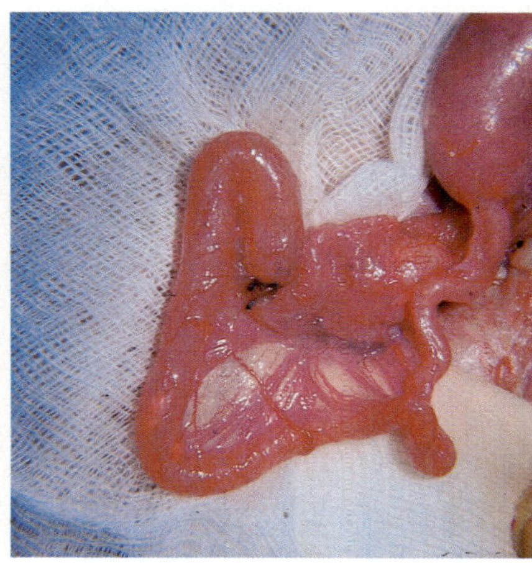

(d)

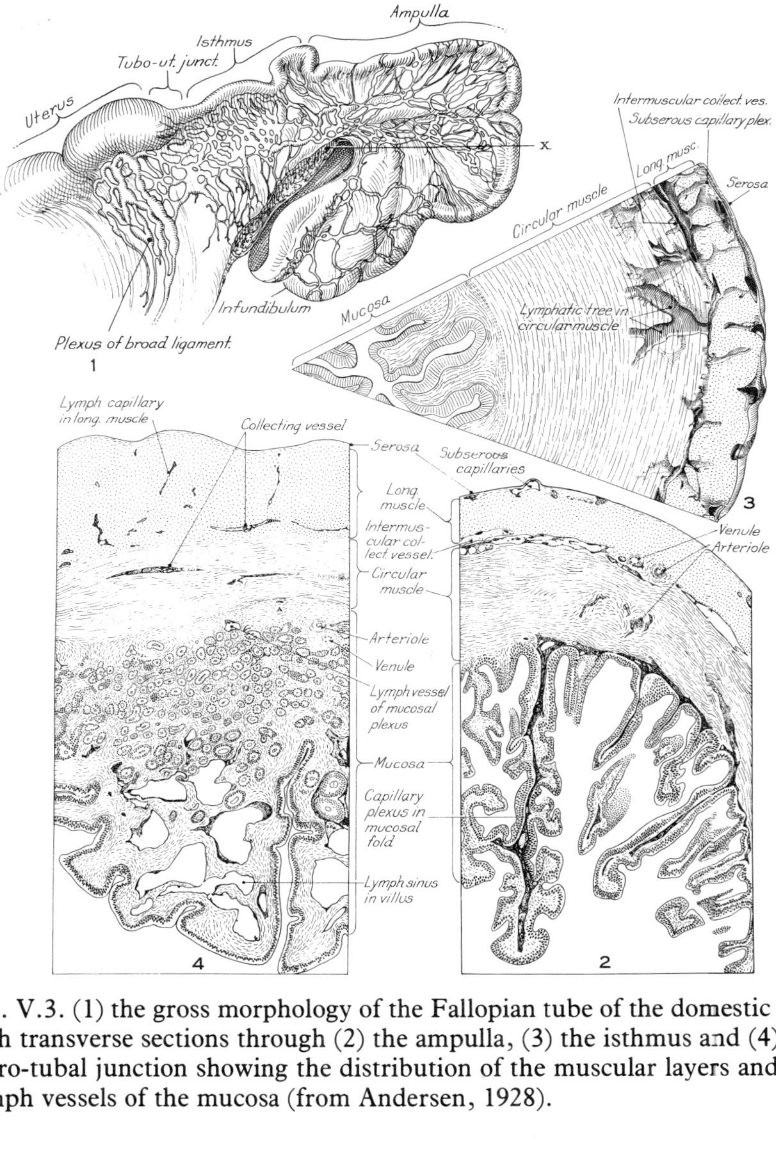

Fig. V.3. (1) the gross morphology of the Fallopian tube of the domestic pig, with transverse sections through (2) the ampulla, (3) the isthmus and (4) the utero-tubal junction showing the distribution of the muscular layers and the lymph vessels of the mucosa (from Andersen, 1928).

would have reached the ampulla before the time of ovulation to be immediately available for penetration of the freshly-ovulated egg(s).

Two lines of evidence suggest that spermatozoa may be largely sequestered in the isthmus in the preovulatory interval, only to be released for further pro-ovarian transport when ovulation has occurred and the products of ovulation have entered the tube. The inference here

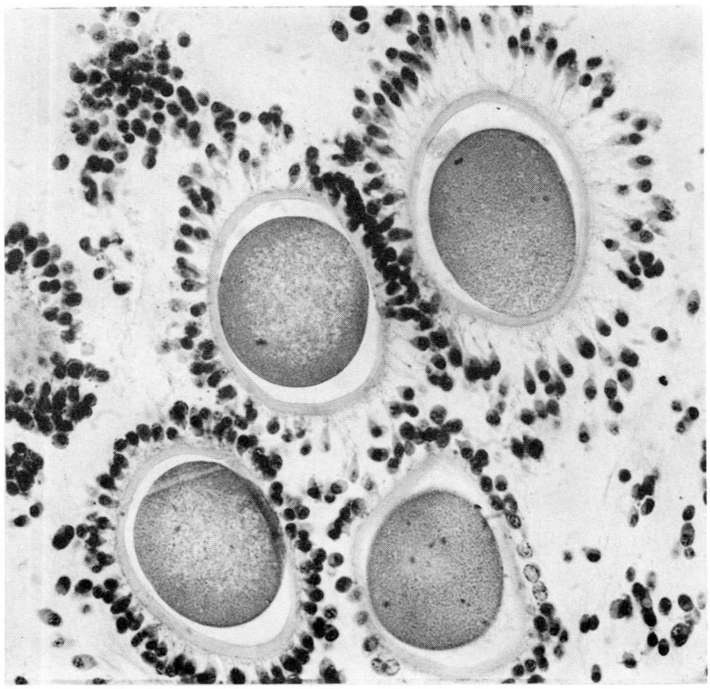

Fig. V.4. The morphological rearrangement of cells of the corona radiata around the rabbit oocyte close to the time of ovulation. Note that extrusion of the first polar body has already occurred. (Courtesy of Professor C. Thibault.)

is that there may be some synchrony between release of the egg from the follicle and a redistribution of spermatozoa from the constricted lumen of the isthmus. First is the evidence from the egg transfer studies of Harper (1973a, b) in rabbits: by transplanting freshly-recovered eggs in cumulus before or after the expected time of ovulation in mated or inseminated recipients, he showed that the products of ovulation do stimulate sperm transport from the proximal isthmus to the site of fertilisation. Second are the observations of Overstreet and Cooper (1975), where spermatozoa accumulating in the isthmus of mated rabbits exhibited diminished motility before ovulation, in contrast to the significant increase in progressive swimming activity near to this event (Fig. V.5). Because of the short fertilisable life of the mammalian egg, mechanisms that co-ordinate and stimulate release of spermatozoa from the lower regions of the tube would have obvious biological advantages, and there appear to be certain analogies with sperm storage and release in the avian oviduct.

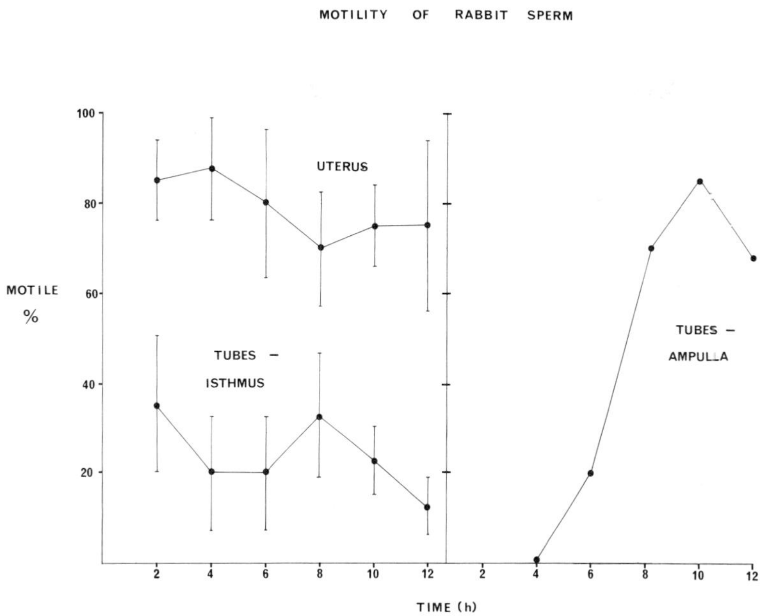

Fig. V.5. Graphical illustration of the motility of rabbit spermatozoa recovered from different portions of the Fallopian tube in the preovulatory interval. Note the progressive increase in motility of spermatozoa in the ampulla as ovulation approaches. (Redrawn from Overstreet and Cooper, 1975.)

Even though a specific attractant substance for spermatozoa has not yet been demonstrated to be released from mammalian eggs or their investments, it seems worthwhile keeping an open mind on whether or not some form of chemotaxis might contribute to the final phase of sperm transport and orientation towards the egg surface.

Capacitation and the acrosome reaction

Before spermatozoa can penetrate between the egg investments (cumulus and corona cells) and particularly through the protective egg membrane (the zona pellucida), they require to undergo a physiological change that confers upon them an increased motility and the ability to release proteolytic enzymes from the acrosomal region of the sperm head. By definition, this physiological change termed capacitation occurs during residence in the female genital tract, although much if not all of the change can nowadays be obtained in *in vitro* culture systems for the spermatozoa of several species of rodents, and also apparently for a number of primates. The phenomenon of capacitation was first specifically identified by Austin (1951) and Chang (1951) working respectively in Sydney, Australia and near Boston, Massachusetts. Surgical deposition of sperm samples into the female tract at various times relative to ovulation indicated that rat spermatozoa needed about 4–5 hours and rabbit spermatozoa some 5–6 hours before acquiring the ability to penetrate recently ovulated eggs (Table V.1).

Table V.1. The fertilising capacity of ejaculated rabbit spermatozoa deposited directly into the Fallopian tubes by means of flank incisions. These results indicate the need for a period of residence in the female tract (capacitation) before spermatozoa finally acquire the ability to fertilise. (Adapted from Chang, 1951.)

Timing of insemination with respect to ovulation	No. of animals	No. of ovulations	Eggs recovered	
			Total	% fertilised
2 hours after	5	58	40	0
1 hour after	2	14	10	0
2 hours before	5	42	31	0
4 hours before	5	39	34	6
6 hours before	4	49	41	78
8 hours before	5	50	47	55
[1]2 hours after	2	20	14	43
[1]2 hours before	1	13	13	100

[1] Control matings were performed in these animals.

Austin (1952) further noticed that mating of rats soon after ovulation reduced the interval until penetration to some 2–3 hours, and it was in this same publication that the term capacitation was first used in this regard. These early results were soon endorsed by Noyes (1953) and Dauzier and Thibault (1956), and further developed in the now classical papers of Chang (1955, 1957, 1958). Moreover, it has since been demonstrated that spermatozoa of other mammalian species such as hamster, guinea-pig, cat, sheep, cow and pig cannot fertilise the eggs immediately upon entering the female tract, even when the problem of sperm transport to the Fallopian tubes is overcome by surgical insemination, but need first to undergo some form of discrete physiological change requiring about $1\frac{1}{2}$–6 hours according to species (Table V.2).

Table V.2. Estimates of the time required for capacitation of mammalian spermatozoa *in vivo*.

Species	Interval in hours
Sheep	1–1·5
Pig	2–3
Cow	5–6
Mouse	<1·0
Rat	2–3
Hamster	3–4
Rabbit	5–6

Although a very considerable research effort since 1951 has been directed towards elucidation of the nature of capacitation, and whilst many points of detail can now be listed and discussed (see comprehensive reviews by Austin, 1967; Bedford, 1970; Chang and Hunter, 1975; Chang *et al.*, 1977), the precise changes that constitute capacitation remain unknown. Nonetheless, there is strong evidence that they are—at least in part—membrane-associated phenomena, expressed particularly as a destabilisation in the anterior region of the sperm head, that function to permit release of the lytic acrosomal enzymes with which the spermatozoon gains access to the vitelline surface of the egg. However, increased respiratory and flagellar activity are also noted after achievement of capacitation, the changes in motility being presumed to facilitate sperm penetration of the egg membranes.

An understanding of the post-testicular changes occurring in spermatozoa in the male reproductive tract has provided some insight into the possible nature of capacitation. During passage along the epididymal duct, spermatozoa undergo a series of modifications that confer increased stability on the head including a compaction of the nuclear chromatin, and a tighter application of the plasma and acrosomal membranes to the sperm nucleus (see Bedford, 1975). Secretions within the epididymal duct also influence the maturing spermatozoa, for the antigenic nature of the sperm surface changes during the one to two weeks or more spent in epididymal transit (Bedford, 1975). There is good evidence that enzyme inhibitors are "added" to the spermatozoa while still in the male, although whether the principal phase of addition occurs in the epididymis, during the pre-ejaculatory mixing with whole seminal plasma, or in the initial passage of spermatozoa within seminiferous tubule and rete testis fluid remains to be clarified. Neutralisation or removal of inhibitors of acrosomal enzymes in the female tract would seem a logical stage in the capacitation process, or an immediate sequel to it, and could explain how these enzymes of fundamental importance in fertilisation are stabilised in the sperm head until spermatozoa are within reach of the eggs.

In fact, there is evidence from at least three mammalian species that different compartments of the female tract act synergistically to promote capacitation, and that sequential exposure to the uterus and then the Fallopian tubes accelerates the process of capacitation (rabbit: Adams and Chang, 1962, Bedford, 1968; golden hamster: Hunter, 1969; pig: Hunter and Hall, 1974a, b). A specific action of the tubal environment is therefore suggested and, in the golden hamster, the effects on stimulated respiration and achievement of the capacitated state are so distinct that they can be seen through the thin wall of the ampulla in terms of a massive increase in sperm motility (Yanagimachi, 1970). Even though hamster sperm reach the tubes in under one hour, another three hours are required for this dramatic change to be accomplished. Precise evidence for the capacitation of primate spermatozoa is not yet available, but a study of sperm incorporation *in vitro* by zona-free hamster eggs or cumulus-free human eggs suggests that 4–5 hours are required for capacitation of ejaculated human spermatozoa (Yanagimachi *et al.*, 1976).

The acrosome reaction follows on from capacitation as a conspicuous structural change in the anterior portion of the sperm head that seems essential to mammalian fertilisation, enabling release of enzymes from this sac-like organelle and thereby assisting penetration of the egg investments, especially the zona pellucida. It entails progressive multiple fusions between the outer acrosomal membrane and overlying plasma (cell) membrane (Fig. V.6), possibly after a phase of swelling in the acrosome itself (Szollosi and Hunter, 1978), such that the anterior portion of the sperm head becomes limited by vesicles. Between these vesicles arise a series of ports or apertures which are presumed to permit escape of lytic enzymes consisting largely of hyaluronidase and a trypsin-like enzyme (Austin, 1972). However, a current interpretation suggests that much of the acrosomal hyaluronidase may diffuse out *before* the acrosome reaction is manifested at the zona surface (Szollosi and Hunter, 1978). On the other hand, the trypsin-like enzyme (acrosin or zona lysin) may be relatively tightly bound to the inner acrosomal membrane (Stambaugh and Buckley, 1969), and require specific activation from a precursor form termed proacrosin (see Hartree, 1977).

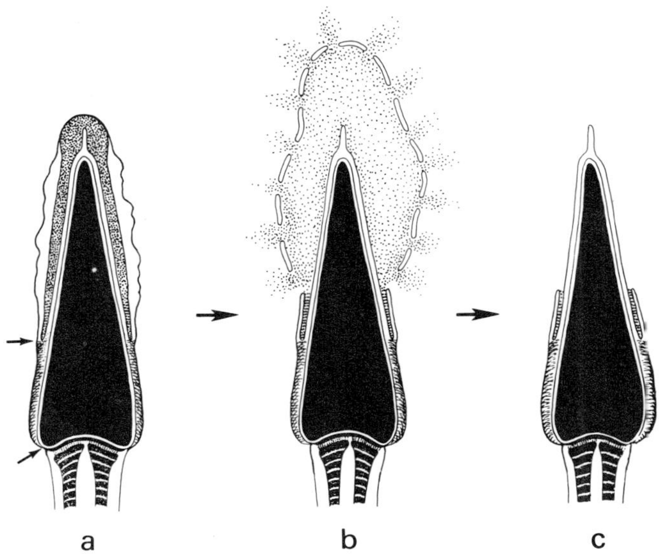

Fig. V.6. Illustration of changes in the arrangement of membranes about the head of a capacitated spermatozoon during the acrosome reaction. Note the multiple fusions between the outer acrosomal membrane and overlying plasma membrane in (b) to form a series of vesicles on the anterior portion of the head. These are displaced as in (c) during penetration of the zona pellucida. (From Bedford and Cooper, 1978.)

The acrosome reaction is nowadays regarded as a process quite distinct from that of capacitation (Austin, 1967; Bedford, 1970; Chang and Hunter, 1975), although it may represent an inevitable sequel to the destabilising changes initiated during capacitation. In this context, some adjustment in the distance separating the plasma and outer acrosomal membranes may be an essential feature of capacitation, enabling this phenomenon to be considered as the physiological process that permits development of a co-ordinated acrosome reaction. After completion of the acrosome reaction, the sperm maintains its surface integrity by a fold at the region of the equatorial segment such that the inner acrosomal membrane becomes the limiting membrane on the anterior portion of the sperm head (Fig. V.6). Because of the metabolic and structural changes that follow upon capacitation, spermatozoa achieving these stages are thought to be in a fragile state with a short lifespan. Loss of the complement of acrosomal enzymes clearly renders the spermatozoon a non-functional cell, as would the associated influence of impending metabolic exhaustion.

During capacitation and the commencement of sperm penetration between the cumulus cell layers that surround the egg, there may be a final phase of maturation in the membranes of the secondary oocyte and a further spatial rearrangement in cells of the corona radiata. Work in laboratory species (Austin and Braden, 1954; Austin, 1961) suggested that the eggs and/or their investments might require a short period of exposure to the luminal contents of the Fallopian tubes before they were penetrable by spermatozoa, in other words a form of postovulatory maturation. Although observations on the time relationships between ovulation and fertilisation have failed to support this notion in domestic animals (Table V.3), exposure of the oocyte to the tubal

Table V.3. Data from mature pigs in which the time of ovulation was controlled precisely and eggs recovered at laparotomy some three hours after insemination. There is no detectable influence of post-ovulatory maturation or ageing of the egg membranes on the proportion of eggs fertilised. (Taken from Hunter and Dziuk, 1968.)

Post-ovulatory age of eggs at insemination (h)	Number of eggs: Examined	With spermatozoa		Fertilised[1]	% fertilised[1] (Mean±S.E.)
			%		
0	104	62	59·6	43	41·4±12·9
4	94	39	41·5	15	16·0± 5·4
8	93	54	58·1	43	46·2±12·8
12	107	53	49·5	42	39·2±10·6
Total	398	208	52·3	143	35·9

[1] Eggs classified as fertilised showed anaphase or telophase of the second meiotic division, or a second polar body together with female chromatin, and contained a spermatozoon in the vitellus.

environment after its prolonged residence in the Graafian follicle undoubtedly does initiate changes at the membranous level, as well as in the granulosa investments. Studies of ionic changes in the tubal fluids by electron probe analysis (Roblero *et al.*, 1976) may indicate the nature of dynamic events at the cell surface. The bicarbonate ion in tubal fluid has already been invoked as a factor prompting dispersal of the corona cell layer from rabbit eggs (Stambaugh *et al.*, 1969).

It remains controversial whether mammalian eggs possess a component associated with the membranes termed fertilisin, which either induces sperm agglutination and possibly the acrosome reaction or requires a period of exposure to a putative antifertilisin in the tubal fluids before spermatozoa can penetrate the zona pellucida. Apart from analogies with the fertilisin reaction in invertebrates (Bishop and Tyler, 1956), the only plausible evidence for this phenomenon in mammals comes from the *in vitro* work of Thibault and Dauzier (1960), in which it was noted that washing rabbit eggs in a physiological medium prior to addition of sperm suspensions resulted in a significantly higher incidence of *in vitro* fertilisation. Conversely, a more recent study in the rabbit failed to find any evidence for a specific fertilisin emanating from the egg surface which might be involved in initiating the acrosome reaction and thereby facilitating a higher incidence of sperm penetration (Overstreet and Bedford, 1975).

There may also be some physiological activity emanating from the zona pellucida upon those sperm heads that confront its surface. A specific binding and attachment of spermatozoa to the zona surface has been noted by Gwatkin's group (Hartmann and Hutchison, 1974a, b, c), based largely on studies in the golden hamster. Even so, this topic of sperm binding raises problems of interpretation, not least because of the loose and spongy structure of the outermost portion of the zona (Fig. V.7) which may function at this time more as a form of trapping mechanism.

Fertilisation

Although capacitation and the acrosome reaction are essential preliminaries, the process of fertilisation involves sperm penetration of the zona pellucida, with detachment and loss of the vesiculated outer acrosomal and plasma membranes at the surface of the zona (Fig. V.8). Digestion of a narrow, curved, pathway through the zona is due in part to a local action of membrane bound acrosin, but the increased motility consequent upon capacitation is undoubtedly important in this phase of penetration. These steps are followed by sperm attachment to the plasma (vitelline) membrane of the egg, cessation of flagellar activity, sperm head incorporation into the ooplasm by a fusion of plasma membranes, decondensation of the chromatin and formation of a male pronucleus (Fig. V.9). Even though the sperm flagellum usually enters the egg in mammals, there is seemingly no contribution from the

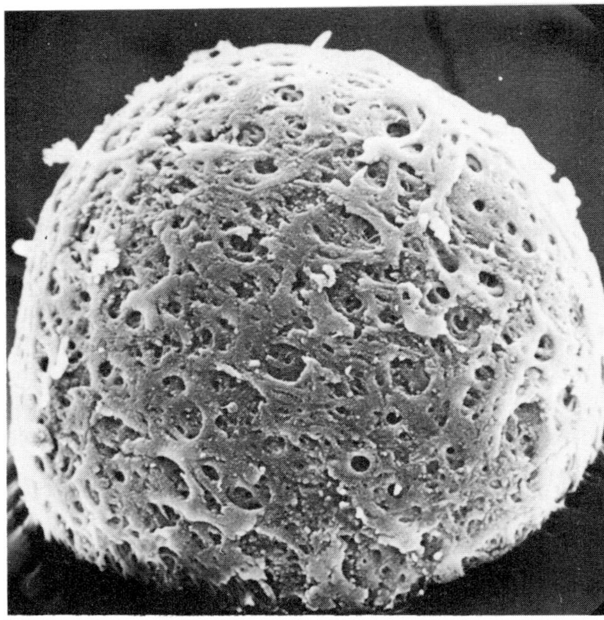

Fig. V.7. Scanning electron micrograph of the zona pellucida of a pig egg. Note the extremely uneven nature of the surface which may facilitate contact reactions between the spermatozoon and egg. (Courtesy of Dr J. E. Fléchon.)

organelles of this structure to the formation of the embryo. The mitochondria of the mid-piece are certainly known to degenerate (Szollosi, 1965).

Specific changes in the egg initiated by sperm contact with the plasma membrane are considered as *activation* changes and are:

(1) Activation of the meiotic division, with resumption from the short resting stage of second metaphase through to telophase and complete abstriction of the second polar body (Fig. V.10).

(2) Activation of a fusion between the cortical granules and the plasma membrane (Fig. V.11) so that release of the contents of the cortical granules into the perivitelline space can instigate a block to polyspermy (see Austin, 1956; Szollosi, 1967; Fléchon, 1970). There is also a conspicuous contraction of the vitellus during this change.

(3) Activation of synthetic processes in the egg, in particular replication of the complement of DNA.

The male and female chromatin groups develop nuclear envelopes, form pronuclei each containing several nucleoli, and the two pronuclei

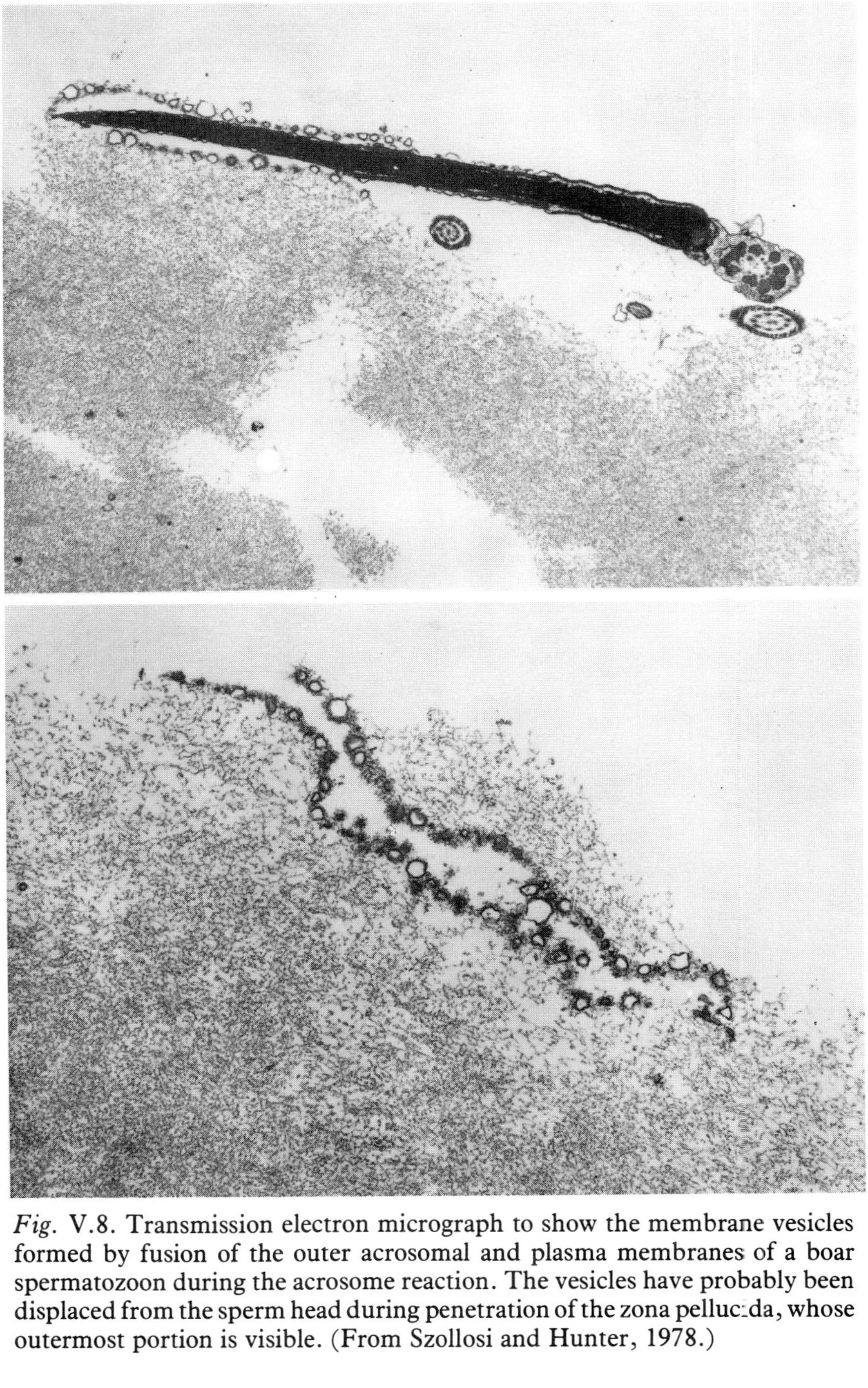

Fig. V.8. Transmission electron micrograph to show the membrane vesicles formed by fusion of the outer acrosomal and plasma membranes of a boar spermatozoon during the acrosome reaction. The vesicles have probably been displaced from the sperm head during penetration of the zona pellucida, whose outermost portion is visible. (From Szollosi and Hunter, 1978.)

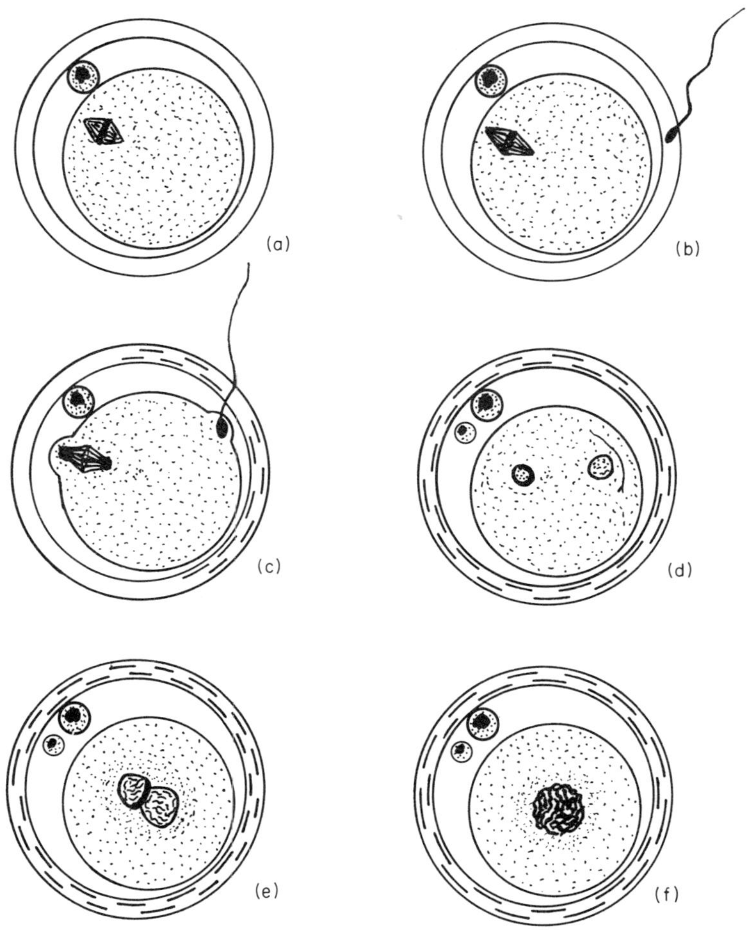

Fig. V.9. Diagrammatic representation of some principal events during fertilisation of the mammalian egg, from the stage of sperm penetration of the zona pellucida and activation of the oocyte (b and c), through pronuclear formation and migration (d) to pronuclear apposition and restoration of the diploid condition (e and f).

then enlarge and migrate to the centre of the egg. It is during this phase of pronuclear migration that DNA replication occurs (Oprescu and Thibault, 1965; Szollosi, 1966). The central meeting of the pronuclei leads to a fading but not fusion of their membranes, followed by condensation of the chromatin and restoration of the diploid condition by assembly of the chromosomes (karyogamy) at metaphase. Anaphase and telophase of the first mitotic division follow almost immediately.

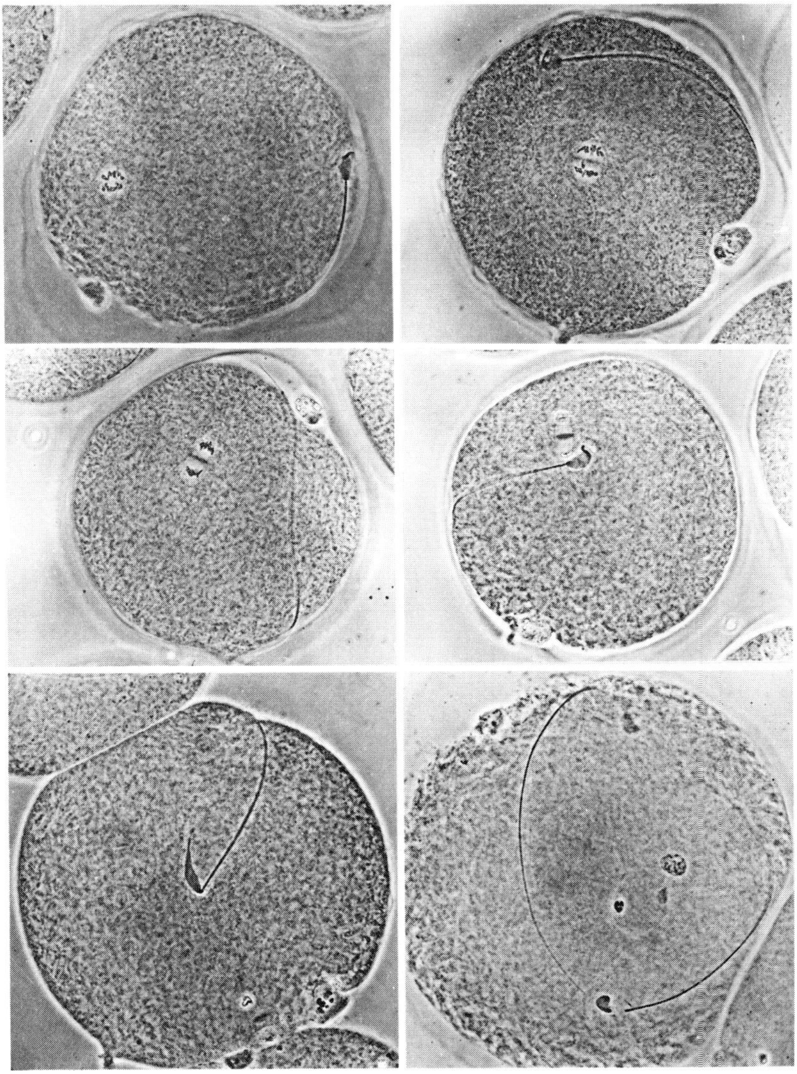

Fig. V.10. Completion of meiosis shortly after activation of eggs of the golden hamster by the fertilising spermatozoon.

The exact mechanism of the block to polyspermy in mammals, that is the prevention of penetration of the egg cytoplasm by more than one sperm, is not fully understood. But the cortical granules are known to play a key rôle, and discharge their contents to influence the penetrability of the zona pellucida or, more rarely, the vitelline surface of the egg

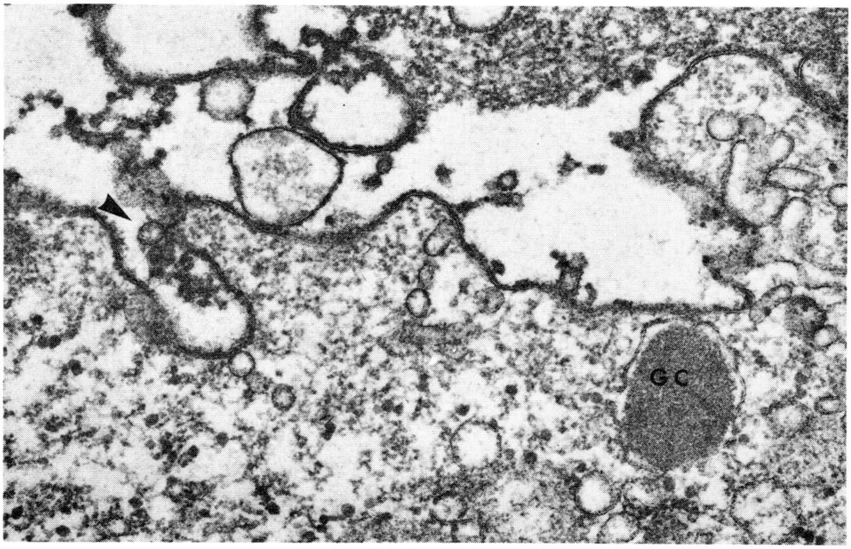

Fig. V.11. Electron micrograph to illustrate the cortical reaction that rapidly follows activation of an oocyte by the fertilising spermatozoon. Note how fusion of the cortical granule membrane with the overlying plasmalemma enables the escape of contents into the perivitelline space. (Courtesy of Dr D. Szollosi.)

(see Braden *et al.*, 1954). Apparently similar changes in the zona can be induced by treating unfertilised eggs with cortical granule material (Barros and Yanagimachi, 1971). The block develops in the form of a propagated change over the egg surface, the point of origin of the change being that at which the sperm makes functional contact with the vitelline membrane. Because of this fact, the greatest chance of dispermic fertilisation arises if a second sperm penetrates the zona at a point diametrically opposite that of the initial penetration (i.e. at the furthest point in the other hemisphere). But when fertilisation occurs after the normal time relationships between mating and ovulation, the Fallopian tube and the egg investments act together very significantly to reduce the chances of two viable spermatozoa arriving simultaneously at the egg surface. On the other hand, under conditions of postovulatory ageing of the egg before mating or insemination, the physiology of tubal function and the condition of the egg membranes both serve to permit an increased chance of polyspermy, added to which the ageing egg itself is more prone to this condition.

Gamete lifespans and abnormalities of fertilisation

The viable life of the gametes in the female reproductive tract has been estimated for a number of mammalian species, from which it is known that the egg commences to degenerate after only a small number of hours, in some contrast to the situation for spermatozoa where the viable lifespan may extend to 24–48 hours or more (Table V.4). Nonetheless, it must be appreciated that the number of eggs is strictly limited, being of course only one or two per cycle in cattle and sheep, whereas there is a relatively vast population of spermatozoa within which there may well be some selection on the grounds of viability (see Chapter IV). In a heterogeneous population of cells such as those contained in the ejaculate, it may therefore be difficult to offer with any precision a meaningful figure for the lifespan of spermatozoa in the female tract. Added to this specific constraint, it is still uncertain to what extent viable lifespans represent an intrinsic property of the cell, be they male or female gametes, and to what degree they may be modified by the tubal milieu. If the latter does indeed play a significant rôle, then it may become possible to develop a treatment for prolonging the lifespan of the gametes, if this should be considered desirable.

Since the fertilisable life of mammalian eggs is generally observed to be quite short (Table V.4), in many species rarely extending beyond 10–12 hours, it is not surprising that lowered fertility frequently occurs when mating or insemination is delayed until after the time of ovulation. In this situation, the age of the egg at penetration will have further

Table V.4. The viable lifespan of gametes in the female reproductive tract, representing the period during which they are able to achieve normal fertilisation and subsequent cleavage. Even so, a proportion of the eggs penetrated at the upper end of their lifespan may succumb to early embryonic loss.

Species	Eggs (hours)	Spermatozoa[1] (hours)
Cow	10–12	24–48
Sheep	10–15	24–48
Pig	8–12	24–42
Horse	8–10	140

[1] Motility is invariably retained for a much longer period than fertilising ability, except in the case of stallion spermatozoa.

increased by the time required for the spermatozoa to have become capacitated and to have reached the surface of the egg. The early studies of Hammond and Walton (1934) in the ferret and Hammond (1934) in the rabbit attributed this postovulatory loss of fertility to an increase in the proportion of eggs failing to undergo fertilisation, but increased embryonic mortality may also result from fertilisation of ageing eggs (i.e. guinea-pig: Blandau and Young, 1939; rat: Blandau and Jordan, 1941; Braden, 1959; rabbit: Chang, 1952). These experimental observations clearly imply that the egg remains fertilisable for longer than it retains the capacity to give rise to a normal embryo. As ageing eggs are more likely to be encountered under conditions of artificial insemination than natural mating, it seems especially important to extend these observations from laboratory rodents to domestic species of livestock. However, any such study requires accurate estimation or control of the time of ovulation, and the only comprehensive work of this nature is in pigs although there are useful data for cattle and sheep.

Several studies in pigs have indicated that matings or inseminations made late in oestrus tend to result in lowered fertility, as measured by the proportion of eggs fertilised (Hancock and Hovell, 1962) and the incidence of embryonic survival (Burger, 1952). These general observations have been extended, since ageing of the gametes could be a contributory factor to the high incidence of embryonic mortality which has been reported for this polytocous species (reviewed by Hanly, 1961). From the fertility curves (Fig. V.12) presented by Willemse and Boender (1967) and the more detailed data of Hunter (1967a, b), it appears that pig eggs commence to deteriorate within 8 hours of ovulation, *even though the animal may still be in behavioural oestrus and thus receptive to the male at this stage*. As recorded in Table V.5, this deterioration does not show up in the proportion of pig eggs fertilised, but it is detectable when embryonic survival is examined 25 days after insemination (Hunter, 1967a, b, 1977b). Fertility curves for cattle relating insemination time to conception rate suggest that the situation in pigs is also true of ruminants (Fig. V.13). Since ovulation occurs after the end of oestrus in cattle and sheep (see Chapter III), the risk of postovulatory egg ageing would normally occur only under conditions of artificial insemination. Specific cytological studies on ageing eggs in ruminants are scarce, probably because of the intrinsically low ovulation rate and the need to predetermine the precise time of ovulation, but Thibault (1967) has recorded the surprisingly large figures of 20–24 hours for the fertilisable life of the cow egg.

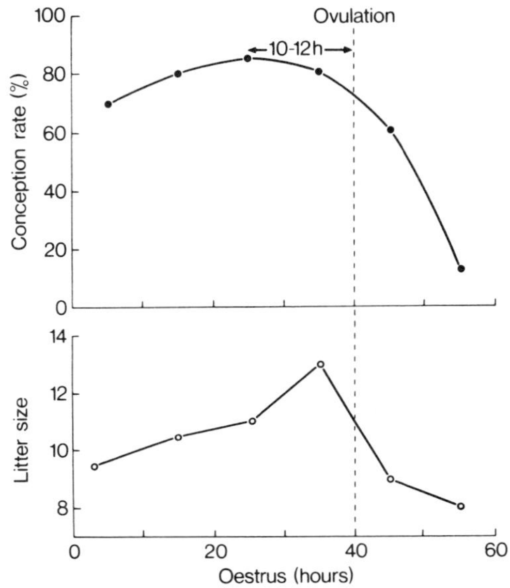

Fig. V.12. Graphical representation of the influence of the time of insemination upon fertility (conception rate and litter size) in pigs. (Redrawn after Willemse and Boender, 1967. From Austin, C. R. and Short, R. V. *Reproduction in Mammals*, Cambridge University Press, Volume 5, Figs 1–7. Courtesy of Dr C. Polge.)

Table V.5. The influence of post-ovulatory ageing of pig eggs in the Fallopian tubes on fertilisation and on embryonic survival at 25 days after insemination. Eighteen animals were inseminated in each group. (Compiled from Hunter, 1967a, b.)

Estimated age of eggs at fertilisation	Eggs fertilised normally		Viable embryos at 25 days		
	%	± (S.E.)	% survival	± (S.E.)	Mean number
0	90·8	(4·5)	87·9	(2·9)	12·0
4	92·1	(2·7)	72·9	(14·9)	11·7
8	94·6	(2·3)	60·5	(13·2)	8·7
12	70·3	(7·8)	53·3	(15·7)	6·8
16	48·3	(8·4)	27·9	(14·5)	4·8
20	50·9	(7·5)	32·3	(15·2)	5·0

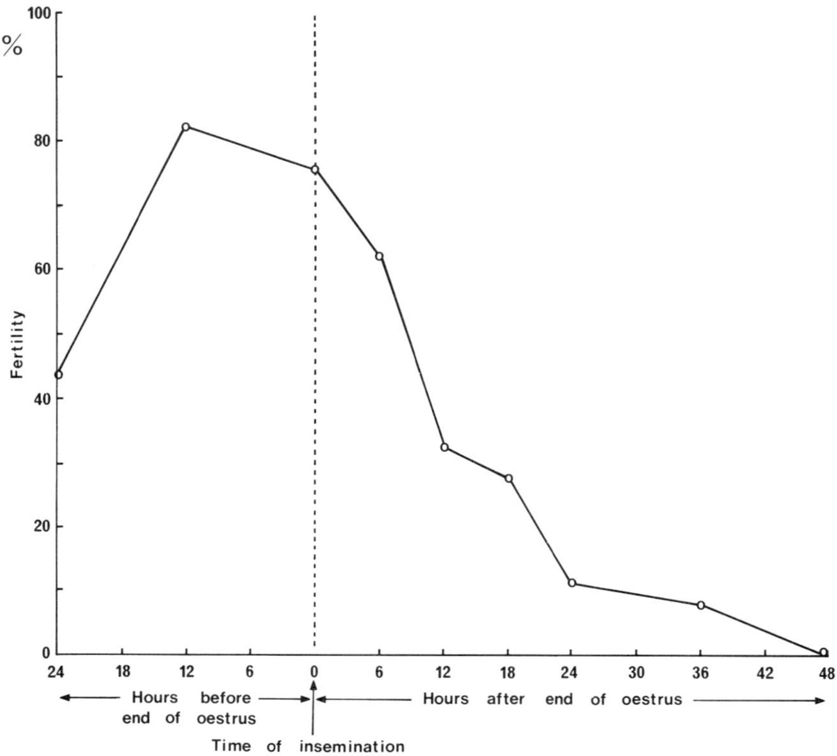

Fig. V.13. To illustrate the deleterious influence of incorrectly timed inseminations upon fertility in cattle. (Redrawn from the data of Trimberger, G. W. 1948; Univ. Nebraska Agric. Exp. Sta. Res. Bull. No. 153.)

After a phase of postovulatory ageing of eggs in the Fallopian tubes, specific anomalies of the process of fertilisation can be detected. These may arise from disorganisation of the microtubules of the meiotic spindle (Fig. V.14) before penetration of the egg by a spermatozoon. The sequel to this condition may involve (1) retention of the second polar body (i.e. digyny; Thibault, 1959) owing to a reorientation of the spindle microtubules relative to the egg's surface; (2) anomalies of the chromosome complements and possibly the formation of triploids (Bomsel-Helmreich, 1961, 1965; McFeely, 1967); or (3) fragmentation of the cytoplasm and degeneration of its organelles (Dziuk and Polge, 1962; Hunter, 1967a); these abnormalities all predispose the egg to early embryonic death. Abnormal fertilisation also occurs when more than one spermatozoon enters the vitellus (Fig. V.15), again found

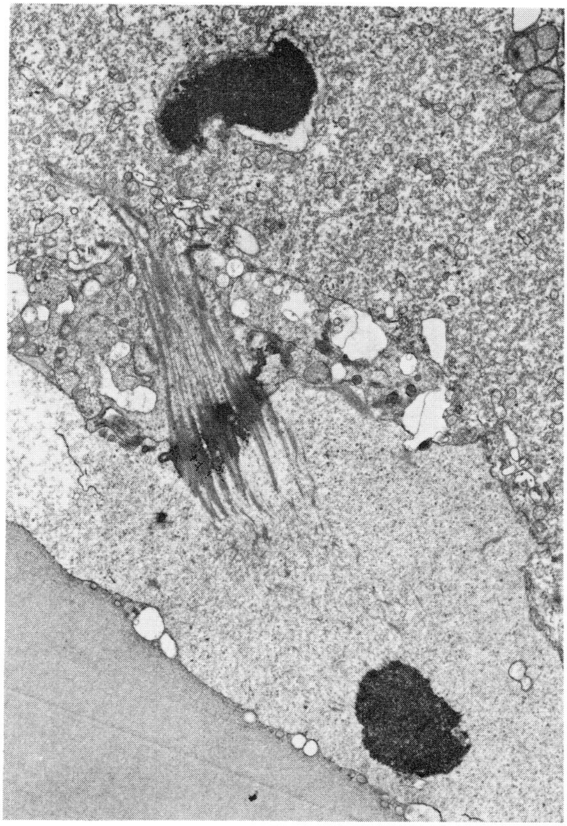

Fig. V.14. Completion of the second meiotic division shortly after activation of a pig egg by the fertilising spermatozoon. In this electron micrograph, the chromatin is no longer associated with the spindle microtubules, and in the vitellus has already developed a nuclear envelope. This is in contrast to its condition in the second polar body whose cytoplasm is largely devoid of organelles. (From Szollosi and Hunter, 1973.)

most commonly after postovulatory mating or insemination (Hancock, 1959; Thibault, 1959; Hunter, 1967a), a situation in which the eggs soon commence to degenerate and regulation of sperm passage by the isthmus is less effective. An anomaly of fertilisation would arise from activation of an egg by a diploid spermatozoon, but experiments in rabbits suggest that the incidence of this occurrence is likely to be very low (Fechheimer and Beatty, 1974; Mortimer, 1978). These and other bizarre forms of fertilisation are discussed in detail by Austin (1961, 1970) and Beatty (1970).

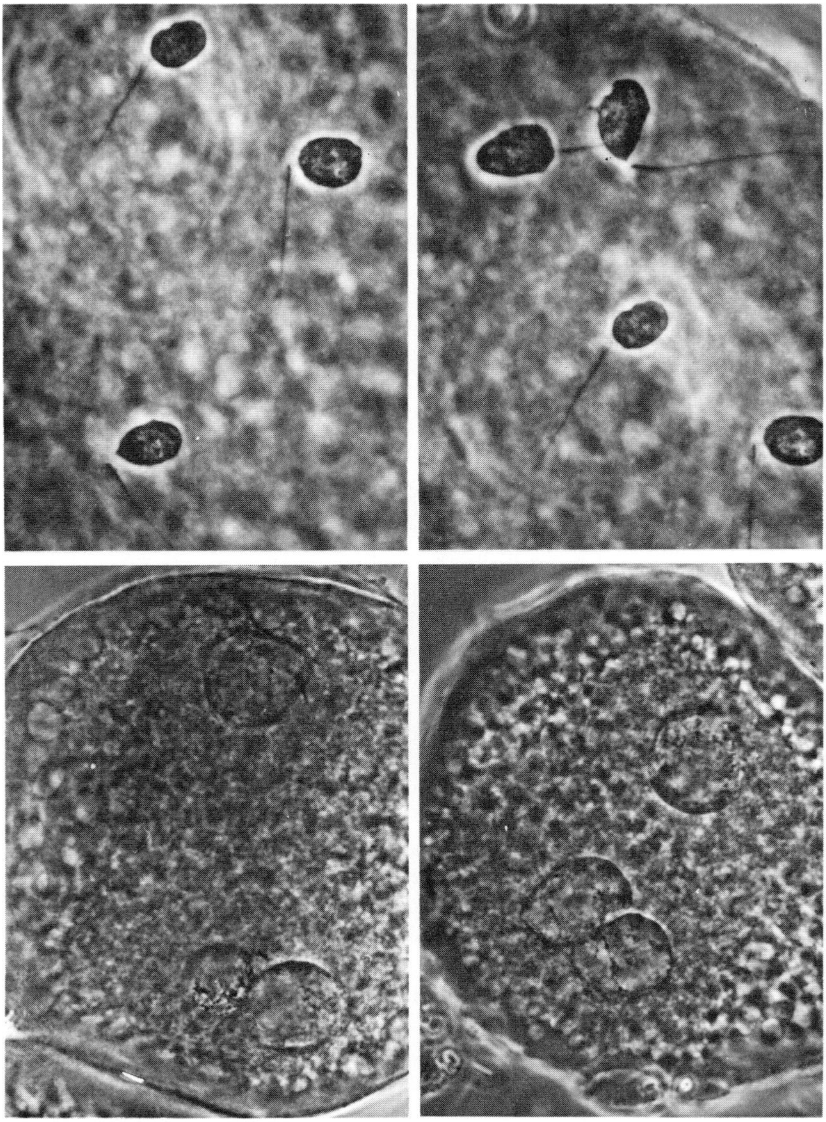

Fig. V.15. Polyspermic fertilisation of pig oocytes obtained in the presence of excessive numbers of spermatozoa in the Fallopian tubes. This condition is also associated with post-ovulatory ageing of the egg.

If insemination is performed too late relative to the time of ovulation, the deleterious influence upon fertility is seen in (1) a reduced conception rate; (2) a lower incidence of normal fertilisation; and (3) a significant embryonic loss. In polytocous species the latter effects are noted ultimately in a reduced litter size (Boender, 1966; Willemse and Boender, 1967), and emphasize the importance of avoiding ageing of gametes, suggesting perhaps that the time of mating or insemination should be predetermined (Dziuk and Polge, 1965; Polge *et al.*, 1968).

In contrast to this situation of delayed insemination and effects of ageing in the eggs, the evidence for a deleterious influence of ageing of mammalian spermatozoa in the female tract on embryonic development is equivocal. As already mentioned, this is doubtless associated in part with the vast number of gametes in the ejaculate, within which there is scope for "selection" of viable cells among the heterogeneous population (Dziuk, 1970; Chang and Hunter, 1975). Also relevant is the fact that artificial insemination is unlikely to be performed at the very onset of oestrus, and rarely sooner than some 20–24 hours before ovulation—an interval during which spermatozoa can retain their fertilising ability in the higher reaches of the female tract (see Table V.4). Moreover, it is worth considering the notion that this relatively prolonged survival of male gametes is occurring in a *preovulatory* environment, whereas postovulatory deterioration of the egg is occurring in an environment under the influence of increasing progesterone secretion from a developing corpus luteum. Nonetheless, most of the published data, especially for pigs in which conception and subsequent embryonic loss can be judged accurately, suggest that best fertility is obtained when insemination takes place 12–16 hours before ovulation (Boender, 1966; Polge *et al.*, 1968; Polge, 1969; Dziuk, 1970). When using diluted semen or frozen-thawed spermatozoa in pigs, there is certainly an optimum time for insemination if litter size is not to be compromised (Polge *et al.*, 1970; Larsson, 1976), and this is definitely closer to ovulation (i.e. 6–8 hours *before*) than with freshly ejaculated spermatozoa.

Fate of gametes in female genital tract

Although there appears to be a clear exception in horses (see below), unfertilised eggs eventually degenerate in the uterus after a gradual softening and distortion of the zona pellucida, and perhaps after a phase of cytoplasmic fragmentation (Dziuk, 1960). Degenerating eggs with these characteristics can be flushed from the uterine lumen as late as 8–10 days after ovulation in the pig, and up to 15 days in sheep (Bindon, 1969), but such eggs are probably expelled from the uterus before the next oestrus. The presence of a zona pellucida seems to protect the degenerating cytoplasm from phagocytosis.

Concerning the fate of spermatozoa not actually involved in fertilisation, there are rather more avenues available than for the eggs and thus there is scope for controversy. The two main routes of disposal of surplus spermatozoa are (1) voiding from the female tract; and (2) engulfment by polymorphonuclear leucocytes. As to those spermatozoa that gain the Fallopian tubes, the controversial question is whether phagocytosis occurs within the lumen of this structure, at least whilst the eggs are in residence. The accepted routes of disposal of tubal spermatozoa include passage into the peritoneal cavity, incorporation into corona cells (Fig. V.16) and, in species such as the pig, accumulation of as many as several hundred spermatozoa in the zona pellucida of fertilised eggs (Hancock, 1961; Hunter and Léglise, 1971). Studies involving intraperitoneal insemination of pigs have also suggested a retrograde passage of the remaining proportion of tubal spermatozoa into the uterus at the time of embryo passage, along with the liberated cumulus and corona cells (Hunter, 1978). Thus, it seems unlikely that incorporation of degenerating spermatozoa into the tubal epithelium (Austin, 1959, 1960), either before or after engulfment, represents a significant route of disposal of surplus tubal spermatozoa.

Transport of eggs and embryos

After liberation from the cumulus oophorus and completion of the events of fertilisation, the developing embryos still encompassed by the zona pellucida remain in the isthmus for three to four days in most of the species examined, entering the uterus as morulae or blastocysts (Chapter VI). In cattle and sheep, the embryo enters the uterus at about three days after ovulation at the 8–16 cell stage. In pigs, entry into the uterus takes about two days when the embryo is at the correspondingly earlier stage of 4-cells (Heuser and Streeter, 1929; Hunter, 1974), but in mares some five to seven days may be required (see Chapter VI).

With the notable exception of findings in the horse, fertilised and unfertilised eggs pass into the uterus on a similar time schedule. In mares, however, unfertilised eggs remain in the Fallopian tubes where they undergo cytoplasmic degeneration (Van Niekerk and Gerneke, 1966; Betteridge and Mitchell, 1974), although they are recognisable within a zona pellucida for many weeks or even months. The reason for

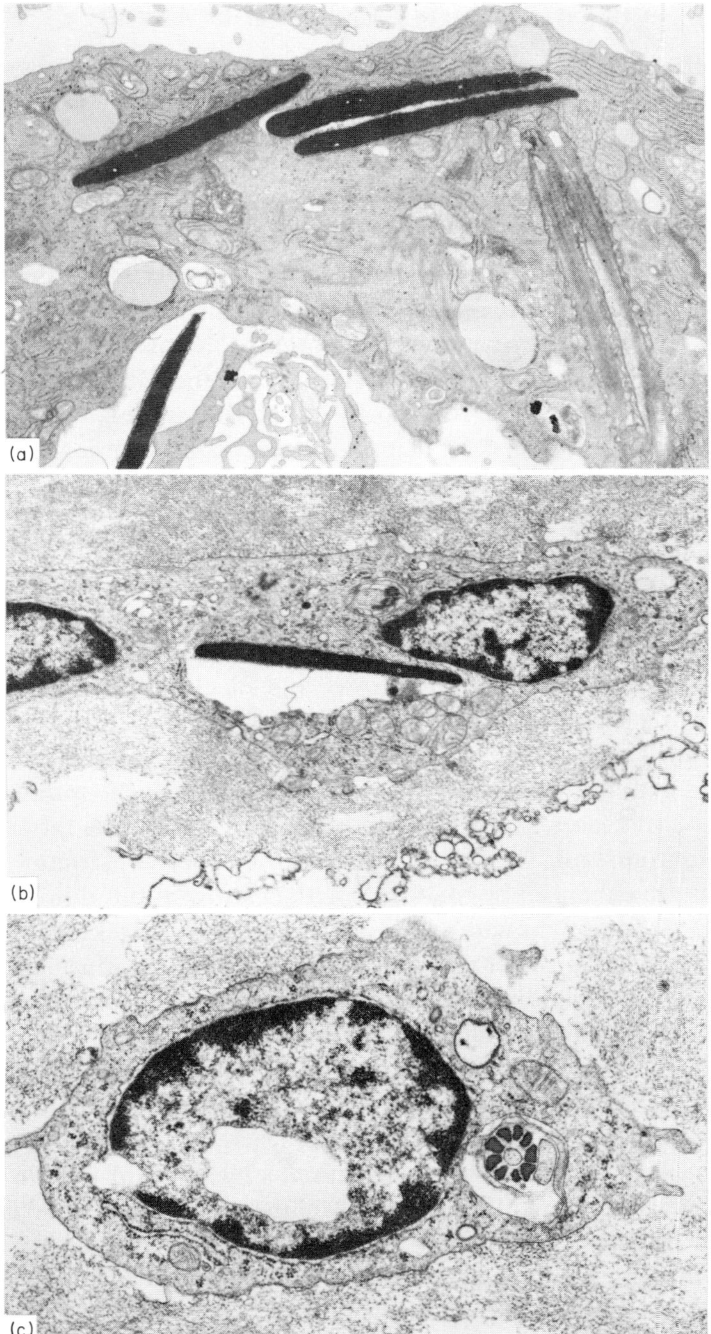

Fig. V.16. Incorporation of washed or preincubated bull spermatozoa by cells of the cumulus oophorus of calf oocytes during a period of culture *in vitro* (a) or *in vivo* incorporation of boar spermatozoa by corona cells of the pig oocyte (b, c). (Courtesy of Dr D. Szollosi.)

this difference is unknown, but secretions from the cleaving egg may act to cause relaxation of the isthmus at the appropriate time, thereby permitting passage of the fertilised egg into the uterus. Another suggestion is that the unfertilised egg is distinguished from the zygote in this species by its surface characteristics (Rowson, 1971), but morphological evidence indicates little difference in the zonae of the two types of egg (Betteridge *et al.*, 1976). An endocrine influence of the embryo on tubal function seems the more plausible explanation, especially in the light of recent evidence on the biosynthetic potential of eggs (Niimura and Ishida, 1976) and blastocysts (Dickmann and Dey, 1974).

The specific pattern of transport through the isthmus varies according to species, but a general feature is a final rapid transit through the distal isthmus into the uterus, in part at least due to relaxation of the musculature. Two components of muscular function in the isthmus have been implicated in the control of egg passage. First, the very rich adrenergic innervation of this portion of the tube enables the isthmus to constrict and act as an adrenergic sphincter, thereby arresting descent of the eggs. Moreover, oestrogens have been shown to potentiate α-adrenergic receptor activity (Howe and Black, 1973; Black, 1974), underlining the rôle of this physiological sphincter close to the time of ovulation. Progesterone, on the other hand, potentiates β-adrenergic receptor activity, mediating in isthmic relaxation as the interval from ovulation increases. After HCG-induced ovulation in the rabbit, both the amplitude and duration of the post-ovulatory constriction of the isthmus are enhanced by treatment with oestrogen, but diminished or blocked by treatment with progesterone (Blair and Beck, 1976).

A second means of control appears to be obtained through a local action of prostaglandins. Much of the recent evidence is controversial due to an apparent contradiction in the action of specific prostaglandins on tubal function *in vivo* when compared with their effect *in vitro* and due also to differential effects of specific prostaglandins on the longitudinal and circular layers of muscle. Nonetheless, prostaglandins of the F series may act to retain eggs in the isthmus by virtue of their occlusive action, whereas those of the E series facilitate egg passage by diminishing the occlusive effects (Spilman and Harper, 1973; Spilman, 1976). There may be regional effects of prostaglandins within the isthmus, with the uterine quarter tending to contract, whilst the remaining three-quarters relax. In both rabbits and humans, however, prostaglandin $F_{2\alpha}$ activates tubal contractility while prostaglandin E induces relaxation (Maia *et al.*, 1976). Thus, in terms of the physiology of isthmic function, a mutual antagonism can be demonstrated between α-adrenergic receptor activity and the effects of prostaglandin E, again seemingly modulated by the prevailing balance of ovarian steroid hormones.

A third mode of controlling the rate of egg transport through the isthmus may involve the condition of the prominent and much-folded mucosa, which is largely dependent upon the balance of gonadal steroids and in the pig, for example, is extremely oedematous at oestrus (Andersen, 1928). The oedema is markedly reduced by the time of egg passage into the uterus and seemingly can be reduced by systemic (Day and Polge, 1968) or, more specifically, by local microinjections (Hunter, 1972) of a solution of progesterone in oil. Overall, therefore, the regulatory influence of ovarian steroids seems to be exerted mainly on the excitability and conduction properties of oviductal smooth musculature and on the morphology of the mucosa.

Anomalies of egg transport

It has been known for many years that the pattern of contractile activity in the Fallopian tube is steroid dependent (Corner, 1923; Seckinger, 1923). Administration of steroid hormones close to the time of ovulation or induction of ovulation in the presence of active corpora lutea disrupts transport of eggs or embryos in several species (Wislocki and Snyder, 1933; Burdick and Pincus, 1935; Greenwald, 1959). This disruption may be in the form of "tube locking" of eggs, or in an accelerated tubal descent so that the eggs or embryos enter the uterus prematurely. Both conditions compromise normal development of the embryo and tend to be lethal, for the time of entry into the uterus is critical for subsequent development and implantation (Chang, 1950).

Between species, it is difficult to state categorically whether exogenous oestrogens cause "tube locking" and progesterone an acceleration of descent, since the underlying processes are influenced by (1) the dose of hormone (which is invariably pharmacological); (2) the actual oestrogenic or progestational compound administered; and (3) most critically the time relationship between treatment and ovulation. As a general rule, however, oestrogen treatment retards tubal transport of eggs, whereas progesterone treatment facilitates a more rapid descent; but treatment with progesterone may only cause such a response when acting on oestrogen-primed tissues of the oviduct. As examples of steroid induced anomalies of tubal function, there are the classical oestrogen-induced "tube locking" of eggs in mice and rabbits (Burdick and Pincus, 1935), and the influence of subcutaneous injections of progesterone shortly before ovulation in pigs causing descent of eggs into the uterus within 8 hours (Day and Polge, 1968). Protracted or excessive responses to superovulation procedures in cattle (Rowson, 1951) or induction of ovulation during the luteal phase in pigs may also lead to accelerated egg transport into the uterus.

Fluid environment of the embryo

As to the rôle of Fallopian tube fluids in the nutritional support of the developing embryo, specific constituents of the fluid appear essential to energy metabolism. Most of the information is derived from *in vitro* studies using mouse embryos (Biggers *et al.*, 1967; Brinster, 1973), which have shown for example that the mouse zygote can divide only when pyruvate or oxaloacetate are present in the culture medium, and that lactate can be used as a substrate only after the 2-cell stage is reached. Pyruvate concentrations in tubal fluid increase considerably after ovulation but, significantly, changes in lactate concentration are the most striking. These are but limited examples, the essential point being that the nature of the substrates available in tubal fluid is closely regulated by the prevailing balance of ovarian hormones and is therefore changing to match the embryo's requirement for and ability to utilise the substrates. A local influence of the embryo on the synthetic activity of the tubal epithelium has been suggested (Stone and Hamner, 1975), but the available evidence is as yet meagre. Nonetheless, the report by Roblero *et al.* (1976) that the ionic compositions of ampullary and isthmic fluid are dissimilar, and that both are very different in composition from blood plasma, could be interpreted in support of such local effects.

Tubal secretion is programmed quantitatively by the balance of ovarian oestrogens and progesterone, the volume of secretion being greatest around oestrus or just before ovulation, and minimal during the mid luteal phase or pregnancy (Table V.6). The influence of ovarian steroids on secretion is well demonstrated after ovariectomy in the oestrous rabbit, when tubal secretion drops from 1·2–0·2 ml per 24 hours, whereas systemic injections of oestradiol will restore fluid production to oestrous levels (Bishop, 1956). Similar observations have

Table V.6. To indicate the range in rate of fluid secretion into the lumen of the Fallopian tube of various species at oestrus and during the luteal phase of the cycle; figures represent ml of fluid per 24 hours. (Taken from Hunter, 1977a.)

Species	Reproductive state of animal	
	Oestrus (or mid-cycle peak)	Luteal phase (or pregnancy)
Rabbit	0·8	0·3
	1·4	–
	2·0	0·4
Sheep	1·3	0.3
	1·0	0·4
	1·2	0·2
	0·9	0·2
Cow	4·1	0·7
	2·0	0·2
Rhesus monkey	2·7	0·2
	2·1	0·3
	1·9	0·7[1]

[1] The latest sample in this study was taken on Day 17 of the menstrual cycle.

been made in sheep (McDonald and Bellve, 1969). As has already been noted, the degree of engorgement of the tubal vasculature also varies cyclically (Fig. V.2).

Concerning direction of flow, the bulk of the fluid passes from the ampulla into the peritoneal cavity at oestrus, but by the time the embryos descend through the utero-tubal junction, the reduced volume of fluid is passing in the other direction into the uterus (see Bellve and McDonald, 1970). The direction of flow appears to be influenced principally by the nature of ciliary currents and the degree of muscular and mucosal stricture in the isthmus and utero-tubal junction. The rôle of tubal fluid in aiding the transport of embryos from the isthmus into the uterus remains to be clarified.

The osmolarity of tubal secretions ranges from 302–310 m-osmols in the oestrous rabbit, but reaches a mean of 388 m-osmols after ovariectomy (Coutinho, 1974). Cyclic ionic changes in the fluid have been characterised, mainly in rabbits, whereas cyclic changes in pH have been noted in many species—fluctuations in bicarbonate secretion being an important factor regulating pH and stimulating glycolysis.

A major point of interest in the protein content of tubal fluid concerns its relationship to the protein profile of blood serum. Gel electrophoresis analysis of proteins from human tubal fluid reveals patterns that are similar to those of the corresponding serum, but some unique proteins such as β-glycoprotein are also observed in women (Moghissi, 1970) and in monkeys (Mastroianni *et al.*, 1970). In rabbits, an oestrogen-modulated protein with the electrophoretic migration properties of uteroglobulin has been described (Feigelson and Kay, 1972), and in cattle a unique tubal protein has also been detected (Roberts *et al.*, 1975). Although the results to date suggest that most proteins in tubal fluid arise from a serum transudate, there may be cyclical variation in the ratio between transudate and specific secretion.

References

Adams, C. E. and Chang, M. C. (1962). Capacitation of rabbit spermatozoa in the Fallopian tube and in the uterus. *J. exp. Zool.* **151**, 159–166.

Andersen, D. H. (1928). Comparative anatomy of the tubo-uterine junction. Histology and physiology in the sow. *Amer. J. Anat.* **42**, 255–305.

Austin, C. R. (1951). Observations on the penetration of the sperm into the mammalian egg. *Aust. J. sci. Res. B.* **4**, 581–596.

Austin, C. R. (1952). The "capacitation" of the mammalian sperm. *Nature, London* **170**, 326.

Austin, C. R. (1956). Cortical granules in hamster eggs. *Expl Cell Res.* **10**, 533–540.

Austin, C. R. (1959). Entry of spermatozoa into the Fallopian tube mucosa. *Nature, London* **183**, 908–909.

Austin, C. R. (1960). Fate of spermatozoa in the female genital tract. *J. Reprod. Fert.* **1**, 151–156.

Austin, C. R. (1961). "The Mammalian Egg". Blackwell, Oxford.

Austin, C. R. (1967). Capacitation of spermatozoa. *Int. J. Fertil.* **12**, 25–31.

Austin, C. R. (1970). Ageing and reproduction: post-ovulatory deterioration of the egg. *J. Reprod. Fert., Suppl*, **12**, 39–53.

Austin, C. R. (1972). Fertilisation. *In* "Reproduction in Mammals" (C. R. Austin and R. V. Short, eds), Book 1, pp. 103–133. University Press, Cambridge.

Austin, C. R. and Braden, A. W. H. (1954). Time relations and their significance in the ovulation and penetration of eggs in rats and rabbits. *Aust. J. biol. Sci.* **7**, 179–194.

Barros, C. and Yanagimachi, R. (1971). Induction of the zona reaction in golden hamster eggs by cortical granule material. *Nature, London* **233**, 268–269.

Beatty, R. A. (1970). The genetics of the mammalian gamete. *Biol. Rev.* **45**, 73–119.

Bedford, J. M. (1968). Importance of the Fallopian tube for capacitation in the rabbit. *Proc. 6th Int. Congr. Anim. Reprod., Paris,* **1**, 35–37.

Bedford, J. M. (1970). Sperm capacitation and fertilization in mammals. *Biol. Reprod., Suppl.* **2**, 128–158.

Bedford, J. M. (1975). Maturation, transport, and fate of spermatozoa in the epididymis. *In* "Handbook of Physiology", Section 7, Endocrinology 5, Male Reproductive System (D. W. Hamilton and R. O. Greep, eds), pp. 303–317. American Physiological Society, Washington.

Bedford, J. M. and Cooper, G. W. (1978). Membrane fusion events in the fertilisation of vertebrate eggs. *In* "Membrane Fusion" (G. Poste and G. L. Nicolson, eds), pp. 65–125. Elsevier-North Holland Biomedical Press, Amsterdam.

Bellve, A. R. and McDonald, M. F. (1970). Directional flow of Fallopian tube secretion in the ewe at onset of the breeding season. *J. Reprod. Fert.* **22**, 147–149.

Betteridge, K. J. and Mitchell, D. (1974). Direct evidence of retention of unfertilized ova in the oviduct of the mare. *J. Reprod. Fert.* **39**, 145–148.

Betteridge, K. J., Flood, P. F. and Mitchell, D. (1976). Possible role of the embryo in the control of oviductal transport in mares. *In* "Symposium on Ovum Transport and Fertility Regulation" (M. J. K. Harper and C. J. Pauerstein, eds), pp. 381–389. Scriptor, Copenhagen.

Biggers, J. D., Whittingham, D. G. and Donahue, R. P. (1967). The pattern of energy metabolism in the mouse oocyte and zygote. *Proc. nat. Acad. Sci. (Wash.)* **58**, 560–567.

Bindon, B. M. (1969). Fate of the unfertilized sheep ovum. *J. Reprod. Fert.* **20**, 183–184.

Bishop, D. W. (1956). Active secretion in the rabbit oviduct. *Amer. J. Physiol.* **187**, 347–352.

Bishop, D. W. and Tyler, A. (1956). Fertilizins of mammalian eggs. *J. exp. Zool.* **132**, 575–601.

Black, D. L. (1974). Neural control of oviduct musculature. *In* "The Oviduct and its Functions" (A. D. Johnson and C. W. Foley, eds), pp. 65–118. Academic Press, New York and London.

Blair, W. D. and Beck, L. R. (1976). A system for measurement of oviductal motility and contractility and chronic changes in luminal diameter. *In* "Symposium on Ovum Transport and Fertility Regulation" (M. J. K. Harper and C. J. Pauerstein, eds), pp. 41–74. Scriptor, Copenhagen.

Blandau, R. J. (1969). Gamete transport—comparative aspects. *In* "The Mammalian Oviduct" (E. S. E. Hafez and R. J. Blandau, eds), pp. 129–162. University of Chicago Press, Chicago.

Blandau, R. J. (1973). Gamete transport in the female mammal. *In* "Handbook of Physiology" Section 7, Endocrinology II, (R. O. Greep and E. B. Astwood, eds), pp. 153–163. American Physiological Society, Washington.

Blandau, R. J. and Jordan, E. S. (1941). The effect of delayed fertilization on the development of the rat ovum. *Amer. J. Anat.* **68**, 275–287.

Blandau, R. J. and Verdugo, P. (1976). An overview of gamete transport—comparative effects. *In* "Symposium on Ovum Transport and Fertility

Regulation" (M. J. K. Harper and C. J. Pauerstein, eds), pp. 138–146. Scriptor, Copenhagen.

Blandau, R. J. and Young, W. C. (1939). The effects of delayed fertilization on the development of the guinea pig ovum. *Amer. J. Anat.* **64**, 303–329.

Boender, J. (1966). The development of AI in pigs in the Netherlands and the storage of boar semen. *Wld Rev. Anim. Prod., Special issue* **2**, 29–44.

Bomsel-Helmreich, O. (1961). Heteroploidie expérimentale chez la truie. *Proc. 4th Int. Congr. Anim. Reprod., The Hague,* **3**, 578–581.

Bomsel-Helmreich, O. (1965). Heteroploidy and embryonic death. *In* "Preimplantation Stages of Pregnancy" (G. E. W. Wolstenholme and M. O'Connor, eds), pp. 246–267. Ciba Fdn Symposium. J. and A. Churchill, London.

Braden, A. W. H. (1959). Are nongenetic defects of the gametes important in the etiology of prenatal mortality? *Fert. Steril.* **10**, 285–298.

Braden, A. W. H., Austin, C. R. and David, H. A. (1954). The reaction of the zona pellucida to sperm penetration. *Aust. J. biol. Sci.* **7**, 391–409.

Brinster, R. L. (1973). Nutrition and metabolism of the ovum, zygote and blastocyst. *In* "Handbook of Physiology" Section 7, Endocrinology II (R. O. Greep and E. B. Astwood, eds), pp. 165–185. American Physiological Society, Washington.

Brundin, J. (1964). The distribution of noradrenaline and adrenaline in the Fallopian tube of the rabbit. *Acta physiol. scand.* **62**, 156–159.

Brundin, J. (1965). Distribution and function of adrenergic nerves in the rabbit Fallopian tube. *Acta physiol. scand.* **66**, *Suppl.* **259**, 1–57.

Burdick, H. O. and Pincus, G. (1935). The effect of oestrin injections upon the developing ova of mice and rabbits. *Amer. J. Physiol.* **111**, 201–208.

Burger, J. F. (1952). Sex physiology of pigs. *Onderstepoort J. vet. Res., Suppl. No.* **2**, 218pp.

Chang, M. C. (1950). Development and fate of transferred rabbit ova or blastocyst in relation to the ovulation time of recipients. *J. exp. Zool.* **114**, 197–216.

Chang, M. C. (1951). Fertilizing capacity of spermatozoa deposited into the Fallopian tubes. *Nature, London* **168**, 697–698.

Chang, M. C. (1952). Effects of delayed fertilization on segmenting ova, blastocysts and fetuses in rabbit. *Fedn Proc. Fedn Am. Socs exp. Biol.* **11**, 24.

Chang, M. C. (1955). Development of fertilizing capacity of rabbit spermatozoa in the uterus. *Nature, London* **175**, 1036–1037.

Chang, M. C. (1957). A detrimental effect of seminal plasma on the fertilizing capacity of sperm. *Nature, London* **179**, 258–259.

Chang, M. C. (1958). Capacitation of rabbit spermatozoa in the uterus with special reference to the reproductive phases of the female. *Endocrinology* **63**, 619–628.

Chang, M. C. and Hunter, R. H. F. (1975). Capacitation of mammalian sperm: biological and experimental aspects. *In* "Handbook of Physiology" Endocrinology V (D. W. Hamilton and R. O. Greep, eds), Chap. 16, pp 339–351. American Physiological Society, Washington.

Chang, M. C., Austin, C. R., Bedford, J. M., Brackett, B. G., Hunter, R. H

F. and Yanagimachi, R. (1977). Capacitation of spermatozoa and fertilisation in mammals. *In* "Frontiers in Reproduction and Fertility Control" (R. O. Greep and M. A. Koblinsky, eds), pp. 435–451. The MIT Press, Cambridge, Mass. and London.

Corner, G. W. (1923). Cyclic variation in uterine and tubal contraction waves. *Amer. J. Anat.* **32**, 345–351.

Coutinho, E. M. (1974). Hormonal control of oviductal motility and secretory functions. *In* "Reproductive Physiology" (R. O. Greep, ed.), pp. 133–153. MTP International Review of Science, Butterworths, London.

Dauzier, L. and Thibault, C. (1956). Recherche expérimentale sur la maturation des gamètes mâles chez les mammifères, par l'étude de la fécondation "in vitro" de l'oeuf de lapine. *Proc. 3rd Int. Congr. Anim. Reprod., Cambridge* **1**, 58–61.

Day, B. N. and Polge, C. (1968). Effects of progesterone on fertilization and egg transport in the pig. *J. Reprod. Fert.* **17**, 227–230.

Dickmann, Z. and Dey, S. K. (1974). Steroidogenesis in the preimplantation rat embryo and its possible influence on morula-blastocyst transformation and implantation. *J. Reprod. Fert.* **37**, 91–93.

Dziuk, P. J. (1960). Frequency of spontaneous fragmentation of ova in unbred gilts. *Proc. Soc. exp. Biol. Med.* **103**, 91–92.

Dziuk, P. J. (1970). Estimation of optimum time for insemination of gilts and ewes by double-mating at certain times relative to ovulation. *J. Reprod. Fert.* **22**, 277–282.

Dziuk, P. J. and Polge, C. (1962). Fertility in swine after induced ovulation *J. Reprod. Fert.* **4**, 207–208.

Dziuk, P. J. and Polge, C. (1965). Fertility in gilts following induced ovulation. *Vet. Rec.* **77**, 236–239.

Fechheimer, N. S. and Beatty, R. A. (1974). Chromosomal abnormalities and sex ratio in rabbit blastocysts. *J. Reprod. Fert.* **37**, 331–341.

Feigelson, M. and Kay, E. (1972). Protein patterns of rabbit oviducal fluid. *Biol. Reprod.* **6**, 244–252.

Fléchon, J. E. (1970). Nature glycoprotéique des granules corticaux de l'oeuf de lapine. *J. de Microscop.* **9**, 221–242.

Greenwald, G. S. (1959). The comparative effectiveness of estrogens in interrupting pregnancy in the rabbit. *Fert. Steril.* **10**, 155–161.

Greep, R. O. (1973). "Handbook of Physiology", Section 7, Endocrinology II, Female Reproductive System, Part 2. American Physiological Society, Washington.

Hafez, E. S. E. and Blandau, R. J. (1969). "The Mammalian Oviduct". University of Chicago Press, Chicago.

Hammond, J. (1934). The fertilization of rabbit ova in relation to time. A method of controlling the litter size, the duration of pregnancy and the weight of the young at birth. *J. exp. Biol.* **11**, 140–161.

Hammond, J. and Walton, A. (1934). Notes on ovulation and fertilization in the ferret. *J. exp. Biol.* **11**, 307–319.

Hancock, J. L. (1959). Polyspermy of pig ova. *Anim. Prod.* **1**, 103–106.

Hancock, J. L. (1961). Fertilisation in the pig. *J. Reprod. Fert.* **2**, 307–331.

Hancock, J. L. and Hovell, G. J. R. (1962). Insemination before and after the onset of heat in sows. *Anim. Prod.* **4**, 91–96.

Hanly, S. (1961). Prenatal mortality in farm animals. *J. Reprod. Fert.* **2**, 182–194.

Harper, M. J. K. (1961a). The mechanisms involved in the movement of newly ovulated eggs through the ampulla of the rabbit Fallopian tube. *J. Reprod. Fert.* **2**, 522–524.

Harper, M. J. K. (1961b). Egg movement through the ampullar region of the Fallopian tube of the rabbit. Proc. 4th Int. Congr. Anim. Reprod., The Hague, p. 375.

Harper, M. J. K. (1973a). Stimulation of sperm movement from the isthmus to the site of fertilization in the rabbit oviduct. *Biol. Reprod.* **8**, 369–377.

Harper, M. J. K. (1973b). Relationship between sperm transport and penetration of eggs in the rabbit oviduct. *Biol. Reprod.* **8**, 441–450.

Hartmann, J. F. and Hutchison, C. F. (1974a). Nature of the pre-penetration contact interactions between hamster gametes *in vitro*. *J. Reprod. Fert.* **36**, 49–57.

Hartmann, J. F. and Hutchison, C. F. (1974b). Contact between hamster spermatozoa and the zona pellucida releases a factor which influences early binding stages. *J. Reprod. Fert.* **37**, 61–66.

Hartmann, J. F. and Hutchison, C. F. (1974c). Mammalian fertilisation *in vitro*: sperm-induced preparation of the zona pellucida of golden hamster ova for final binding. *J. Reprod. Fert.* **37**, 443–445.

Hartree, E. F. (1977). Spermatozoa, eggs and proteinases. *Biochem. Soc. Trans.* **5**, 375–394.

Heuser, C. H. and Streeter, G. L. (1929). Early stages in the development of pig embryos. *Contrib. Embryol. Carneg. Instn.* **20**, 1–30.

Howe, G. R. and Black, D. L. (1973). Autonomic nervous system and oviduct function in the rabbit. *J. Reprod. Fert.* **33**, 425–430.

Hunter, R. H. F. (1967a). The effects of delayed insemination on fertilization and early cleavage in the pig. *J. Reprod. Fert.* **13**, 133–147.

Hunter, R. H. F. (1967b). Effect of ageing eggs on embryonic survival. *J. Anim. Sci.* **26**, 945.

Hunter, R. H. F. (1969). Capacitation in the golden hamster, with special reference to the influence of the uterine environment. *J. Reprod. Fert.* **20**, 223–237.

Hunter, R. H. F. (1972). Local action of progesterone leading to polyspermic fertilization in pigs. *J. Reprod. Fert.* **31**, 433–444.

Hunter, R. H. F. (1974). Chronological and cytological details of fertilization and early embryonic development in the domestic pig, *Sus scrofa*. *Anat. Rec.* **178**, 169–186.

Hunter, R. H. F. (1977a). Function and malfunction of the Fallopian tubes in relation to gametes, embryos and hormones. *Eur. J. Obstet. Gynec. Reprod. Biol.* **7**, 267–283.

Hunter, R. H. F. (1977b). Physiological factors influencing ovulation, fertilisation, early embryonic development and establishment of pregnancy in pigs. *Brit. vet. J.* **133**, 461–470.

Hunter, R. H. F. (1978). Intraperitoneal insemination, sperm transport and capacitation in the pig. *Anim. Reprod. Sci.* **1**, 167–179.

Hunter, R. H. F. and Dziuk, P. J. (1968). Sperm penetration of pig eggs in relation to the timing of ovulation and insemination. *J. Reprod. Fert.* **15**, 199–208.

Hunter, R. H. F. and Hall, J. P. (1974a). Capacitation of boar spermatozoa: synergism between uterine and tubal environments. *J. exp. Zool.* **188**, 203–214.

Hunter, R. H. F. and Hall, J. P. (1974b). Capacitation of boar spermatozoa: the influence of post-coital separation of the uterus and Fallopian tubes. *Anat. Rec.* **180**, 597–604.

Hunter, R. H. F. and Léglise, P. C. (1971). Polyspermic fertilization following tubal surgery in pigs, with particular reference to the rôle of the isthmus. *J. Reprod. Fert.* **24**, 233–246.

Johnson, A. D. and Foley, C. W. (1974). "The Oviduct and its Functions". Academic Press, New York.

Larsson, K. (1976). Fertility of deep frozen boar spermatozoa at various intervals between insemination and induced ovulation. *Acta vet. Scand.* **17**, 63–73.

McDonald, M. F. and Bellve, A. R. (1969). Influence of oestrogen and progesterone on flow of fluid from the Fallopian tube in the ovariectomized ewe. *J. Reprod. Fert.* **20**, 51–61.

McFeely, R. A. (1967). Chromosome abnormalities in early embryos of the pig. *J. Reprod. Fert.* **13**, 579–581.

Maia, H., Barbosa, I. and Coutinho, E. M. (1976). Relationship between cyclic AMP levels and oviductal contractility. *In* "Symposium on Ovum Transport and Fertility Regulation" (M. J. K. Harper and C. J. Pauerstein, eds), pp. 168–181. Scriptor, Copenhagen.

Mastroianni Jr., L., Urzua, M. and Stambaugh, R. (1970). Protein patterns in monkey oviductal fluid before and after ovulation. *Fert. Steril.* **21**, 817–820.

Moghissi, K. S. (1970). Human Fallopian tube fluid. I. Protein composition. *Fert. Steril.* **21**, 821–829.

Mortimer, D. (1978). The distribution of diploid rabbit spermatozoa in the female tract after artificial insemination. *Anim. Reprod. Sci.* **1**, 245–250.

Niimura, S. and Ishida, K. (1976). Histochemical studies of Δ^5–3β, 20α and 20β hydroxysteroid dehydrogenases and possible progestagen production in hamster eggs. *J. Reprod. Fert.* **48**, 275–278.

Noyes, R. W. (1953). The fertilizing capacity of spermatozoa. *Western J. Surg. Obstet. Gynecol.* **61**, 342–349.

Oprescu, St. and Thibault, C. (1965). Duplication de l'ADN dans les oeufs de lapine après la fécondation. *Annls Biol. anim. Biochim. Biophys.* **5**, 151–156.

Overstreet, J. W. and Bedford, J. M. (1975). The penetrability of rabbit ova treated with enzymes or anti-progesterone antibody: a probe into the nature of a mammalian fertilizin. *J. Reprod. Fert.* **44**, 273–284.

Overstreet, J. W. and Cooper, G. W. (1975). Reduced sperm motility in the isthmus of the rabbit oviduct. *Nature, London* **258**, 718–719.

Polge, C. (1969). Advances in reproductive physiology in pigs. *J. Aust. Inst. agric. Sci.* **35**, 147–153.

Polge, C., Day, B. N. and Groves, T. W. (1968). Synchronisation of ovulation and artificial insemination in pigs. *Vet. Rec.* **83**, 136–142.

Polge, C., Salamon, S. and Wilmut, I. (1970). Fertilising capacity of frozen boar semen following surgical insemination. *Vet. Rec.* **87**, 424–428.

Roberts, G. P., Parker, J. M. and Symonds, H. W. (1975). Proteins in the luminal fluid from the bovine oviduct. *J. Reprod. Fert.* **45**, 301–313.

Roblero, L., Biggers, J. D. and Lechene, C. P. (1976). Electron probe analysis of the elemental microenvironment of oviducal mouse embryos. *J. Reprod. Fert.* **46**, 431–434.

Rowson, L. E. (1951). Methods of inducing multiple ovulation in cattle. *J. Endocr.* **7**, 260–270.

Rowson, L. E. A. (1971). The role of reproductive research in animal production. *J. Reprod. Fert.* **26**, 113–126.

Seckinger, D. L. (1923). Spontaneous contractions of the Fallopian tube of the domestic pig with reference to the oestrous cycle. *Bull. Johns Hopk. Hosp.* **34**, 236–239.

Spilman, C. H. (1976). Prostaglandins, oviductal motility and egg transport. *In* "Symposium on Ovum Transport and Fertility Regulation" (M. J. K. Harper and C. J. Pauerstein, eds), pp. 197–211. Scriptor, Copenhagen.

Spilman, C. H. and Harper, M. J. K. (1973). Effect of prostaglandins on oviduct motility in estrous rabbits. *Biol. Reprod.* **9**, 36–45.

Stambaugh, R. and Buckley, J. (1969). Identification and subcellular localization of the enzymes effecting penetration of the zona pellucida by rabbit spermatozoa. *J. Reprod. Fert.* **19**, 423–432.

Stambaugh, R., Noriega, C. and Mastroianni, L. (1969). Bicarbonate ion: the corona cell dispersing factor of rabbit tubal fluid. *J. Reprod. Fert.* **18**, 51–58.

Stone, S. L. and Hamner, C. E. (1975). Biochemistry and physiology of oviductal secretions. *Gynec. Invest.* **6**, 234–252.

Szollosi, D. (1965). The fate of sperm middle-piece mitochondria in the rat egg. *J. exp. Zool.* **159**, 367–377.

Szollosi, D. (1966). Time and duration of DNA synthesis in rabbit eggs after sperm penetration. *Anat. Rec.* **154**, 209–212.

Szollosi, D. (1967). Development of cortical granules and the cortical reaction in rat and hamster eggs. *Anat. Rec.* **159**, 431–446.

Szollosi, D. and Hunter, R. H. F. (1973). Ultrastructural aspects of fertilisation in the domestic pig: sperm penetration and pronucleus formation. *J. Anat.* **116**, 181–206.

Szollosi, D. and Hunter, R. H. F. (1978). The nature and occurrence of the acrosome reaction in spermatozoa of the domestic pig, *Sus scrofa*. *J. Anat.* **127**, 33–41.

Thibault, C. (1959). Analyse de la fécondation de l'oeuf de la truie après accouplement ou insémination artificielle. Annls Zootech. Sér. D., Suppl., pp. 165–177.

Thibault, C. (1967). Analyse comparée de la fécondation et de ses anomalies chez la brebis, la vache et la lapine. *Annls Biol. anim. Biochim. Biophys.* **7**, 5–23.

Thibault, C. and Dauzier, L. (1960). "Fertilisines" et fécondation in vitro de l'oeuf de lapine. *C. R. Acad. Sci. (Paris)* **250**, 1358–1359.

Van Niekerk, C. H. and Gerneke, W. H. (1966). Persistence and parthenogenetic cleavage of tubal ova in the mare. *Onderstepoort J. vet. Res.* **33**, 195–231.

Willemse, A. H. and Boender, J. (1967). The relation between the time of insemination and fertility in gilts. *Tijdschr. Diergeneesk.* **92**, 18–34.

Wintenberger-Torres, S. (1961). Mouvements des trompes et progression des oeufs chez la brebis. *Annls Biol. anim. Biochim. Biophys.* **1**, 121–133.

Wislocki, G. B. and Snyder, F. F. (1933). The experimental acceleration of the rate of transport of ova through the Fallopian tube. *Bull. Johns Hopk. Hosp.* **52**, 379–386.

Yanagimachi, R. (1970). The movement of golden hamster spermatozoa before and after capacitation. *J. Reprod. Fert.* **23**, 193–196.

Yanagimachi, R., Yanagimachi, H. and Rogers, B. J. (1976). The use of zona-free animal ova as a test system for the assessment of fertilizing capacity of human spermatozoa. *Biol. Reprod.* **15**, 471–476.

Chapter VI

Uterine Function in Relation to Development and Distribution of Embryos, Implantation, and Establishment of Pregnancy

Introduction

In simple terms, the rôle of the uterus with regard to the embryo is frequently presented as one of nutrition, implantation, differentiation of the embryo proper from its surrounding membranes, and then support of the foetus until the normal time for the events of parturition. Little emphasis is sometimes given to the fact that the preimplantation phase of embryonic development in the uterus may occupy a week or more in many domestic animals (Table VI.1), even though the rarer phenomenon of delayed implantation has been recognised for many years in species such as roe deer, badger, certain bears and mink (Enders, 1963; Asdell, 1964; Aitken, 1977). It is therefore worth stressing at the beginning of this chapter that during the preimplantation phase in the uterus, major developmental changes are occurring in the embryo as it progresses from a 4- or 8-cell aggregate to a morula and then a blastocyst. In the early part of this transition, the embryo is still protected by an accessory membrane, the zona pellucida, and may rely quite extensively upon cytoplasmic reserves of nutrients (there is a relatively large amount of yolk in the egg cytoplasm of species such as the pig), but upon shedding of the zona, the rapidly growing preimplantation embryo relies heavily on constituents of uterine fluids (the so called uterine milk or histotrophe) for nutritional support; protein components of uterine fluid may be especially involved in promoting growth at this stage.

A spatial relationship between embryo and uterus is also achieved during this phase, and in polytocous species this may involve a substantial redistribution of embryos throughout the cornua. During these preimplantation events, or in some species close to the time of commencement of implantation, a critical decision is imposed on the mother as to whether to continue oestrous cycles or, in contrast,

Table VI.1. Chronology of preimplantation development, passage into the uterus from the Fallopian tubes, and commencement of attachment (implantation). Preovulatory mating or insemination has been assumed, and the stages are therefore presented as times after ovulation or sperm penetration of the eggs.

Species	First cleavage (hours)	Second cleavage (hours)	Entry into uterus		Loss of zona pellucida (days)	Commencement of attachment (days)
			Time (hours)	Stage of development		
Cow	20–24	32–36	72–84	8–16 cells	9–11	22
Sheep	16–18	28–30	66–72	8–16 cells	7–8	15
Pig	14–16	20–24	46–48	4–cells	6	13
Horse	24	30–36	140–144[1]	Blastocysts	8?	37

[1] Approximate figure for fertilised eggs; unfertilised eggs remain and degenerate in the Fallopian tube.

whether to prolong the lifespan of the corpus luteum to provide endocrine support for the maintenance of pregnancy. Even with a successful establishment of the embryo at this sensitive stage, it cannot be assumed that all pregnancies will go to term, for a variety of factors can contribute to prenatal loss. In a polytocous species such as the pig, prenatal loss is usually expressed as a smaller litter size, but in a predominantly monotocous species such as the cow, prenatal loss will lead to the failure of pregnancy.

Entry of embryos into uterus and preliminary differentiation

As noted in the previous chapter, fertilisation takes place in the Fallopian tubes of eutherian mammals, and the fertilised eggs do not descend to the uterus for some 3–4 days or more in most species so far examined. Thus, cow and sheep embryos enter the uterus 66–72 hours after ovulation at the 8 to 16-cell stage of development. However, the pig is an exception in this respect with embryos entering the uterus 46–48 hours after ovulation at the 4-cell stage, whereas horse embryos may not enter the uterus for 5–7 days by which time they are blastocysts (Table VI.1). Compared with the situation in cattle and sheep, the precocious transport of pig embryos into the uterus is also reflected in the timing of subsequent events such as emergence from the zona pellucida (hatching), expansion of the blastocyst, and commencement of a specific morphological interaction between the trophoblast and the endometrium. This latter event varies in time from about two weeks in the pig up to as late as three to four weeks after fertilisation in cattle.

The rate of passage of embryos along the tubes into the uterus is principally under the control of ovarian steroid hormones (see Chapter V), and the timing of entry into the uterus has been shown to be quite critical if embryonic development and a successful pregnancy are to ensue. This has been established experimentally by the two different approaches of disturbing the normal rate of tubal transport by treatment with gonadal steroids (Burdick and Pincus, 1935; Greenwald, 1959; Chang, 1966) and by the more elegant procedure of embryo transplantation (Chang, 1950; Noyes and Dickmann, 1960; Rowson and Moor, 1966a). With the possible exception of the situation in primates (see Hunter, 1977a), the latter approach has demonstrated that the stage of embryonic development needs to be closely synchronised with that of the hormonally regulated endometrial proliferation if a high incidence of pregnancy is to result. Among the larger domestic species with their longer oestrous cycles and gestation periods, there appears to be some tolerance in the synchronisation requirements: Rowson and his colleagues have shown that cow and sheep embryos can tolerate an asynchrony *vis à vis* the stage of uterine development of ± 1 and ± 2 days, respectively (Rowson and Moor, 1966a; Rowson *et al.*, 1969; Rowson, 1971; Rowson *et al.*, 1972). Even so, the highest conception rates following transplantation have usually been recorded when synchronisation has been precise.

Whilst the embryos are held and sustained in the Fallopian tubes, the uterus is prepared for their arrival by phagocytic removal of post-coital débris such as dead spermatozoa and sperm remnants, other seminal products, and bacteria that may have been introduced at the time of mating. This phase of preparation also includes a more positive form of stimulation arising from the influence of increasing progesterone output by the developing corpus luteum: the latter leads to endometrial proliferation with increased glandular activity and secretion of fluid into the uterine lumen. Such uterine fluids together with certain cellular components combine to form the histotrophe.

Once the dividing blastomeres have formed a substantial group of cells (i.e. 8–16, but sometimes more), the embryo is called a morula. In terms of their developmental potential, the first few blastomeres are probably equipotent in the rabbit and farm species, at least until the 4-celled stage (see Moore *et al.*, 1968), and in due course seem to differentiate according to their position in the morula. Formation of a blastocyst occurs when cells of the morula commence to secrete fluid and rearrange themselves around a central fluid-filled cavity termed the blastocoele (Fig. VI.1). In the blastocysts of the farm species, as judged particularly by observations with the scanning electron microscope, there seems to be a distribution of blastomeres according to size—the larger to one pole, the smaller to the other (Fléchon and Renard, 1978). The blastocyst rapidly differentiates into the group of cells that will proliferate to form the embryo proper, namely the inner cell mass, and those that will give rise to the extra-embryonic membranes, the trophectoderm or trophoblast. The latter is thus the tissue specialised for interaction with the uterus and which, despite its foreign (paternal) complement of antigens, is not rejected by the mother as implantation proceeds.

A layer of endoderm originates from cells of the inner cell mass,

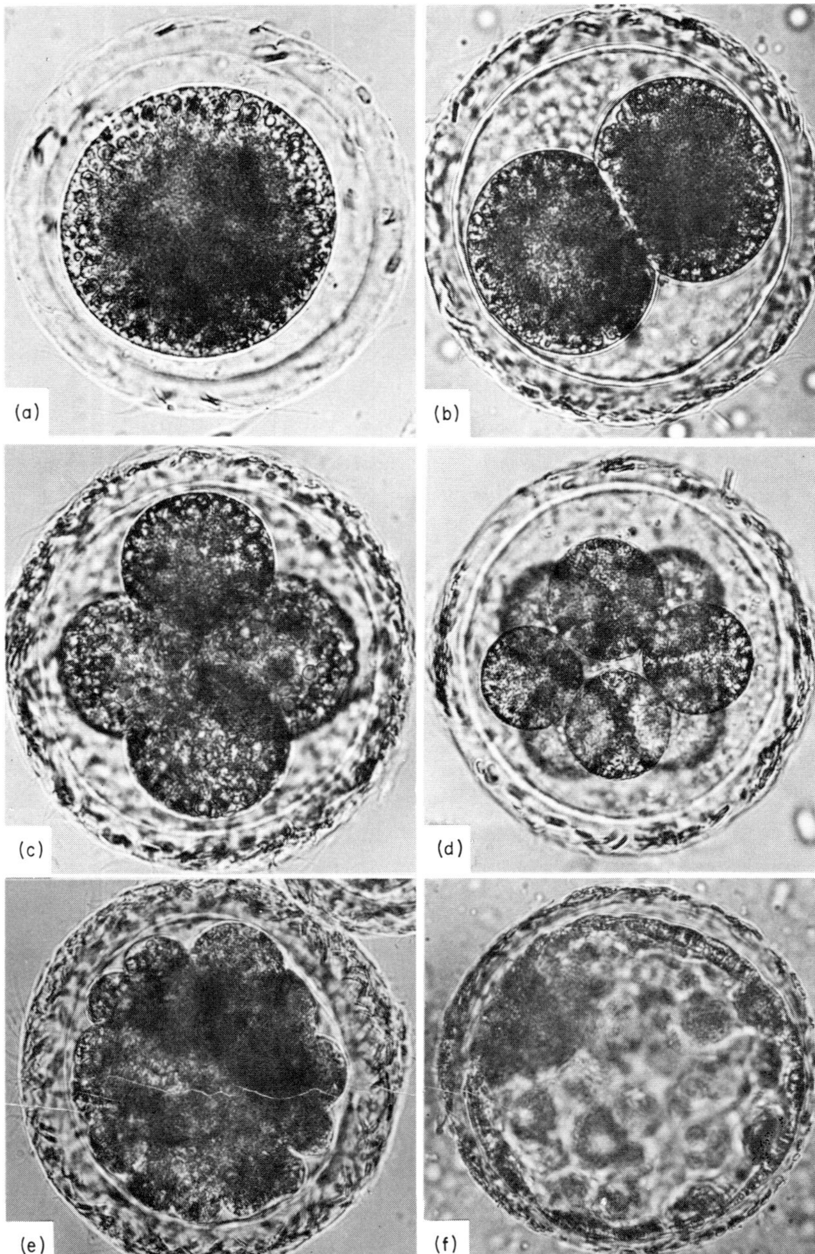

Fig. VI.1. Fertilisation and development of pig eggs from the pronucleate (a) and early cleavage stages (b, c, d) to a morula (e) and blastocyst (f). Note the increasing numbers of accessory spermatozoa trapped in the zona pellucida; this will be shed shortly after the stage depicted in (f). (Reproduced from Hunter, 1974.)

giving rise to a bilaminar blastocyst, and this is followed by formation of the mesoderm (Fig. VI.2). At the stage of initial blastocyst formation, the embryo is still supported and protected within the zona pellucida, but as development in the uterine environment proceeds, the zona pellucida is shed and the ungulate embryo undergoes further growth and differentiation before developing an intimate association with the endometrium. At this stage in the large farm species, the much elongated embryo is frequently termed a conceptus.

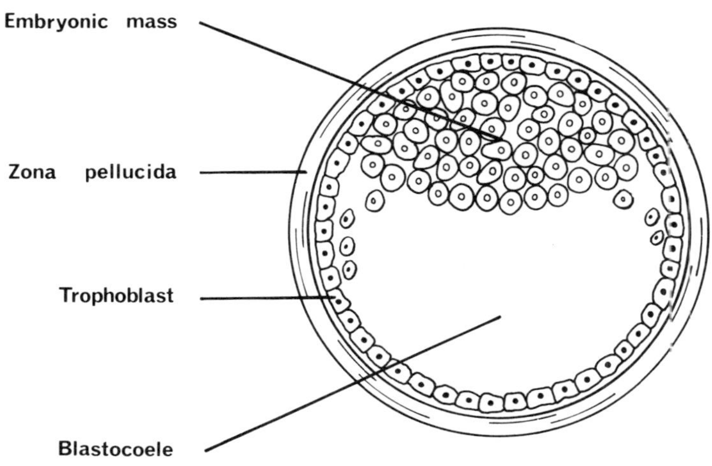

Fig. VI.2. Diagrammatic representation of a blastocyst still within the zona pellucida to show a simplified distinction between cells of the trophoblast and those of the inner cell mass (embryonic mass).

Emergence from the zona pellucida

The mechanisms involved in release of the embryo from the zona pellucida are still the subject of research, and may well differ in important respects between species. But apart from the physical effect of growth and expansion of the blastocyst causing a stretching and thinning of the zona (Fig. VI.3), and thereby contributing to its rupture (see Fléchon and Renard, 1978), there is also a possible involvement of locally released enzymes giving a weakening in one portion of the zona. Once an aperture has been formed, an active pumping of fluid into the blastocoele causing expansion of the embryo seems to be a major factor

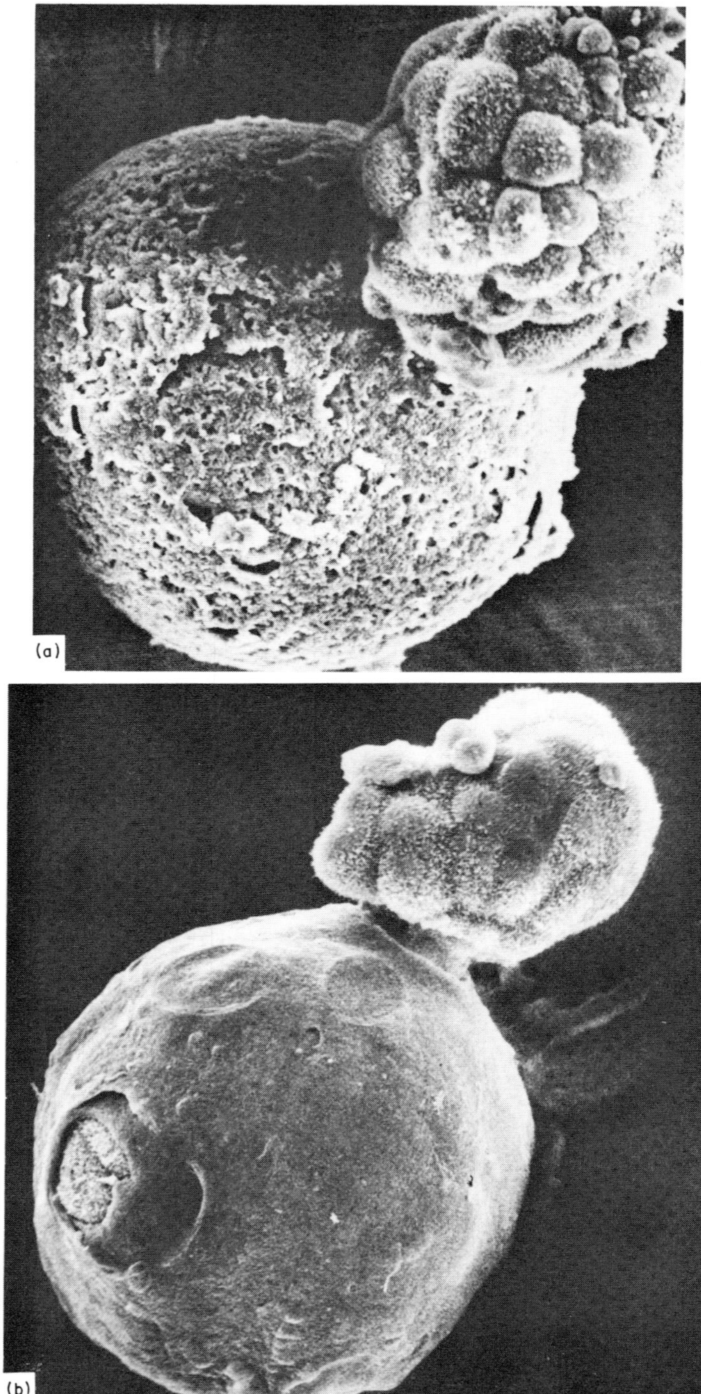

Fig. VI.3. Scanning electron micrographs of cow (a) and pig (b) blastocysts during the process of escape of the embryo from the zona pellucida. The cow blastocyst had been cultured *in vitro* from Days 7 to 9 whereas the pig embryo was recovered and fixed six days after ovulation. (Courtesy of Dr J. E. Fléchon.)

involved in hatching. A relative movement between and within blastomeres, together with an action of their conspicuous microvilli, may also facilitate escape of the embryo. Some degree of morphological distortion is clearly inevitable during these events.

Secretion of proteolytic enzymes by cells of the trophoblast has been suggested as a means of weakening and lysing the zona (see Dickmann, 1969) and could explain the aperture sometimes seen in the zona shortly before blastocyst emergence; some form of secretion of lytic enzymes by the uterine epithelium has also been invoked in this respect (McLaren, 1970). Both physical and biochemical factors will doubtless eventually be found to be involved in the liberation of the blastocyst from its zona pellucida, and in fact the embryo may be dictating in some trophic manner a local release of lytic enzymes from that portion of the endometrium with which it is in contact. However, having described the possible involvement of physical and chemical factors, it is important to realize that emergence of an intact blastocyst from its zona is not the only means whereby the trophoblast establishes a functional contact with the endometrium. In the guinea pig, for example, quite extensive cytoplasmic processes are seen to pass from the abembryonic pole of the blastocyst *through* the zona pellucida (Blandau, 1949; Blandau and Rumery, 1957).

As to the timing of hatching, pig blastocysts emerge from the zona on the sixth day of embryonic life, whereas sheep embryos are liberated between the seventh and eighth days and cow embryos between the ninth and eleventh days of development, usually the tenth (Table VI.1). The fate of the zona pellucida after liberation of the embryo has not been specifically examined, but it may undergo further degeneration followed by lysis in the uterus or be incorporated by elements of the trophoblast or endometrium. Either process would need to account not only for the substance of the zona but also for sperm head material which may be associated with this membrane. As already noted (Chapter V), several hundred sperm heads may be embedded in the zona pellucida of fertilised pig eggs.

Elongation of the embryo

As appears to be the case for a variety of ungulates, the preimplantation phase in cattle, sheep and pigs includes a remarkable expansion of the blastocyst by extensive proliferation of the trophoblast shortly after hatching from the zona. This proliferation is accompanied by a morphological transition from a spherical, fluid-filled blastocyst to a much extended although somewhat wrinkled conceptus in the cow (Winters *et al.*, 1942; Chang, 1952) and sheep (Green and Winters, 1945; Rowson and Moor, 1966b) or in the case of the pig, to an extremely long thread-like conceptus (Heuser and Streeter, 1929; Green and Winters, 1946; Perry and Rowlands, 1962). Because the surface of the conceptus is wrinkled in this post hatching phase (Figs VI.4 and 5), the rate of accumulation and/or secretion of fluid into the trophoblastic cavity clearly does not keep pace with the increase in surface area of the trophoblast.

Since bovine blastocysts commence their implantation reaction considerably later than those of sheep and pigs (see below), there is a somewhat slower initial phase of preimplantation growth that follows hatching on about the tenth day. By the thirteenth day, the blastocyst still shows only a relatively small expansion, measuring about 5 mm in diameter (Perry *et al.*, 1976), but by the sixteenth day it has grown to many centimetres and may occupy two thirds the length of the uterine horn (Winters *et al.*, 1942; Betteridge, 1977).

In sheep, blastocysts formed on the fifth day after ovulation develop quite slowly until the seventh, when they are still spherical and contain approximately 300 cells. Growth then begins to accelerate, the blastocyst having some 3000 cells on the ninth day and by the eleventh it has attained a length of 10–22 mm (Wintenberger-Torrès and Fléchon, 1974). A tubular conceptus some 10 cm in length has usually formed by the thirteenth day (Fig. VI.4), and by the fifteenth the conceptus may measure up to 15–19 cm in length (Chang and Rowson, 1965; Rowson and Moor, 1966b).

After hatching from the zona, considerable variation in size is soon seen between pig blastocysts from the same uterus, these differences being magnified progressively in 7- and 8-day blastocysts, during which time the spherical form is lost and a wrinkled appearance develops (Heuser and Streeter, 1929; Hunter, 1974). Thereafter, pig blastocysts expand more rapidly than those of sheep, occasionally measuring 4–5 mm in diameter on the ninth day of embryonic life and then elongating into a tubular conceptus of some 5 cm or more by the eleventh or twelfth day (Fig. VI.5). One day later, a massive expansion in length has occurred in several but rarely all of the blastocysts which, when rearranged after flushing from the uterus, may extend to lengths of 80–110 cm (see Perry

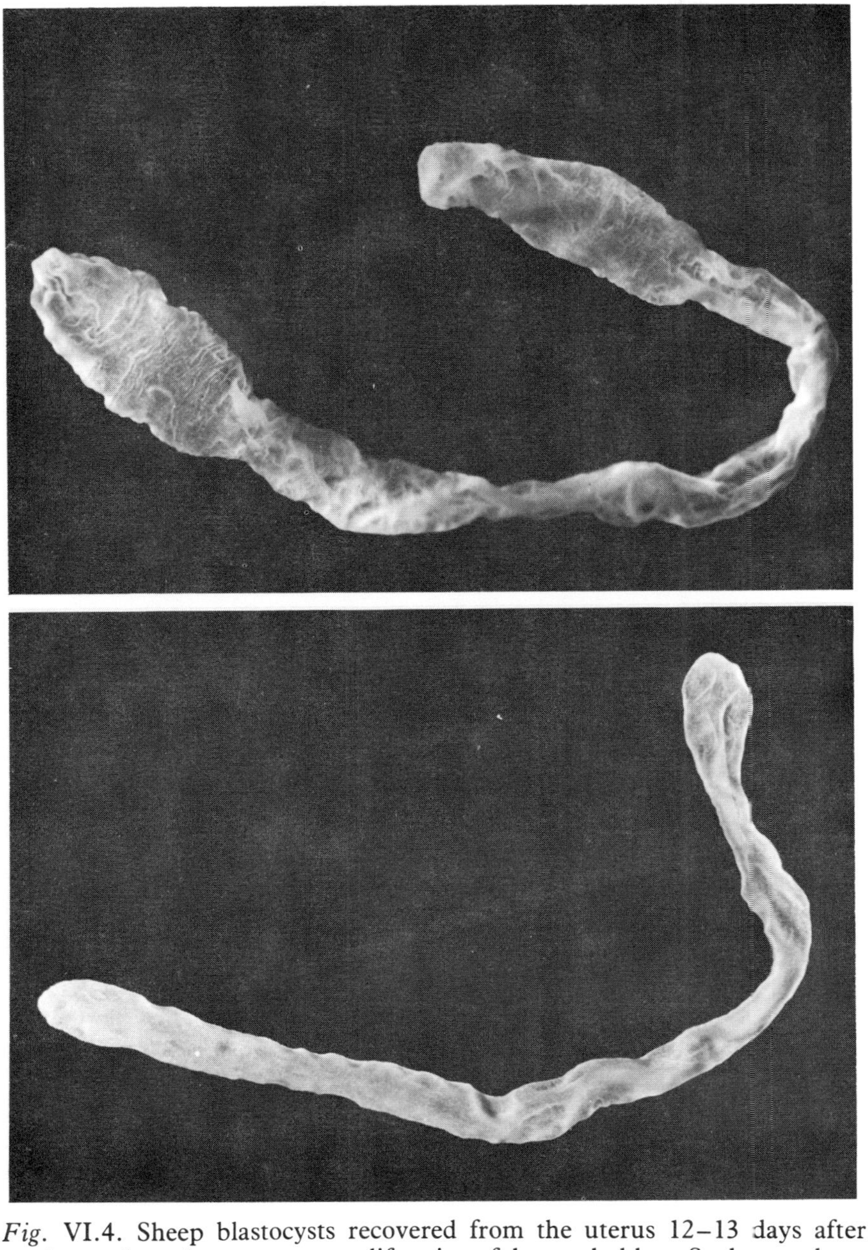

Fig. VI.4. Sheep blastocysts recovered from the uterus 12–13 days after mating to show the enormous proliferation of the trophoblast. Such an embryo is more frequently termed a conceptus. (Courtesy of Dr R. M. Mcor.)

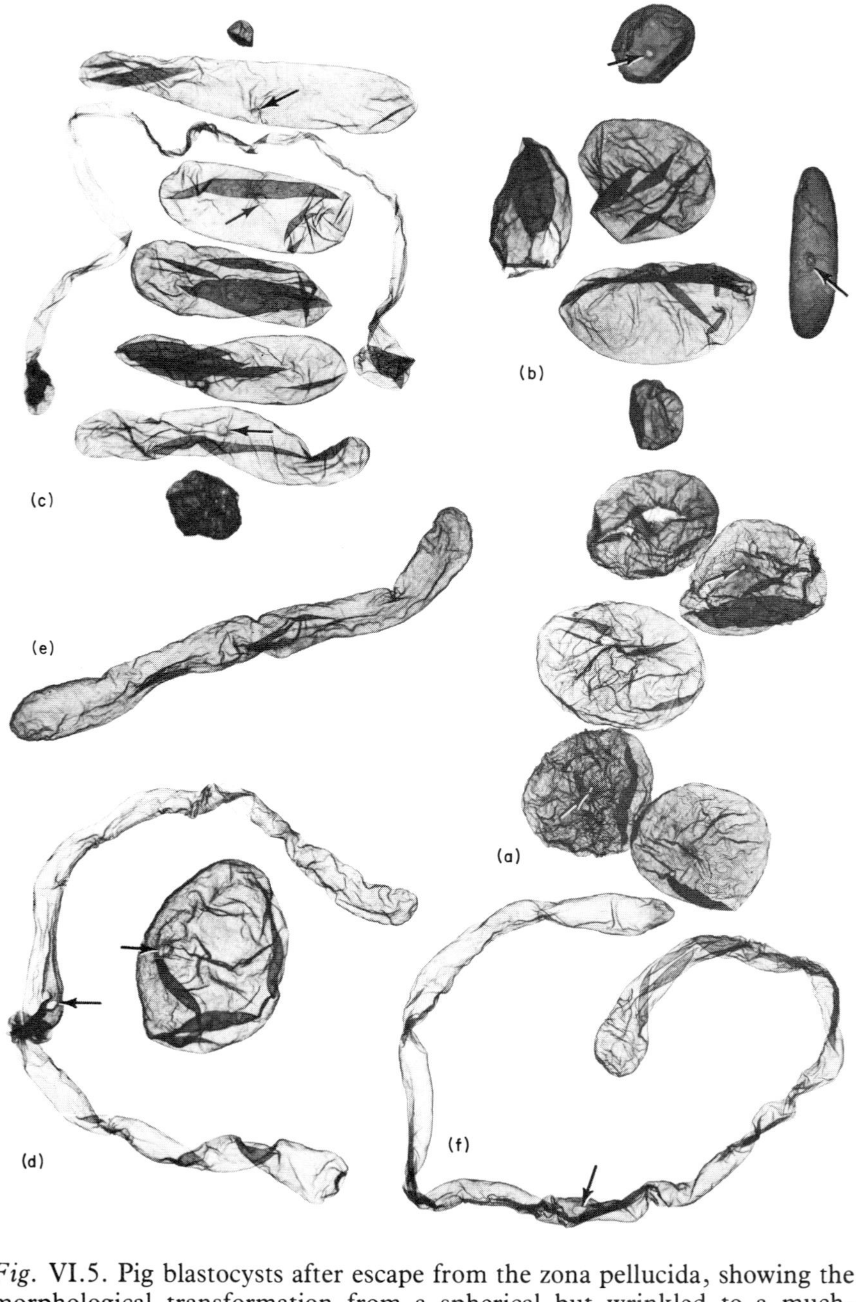

Fig. VI.5. Pig blastocysts after escape from the zona pellucida, showing the morphological transformation from a spherical but wrinkled to a much-elongated conceptus due to proliferation of the trophoblast. (Reproduced from Anderson (1978). Courtesy of the Editor of *Anatomical Record*.)

and Rowlands, 1962; Anderson, 1978). The longest individual pig conceptus dissected from the uterus by Perry and Rowlands—one of 14 days—apparently measured some 306 cm but, as is described below, only a small fraction of this length of uterine horn is occupied by an individual conceptus of this age.

Later stages of embryonic development in the horse have been described by Ewart (1897) and Van Niekerk and Allen (1975). The latter authors noted that the conceptus measured 7·5 cm in diameter by Day 30 of gestation, 9·0 cm in diameter by Day 36 and 11·0 cm by Day 38. However, specific details of the earlier phases of post-hatching enlargement still need to be accurately chronicled, although variation in the moment of ovulation during the 5–7 day period of oestrus may frustrate this objective.

Cow, sheep, pig and horse embryos thus undergo a prolonged free-living stage in the uterine lumen before commencement of the attachment process. The alteration in size and shape of the embryo may be associated in part with constituents of the uterine fluid (histotrophe), which are known to be closely regulated by the prevailing balance of ovarian steroids. There is accumulating evidence for the pig in particular that a progesterone-dependent secretion of uterine proteins may be essential to this growth (Daniel, 1972; Murray *et al.*, 1972; Knight *et al.*, 1973), but analogous aspects of uterine protein secretion have yet to be analysed systematically in sheep and cattle.

It remains unknown to what extent a retarded growth of the trophoblast at this post-hatching phase influences the future development and viability of the embryo.

Redistribution and spacing of embryos

During the free-living stage, a redistribution of embryos between the horns of the bicornuate pig uterus or bipartite ruminant uterus may occur, this phenomenon being traditionally referred to as intra-uterine migration of the embryo (Corner, 1921; Warwick, 1926; Dziuk *et al.*, 1964). Such migration has been noted at only a low incidence in cattle and sheep with a single corpus luteum, although the placental membranes extend around the septum and body of the uterus as gestation progresses. But in 31 of 32 ewes which had two corpora lutea in the same ovary and in which twins were being carried, one embryo had always migrated to the contralateral horn (Reimers *et al.*, 1973). Similar results have been reported for cows (Scanlon, 1972). However, intra-uterine migration and mixing of embryos is an invariable feature of reproduction in the domestic pig. Indeed, if it were not found in this polyovular species in which the disparity in the number of eggs released from the two ovaries may exceed 14 to one, competition between embryos would undoubtedly elevate the already significant level of prenatal mortality. In a recent review of reproduction in the pig, it has been suggested that intra-uterine migration should more correctly be viewed as a redistribution (rather than migration) since the embryos are probably passive in the overall physical sense, although their biosynthetic activity in the form of oestrogen and/or prostaglandin secretion may contribute to a local enhancement of myometrial contractions (Hunter, 1977b; Cook and Hunter, 1978).

As can be seen from Fig. VI.6, representing the results of an experiment in which different numbers of genetically marked embryos from homozygous donors were transferred to the uterine horns of unmated recipient pigs, and the foetal skin pigment noted at autopsy after Day 75 of gestation, the redistribution of embryos is seemingly at random and

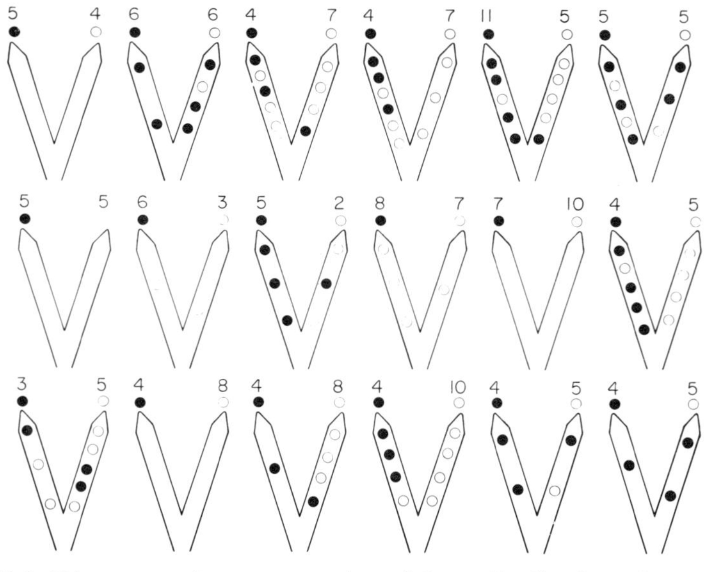

Fig. VI.6. Diagrammatic representation of the redistribution of transplanted embryos within the uterine horns of unmated recipient pigs. Genetically marked "black" or "white" embryos were transferred to the tip of each uterine horn in the numbers shown, and the position of those developing was noted when the foetal skin pigment became detectable at about 75 days. (Redrawn from Dziuk *et al.*, 1964.)

leads to a substantial mixing both within and between the two horns (Dziuk *et al.*, 1964). Much earlier evidence for the phenomenon of intra-uterine redistribution of embryos derives from the unilateral ovariectomy experiments of Hunter (1787), in which it was found that compensation in the number of ovulations occurred in the remaining ovary after removal of one gonad. In this experimental situation, normal pregnancies could be established with embryos developing in both uterine horns. The actual stage of redistribution of embryos between the two horns of the pig uterus occurs principally during the ninth to eleventh days of embryonic life (Dhindsa *et al.*, 1967; Polge and Dziuk, 1970), such that an approximately even spacing of embryos without overlap is achieved by the time attachment is initiated on Day 13 of embryonic life. Distribution of embryos is therefore occurring largely during the phase of rapid trophoblastic enlargement, and pig blastocysts still encompassed by the zona pellucida have rarely been noted more than a third of the distance along the horn from the utero-tubal junction (see Dhindsa and Dziuk, 1968).

The manner whereby an efficient spacing of embryos is achieved before the commencement of attachment is still not understood. However, a conclusion that might be drawn from observations in the pig is that some form of repulsion is exerted if embryos are too closely positioned just before attachment commences, although how the embryos would interact with the uterus to achieve this end remains purely speculative. Nonetheless, a local physical or humoral influence by the embryo on the contractile activity of the myometrium may be involved in preventing overlap, followed by a regional influence of the embryos on growth of the uterine tissues. As the implantation process proceeds, the extremely long and tenuous pig embryo is accommodated in the folds of the endometrium in the zig-zag arrangement illustrated in Fig. VI.7. The degree of folding assumed by the embryos at this stage could certainly contribute to the spacing mechanism by regulating the extent of uterine cornua occupied by an individual, subsequent growth of the uterine tissues being in step with the requirements of the embryos.

Commencement of implantation

Association of the embryo with the uterine epithelium, either by superficial attachment or specific embedding in or beneath the

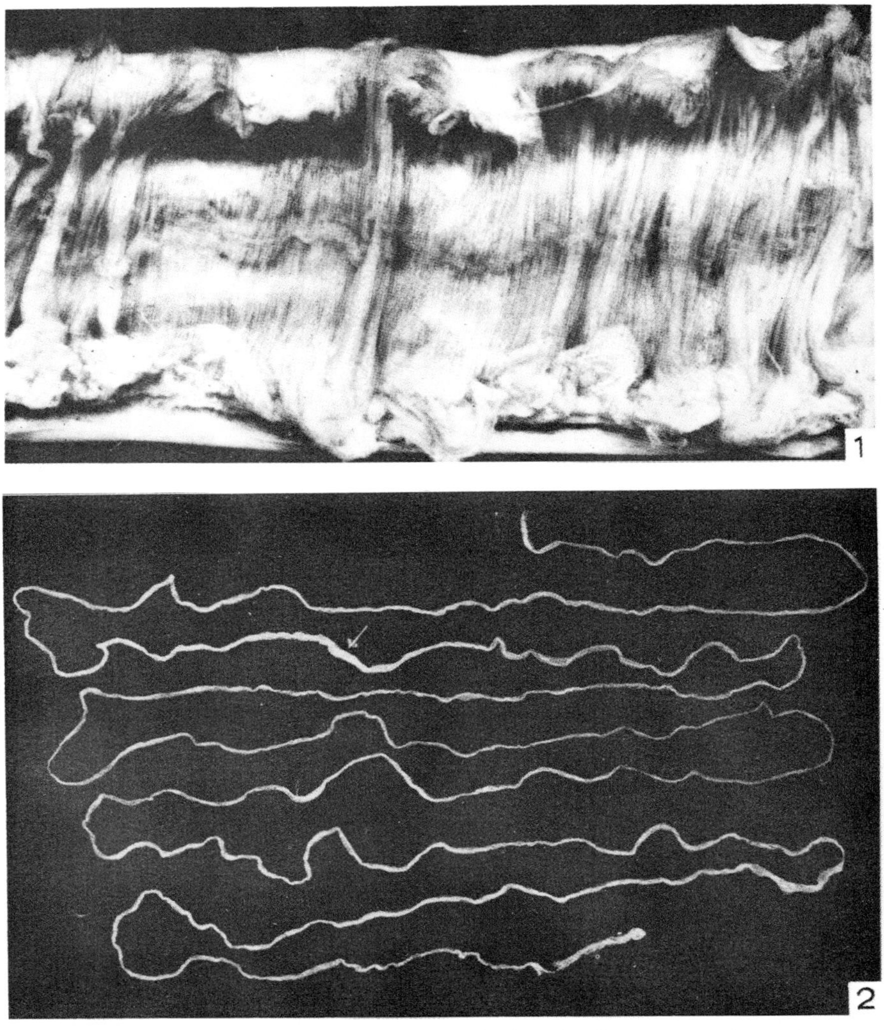

Fig. VI.7. The form of the pig conceptus during the first stages of attachment showing the manner in which the enormously elongated trophoblastic sac is accommodated in folds of the uterine mucosa. (Reproduced from Perry and Rowlands, 1962. Courtesy of the Editor of *Journal of Reproduction and Fertility*.)

endometrium, leads in due course to the formation of a placenta and complete dependence of the differentiating embryo upon metabolic support from the mother. Implantation and placentation exhibit a variety of forms, but in all instances the hormonal status of the mother is of great importance in determining whether or not implantation can proceed (see Psychoyos, 1973). In several species of rodent such as rats and mice, implantation is held in abeyance during lactation, whilst in other mammals such as roe deer, badger, seals and certain bears, there is an obligatory period of delayed implantation termed embryonic diapause; the inappropriate endocrine conditions associated with delayed implantation render the blastocyst metabolically dormant (see Enders, 1963). Ovariectomy shortly before the expected time of implantation can prevent implantation from occurring, whereas ovariectomy plus replacement therapy with a suitable ratio of progesterone and oestrogen permits implantation.

Despite this dominance by maternal hormonal factors, several studies have indicated that the embryo itself may be modifying the local environment between its trophoblastic surface and the endometrium by means of steroid hormone secretion (Dickmann, *et al.*, 1976) and perhaps also by CO_2 production (Menke and McLaren, 1970). The cells of the trophoblast possess enzymes capable of synthesizing "ovarian" hormones, and embryonic membranes supplied in culture with suitable biochemical substrates will secrete oestrogens and progesterone (Perry *et al.*, 1973). Such a pattern of synthetic activity in the late preimplantation embryo, which has also been indicated by histochemical studies, would explain in part the regional changes in the endometrium corresponding specifically to the future implantation sites, and also the local increase in capillary permeability of the rodent uterus as demonstrated by the Psychoyos pontamine-blue reaction (Psychoyos, 1960); the implantation sites appear as bands showing a concentration of the macro-molecular dye. But a recent study in the hamster suggests that prostaglandins may well be mediators in the local vascular response that precedes implantation (Evans and Kennedy, 1978).

The physical relationship between the implanting embryo and the uterus is best summarized according to whether the embryo remains in the lumen of the uterus, referred to as central implantation as in cattle, sheep, pigs and horses, or colonises the glandular crypts and maternal tissues leading to eccentric or interstitial implantation, as in rodents and man (see Wimsatt, 1975). Implantation in cattle, sheep, pigs and horses is essentially superficial, the placenta that is subsequently formed being of the epithelio-chorial (pig and horse) or syndesmo-chorial (cow and sheep) type. Bearing in mind the earlier remarks concerning expansion of the trophoblast and the long free-living stage, it is scarcely surprising that the embryo does not invade deeply into the maternal tissues or initiate a uterine reaction of the type characteristic of rodents. Use of the term nidation as a synonym for implantation is therefore inappropriate for the domestic ungulates, whereas the word attachment would seem more accurate and descriptive. The process of invasion by the embryonic trophoblast in the mouse, on the other hand, promotes characteristic changes in the uterine stroma termed the decidual reaction (Finn, 1971). This cellular proliferation occurs on the fourth day of pregnancy in mice in response to the presence of a blastocyst, but the physiological nature of the stimulus leading to these changes is not understood. One other feature of the implantation process that influences the form of the placenta is the extent to which cell layers in the uterine wall remain interposed between the embryonic tissues and maternal blood. In species where the embryo is invasive and penetrates the maternal tissues to reach the wall or lumen of the blood vessels, an endothelio-chorial (carnivores) or haemo-chorial (rodents, primates) type of placenta is formed. In other words, cells of the maternal epithelium no longer interpose between the maternal vascular supply and the developing embryo, a situation which gives direct access to the haemotrophe (i.e. the blood) and therefore facilitates transfer of nutrients to the conceptus.

In pigs and horses, there are no specialised and predetermined areas of the uterine epithelium with which the embryo will react at attachment. By contrast, in cattle, sheep and other ruminants, there are uterine caruncles which interact with the embryonic membranes in due course to give the placental cotyledons (Fig. VI.8). The commencement of a tenous attachment to the endometrium is seen in 13-day pig conceptuses in the form of short microvilli extending from the surface of the trophoblast, microvilli also becoming a feature of the corresponding portions of the endometrial surface at this time (Crombie, 1970). Even though contact between maternal and embryonic surfaces remains diffuse, attachment of the pig embryo is well advanced by the 18th day and by the 21st or 22nd day great care must be taken if the conceptuses are to be "stripped" from the endometrium without rupturing the chorionic sacs.

Fig. VI.8. Freshly-dissected gravid uteri to illustrate the mode of attachment of the foetal membranes to the uterine wall in cattle (a and b) and sheep (c and d). The foeto-placental specialisations (cotyledons) interact most intimately with the maternal layers within the distinctive caruncles. Note that the latter structures have an overall convex form in the cow (e) in contrast to the concave arrangement in sheep (f).

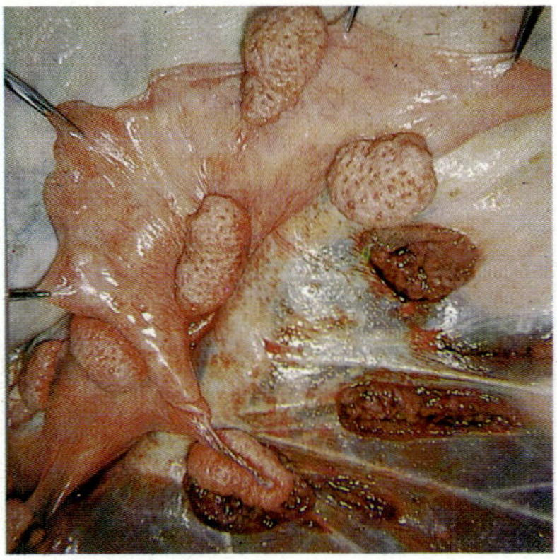

(a)

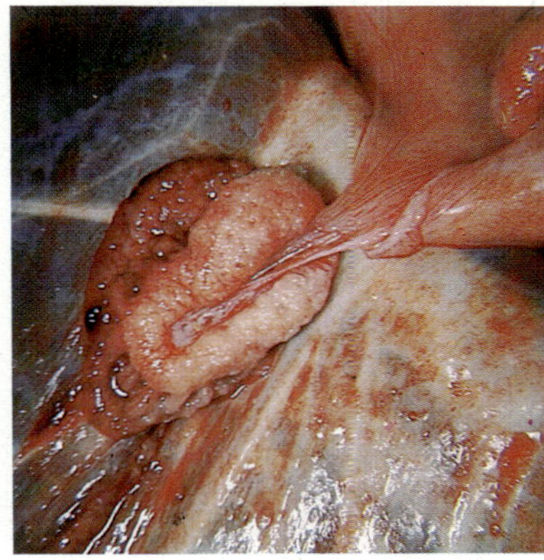

(b)

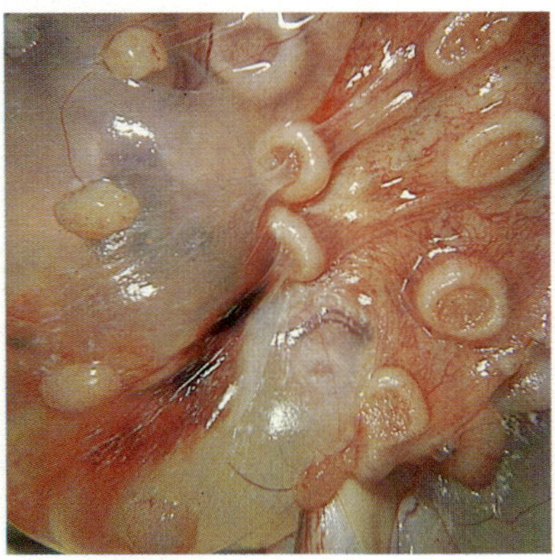

(c)

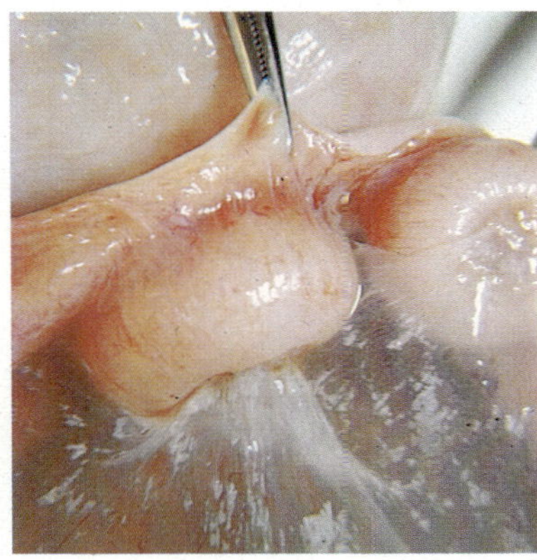

(d)

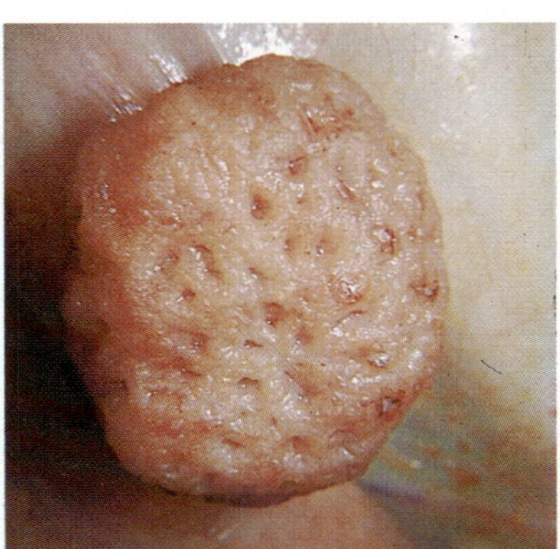

(e)

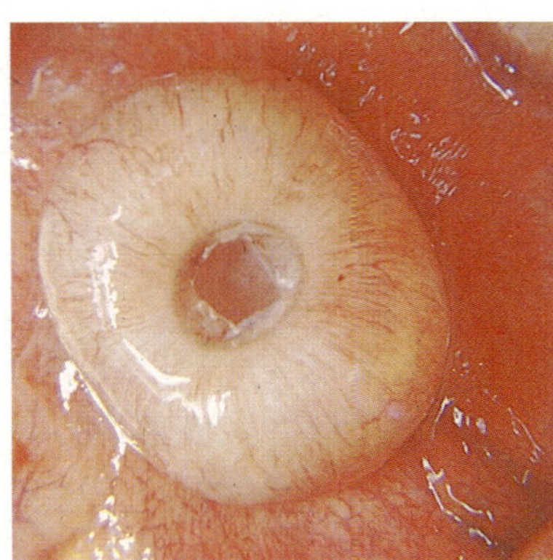

(f)

In sheep, the more specialised caruncular regions of the epithelium show the earliest morphological signs of attachment on about the 15th day (Boshier, 1969), with a concomitant increase in capillary permeability. At this time, the conceptus may measure up to 19 cm in length. By the 16th day, endometrial microvilli interact with the trophoblastic surface (Fléchon *et al.*, 1978), and the embryonic membranes may be commencing to elongate around the septum into the opposite horn of the uterus (Green and Winters, 1945; Rowson and Moor, 1966b). In both pigs and sheep, the timing of the initial embryonic attachment is such that the lifespan of the corpus luteum could be influenced by this event. Classical morphological evidence in the cow suggested that attachment was not commencing until after Day 30 (Winters *et al.*, 1942; Chang, 1952), when chorionic (trophoblastic) villi make specific contact with maternal tissues in the crypts of the caruncles, in due course to form the functional unit of the placentome (Fig. VI.8). However, a more recent ultrastructural study shows that the first phase of adhesion of the bovine embryo has commenced by the 22nd day after insemination (Fig. VI.9), and that this phase continues to the 27th day (Leiser, 1975). Trophoblastic activity may influence areas of the uterine epithelium in ruminants other than the caruncles by both invasion and destruction.

Compared with the above species, the attachment relationship of the horse embryo is even further delayed, the conceptus remaining essentially free in the lumen for more than one month, although it is of course in close contact with the uterine epithelium since it has swollen by fluid accumulation. By Day 37, however, an intimate attachment develops between the cells of the chorionic girdle and the underlying endometrial epithelium (Moor *et al.*, 1975), and by Day 45 rudimentary villi occur over the entire surface of the allanto-chorion and interdigitate with the maternal crypts (Van Niekerk and Allen, 1975).

The process of delayed implantation of the blastocyst referred to above has not been documented as a feature of reproduction in domestic cattle, sheep or pigs.

Establishment and maintenance of pregnancy–endocrine aspects

A response of major biological significance to the presence of a developing embryo in the uterus is the prolongation of the lifespan of the corpus

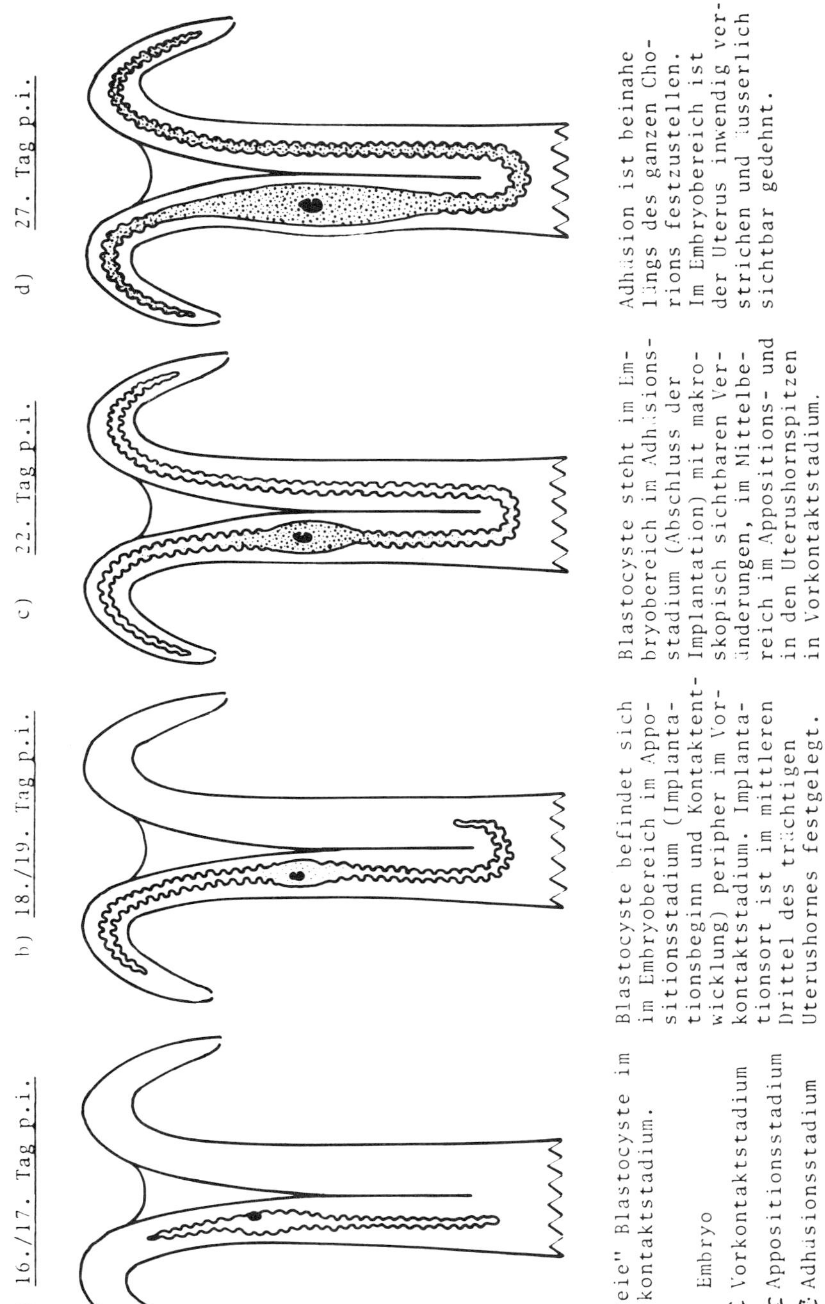

Fig. VI.9. Diagrammatic representation of growth and the first stages of attachment of the bovine conceptus. Note that the embryonic membranes gradually extend around the septum of the uterus, and that an attachment reaction can be detected with the electron microscope as early as 22 days after insemination. (Reproduced from Leiser, 1975. Courtesy of Paul Parey, Berlin and Hamburg.)

luteum (or corpora lutea) with a continuing secretion of progesterone. This is frequently and somewhat misleadingly referred to as conversion of the corpus luteum of the oestrous cycle into one of pregnancy, but no dramatic form of conversion in fact occurs; rather, the physiological events are more concerned with the prevention of regression of the corpus luteum, although the synthesis of progesterone by the luteal structures may increase after this critical stage is passed, due to a luteotrophic influence of the embryo and its membranes. Ovarian secretion of progesterone is required for the support of early pregnancy in most of the eutherian mammals so far studied, but the placenta may take over in some species as a major site of progesterone biosynthesis in the later stages of gestation (see below).

A question that for long perplexed workers in this field is the manner in which an animal discerns that it has successfully mated and contains one or more viable embryos in the uterus. In other words, how does the female decide whether to prolong the lifespan of the corpus luteum to establish and give endocrine support to the pregnancy or, alternatively, to effect rapid regression of the corpus luteum, maturation of another population of Graafian follicles, the onset of behavioural oestrus, and once more the opportunity for initiating pregnancy. Before discussing this problem, it is worth noting that in terms of the length of the oestrous cycle, the animal devotes a major portion to the luteal phase and thereby delays making a decision as to the fate of the corpus luteum for as long as possible, giving the embryo the maximum chance to express its presence. However, in relation to the length of gestation, the decision is taken very early (Table VI.2).

Table VI.2. The duration of the luteal phase of the oestrous cycle in relation to overall cycle length and to the duration of pregnancy.

Species	Duration in days			
	Luteal phase	Oestrous cycle	Pregnancy Mean	Range
Cow	17–18	21	282[1]	275–290
Sheep	14–15	16–17	148[1]	144–153
Pig	15–16	21	114	111–119
Horse	14–15	21–22	336[2]	320–341

[1] Breed differences in gestation length are well established in cattle and sheep.
[2] Seasonal influences on gestation length are found in mares.

Luteotrophins and luteolysins

Reasoning from first principles and with reference to species that show characteristic luteal and follicular phases during the oestrous cycle, luteal lifespan might be supposed to be prolonged because the embryo in some manner exerts a trophic influence on the corpus luteum, i.e. the embryo is directly or indirectly luteotrophic. Alternatively, the luteal lifespan might be curtailed because a non-gravid uterus elaborates a lytic factor to bring about the demise of the corpus luteum—a uterine luteolysin. It does not follow, however, that such influences need necessarily be exerted directly on the ovary, and it is possible that a uterus containing a developing embryo might also transmit appropriate information by nervous pathways, although findings that have accumulated in the last 15 to 20 years strongly suggest that humoral factors are the principal messengers in the present context. But establishment of pregnancy in farm and laboratory species, if not for primates, is not a simple switching on or off of luteal maintenance, but instead represents a complex relationship that depends on a delicate and dynamic balance between luteotrophic and luteolytic factors. Furthermore, establishment of pregnancy involves not only a relationship between the ovaries and uterus, but also one between these organs and the pituitary gland with at least partial mediation by the hypothalamus. The following paragraphs are intended to give a simplified understanding of some of the mechanisms involved, and do not attempt a comprehensive coverage of this extremely complex field of research. Readers requiring greater depth are referred to the reviews of Moor (1968), Anderson *et al.* (1969), Heap (1972), Hansel *et al.* (1973), Heap *et al.* (1973), Perry *et al.* (1976) and Cook and Hunter (1978).

Hysterectomy

The most straightforward and compelling evidence for the presence of luteolytic and luteotrophic factors comes from studies involving, respectively, hysterectomy (i.e. removal of the uterus) and embryo transplantation. With the exception of results in certain primates, hysterectomy performed during the luteal phase of the oestrous cycle invariably leads to a prolongation of the lifespan of the corpus luteum, and in farm animals subjected to this operation, the corpora lutea may be maintained for a period at least equivalent to that of gestation (reviews by Anderson, 1973; Wintenberger-Torrès, 1977). The conclusion to be drawn from such observations is that the non-gravid uterus at the end of the luteal phase acts to terminate the lifespan of the corpus luteum, and that removal of the uterine tissues prevents the termination mechanism from being called into action.

Embryo-transfer

On the other hand, transplantation of blastocysts from a donor animal into a synchronised but unmated recipient, even at a late stage in the luteal phase of the recipient, can lead to maintenance of the corpora lutea. For example, normal pregnancies have been established after the transfer of 13-day sheep embryos into the uterus of unmated $12\frac{1}{2}$-day recipients (Table VI.3: Moor and Rowson, 1964), and after the transfer

Table VI.3. The influence of transferring sheep embryos to the uterus of unmated recipients in the late luteal phase of the oestrous cycle. (Adapted from Moor and Rowson, 1964.)

Post-oestrum stage (days)		Number of ewes	
Donor	Recipient	per group	pregnant
12	12	9	5
13	12	9	6
12	13	3	0
13	13	14	2
13	14	4	0
14	14	9	0

of 16-day cow embryos into unmated synchronised recipients (Betteridge, 1977). Conversely, removal of sheep embryos from the uterus before the twelfth day leads to normal regression of the corpus luteum (Moor and Rowson, 1966). Two conclusions can be drawn from these observations. First, that an irrevocable decision on the lifespan of the corpus luteum is delayed, as mentioned above, until as late a stage as possible. Second, that under appropriate conditions of transplantation, the embryo(s) prevents luteal regression by overriding any luteolytic influence of the uterus, and the embryonic influence in effect is therefore anti-luteolytic and possibly also luteotrophic.

Pituitary factors

Whilst discussion so far has focussed on "uterine" factors that may influence the lifespan of the corpus luteum, there is much evidence, especially from hypophysectomy studies (i.e. removal of the pituitary gland), to suggest that a full physiological function of the corpus luteum depends on trophic hormone secretion by the anterior pituitary gland. Accordingly, it should be realized that the luteotrophic effect of an embryo might be mediated in part by anterior pituitary hormones, whilst the luteolytic effect of a non-gravid uterus could, in theory, involve a withdrawal of pituitary trophic support. Surgical or pharmacological hypophysectomy performed in hysterectomised animals leads to rapid regression of the corpus luteum, indicating a continuing involvement of pituitary support in maintaining the corpus luteum of hysterectomy. Hypophysectomy performed in mated animals late in the luteal phase or shortly thereafter indicates quite clearly that luteotrophic factors secreted by the pituitary are required for the maintenance of pregnancy in domestic animals (Table VI.4). In the absence of this support, a rapid decline in the concentration of systemic progesterone reveals that regression of the corpus luteum underlies the termination of pregnancy. Hormonal replacement therapy in surgically hypophysectomised animals has clarified the nature of the pituitary luteotrophic support that is sufficient to maintain a pregnancy (Table VI.4). In extensive work in hypophysectomised sheep, it was initially thought that such support might reside in the properties of a single anterior pituitary hormone (Kaltenbach *et al.*, 1968; Denamur *et al.*, 1973), but it is now agreed that a luteotrophic complex is secreted by this gland in sheep and certain other species (Table VI.4).

Uterine luteolysin

Although some of the key enquiries into the mechanisms of luteal maintenance involving the approaches of hysterectomy, embryo transplantation and hypophysectomy were made during the early 1960s, it was not until the luteolytic rôle of prostaglandin $F_{2\alpha}$ was first proposed (Pharriss and Wyngarden, 1969) that the principal uterine event underlying regression of the corpus luteum commenced to be understood. Successive experiments, especially in sheep, showed that $PGF_{2\alpha}$ was elaborated in the uterine tissues, released in elevated concentrations

Table VI.4. Two examples of the rôle of the pituitary gland in the maintenance of pregnancy, as demonstrated by hypophysectomy and replacement therapy with hypophyseal hormones.

Species	Influence of hypophysectomy		Nature of pituitary luteotrophic factors in early pregnancy
	during first half of pregnancy	during second half of pregnancy	
Sheep	PF	PM	LH and prolactin
Pig	PF	PF	LH[1]

PF indicates failure of pregnancy; PM indicates maintenance of pregnancy.
[1] Prolactin also becomes involved in luteotrophic support in later stages of pregnancy.

into the uterine venous drainage close to the end of the luteal phase (probably under the influence of ovarian follicular oestrogens), and acted in some local manner on the corpus luteum in the adjacent (ipsilateral) ovary to cause its regression. A local action of the postulated luteolysin had been recognised for some time from the results both of hemi-hysterectomy experiments and from those of unilateral embryo transfer; and after the identification of the involvement of $PGF_{2\alpha}$, pharmacologists were able to explain this requirement on the basis of a very high proportion (> 95%) of physiological concentrations of $PGF_{2\alpha}$ in the blood being deactivated by oxidation during one passage through the pulmonary circulation (Piper *et al.*, 1970). In other words, for elevated uterine venous levels of $PGF_{2\alpha}$ to cause regression of the corpus luteum at the end of the luteal phase, they would have to reach the ovary via some local route and not through the systemic circulation.

The postulated route of transfer invoked passage of $PGF_{2\alpha}$ from the uterine vein(s) into the closely apposed ovarian artery by a counter-current mechanism. Whether transfer involves simple diffusion or a more active form of transport remains unknown, although from experiments using radio-actively labelled prostaglandin it has been noted that the efficiency of the transfer mechanism is extremely low—maybe only 1–2% (McCracken, 1971; McCracken *et al.*, 1971). Nonetheless, the intimate vascular morphology between uterine and ovarian vessels (Fig. VI.10) would seem able to facilitate such a transfer; and surgical dissociation of these vessels in both sheep and guinea pigs leads to extension of the lifespan of the corpus luteum.

Clarification of the involvement of $PGF_{2\alpha}$ in the luteolytic process was greatly aided by the technique of transplanting the ovaries and/or

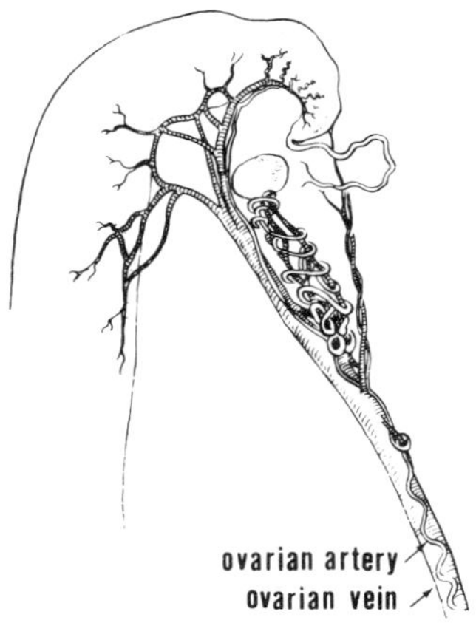

Fig. VI.10. Diagram to show the intimate relationship between the utero-ovarian vein and ovarian artery in sheep. (Reproduced from Del Campo and Ginther, 1973. Courtesy of the Editor of *American Journal of Veterinary Research.*)

the uterus of the sheep from the abdominal site to a more accessible one created in a skin fold of the neck. This sophisticated approach was first developed by Goding *et al.* (1967) and involved anastomosing the ovarian vasculature with the carotid and jugular vessels (Fig. VI.11), thereby providing a straightforward means of sampling ovarian venous blood after local or systemic treatment with lytic or trophic hormones. Incidentally, it was also found during development of this technique that the surgical separation of ovaries and uterus that necessarily occurred during relocation of the ovaries in the neck acted in a manner analogous to hysterectomy experiments, causing the lifespan of the corpus luteum to be extended. Use of the transplant preparation has greatly aided quantitative studies into the relationship between concentrations of $PGF_{2\alpha}$ and ovarian steroid secretion (McCracken *et al.*, 1971), but this particular technique remains essentially restricted to the sheep. By the early 1970s, $PGF_{2\alpha}$ was somewhat dogmatically being termed *the uterine luteolysin* in sheep (e.g. Goding, 1974), but studies involving active immunisation against a $PGF_{2\alpha}$ protein conjugate have supported this claim since they resulted in cessation of oestrous cycles and gave persistence of the corpora lutea with appropriate levels of circulating progesterone (Scaramuzzi and Baird, 1976). Even so, several other experienced physiologists have suggested caution and sought further components of the luteolytic mechanism (e.g. Hansel, 1975;

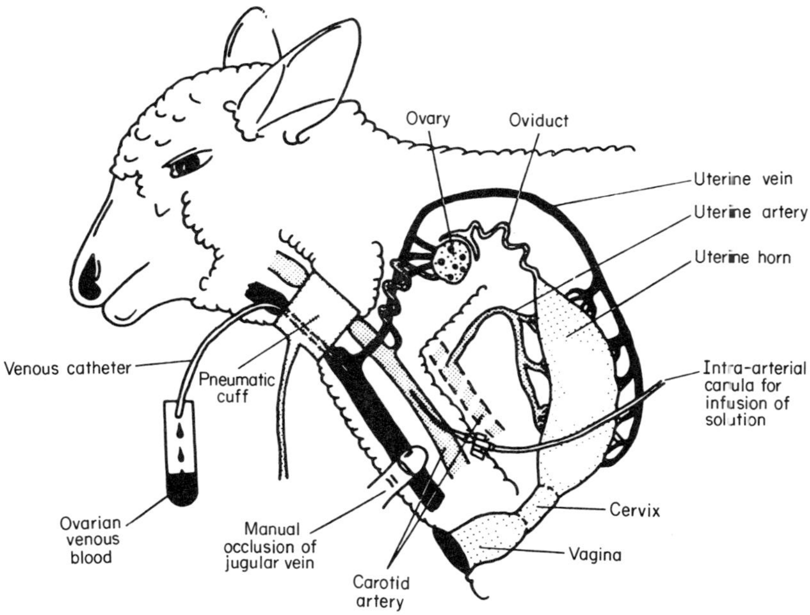

Fig. VI.11. A utero-ovarian transplant preparation used in endocrine studies of ovarian function in sheep. Note the manner in which the "reproductive blood vessels" are anastomosed with the carotid artery and jugular vein enabling procedures of infusion and blood sampling without recourse to further abdominal surgery. (Courtesy of Professor D. T. Baird.)

Hansel and Fortune, 1978). The apparent quantitative inefficiency of the counter-current transfer mechanism for $PGF_{2\alpha}$ has been one of the causes of worry, but evidence from species such as the pig suggests that "transferred" uterine $PGF_{2\alpha}$ may be only a portion of the lytic mechanism, and that the corpora lutea may expedite their own regression by themselves synthesizing $PGF_{2\alpha}$ (Patek and Watson, 1976). The precise manner in which regression of corpora lutea is brought about by $PGF_{2\alpha}$ is still a subject of debate, although alteration in the vascular supply to the luteal structures seems to be involved.

Embryonic factors

Turning from uterine luteolysins to the actual nature of embryonic luteotrophins, an early study in sheep showed that daily infusion of homogenates of 14- and 15-day old sheep embryos in saline directly into the uterine lumen from Day 12 of the cycle resulted in prolongation of luteal lifespan to at least Day 23 (Rowson and Moor, 1967). A similar experiment in pigs also prevented the expected regression of the corpora lutea (Longenecker and Day, 1972). More recent work in the pig has strongly suggested that oestrogen biosynthesis by elements of the trophoblast as soon as Day 12 contributes to the maintenance of the corpora lutea (Perry *et al.*, 1973; Robertson and King, 1974; Perry *et al.*, 1976), but again it remains to be spelled out whether trophoblastic oestrogen acts locally on the uterus (Bazer and Thatcher, 1977), ovaries or via a central (i.e. hypothalamic and pituitary) mechanism. Whilst the luteotrophic rôle of injected oestrogens in intact unmated pigs is a long-standing observation (Nishikawa and Waide, 1958; Gardner *et al.*, 1963), at least two studies suggest that oestrogens are here acting in conjunction with pituitary luteotrophins and cannot function to maintain the corpora lutea in hypophysectomised animals (Du Mesnil du Buisson, 1966; Anderson *et al.*, 1967). An anti-luteolytic action of trophoblastic steroids may also contribute to the mechanism of pregnancy maintenance in cattle and sheep (see Gadsby *et al.*, 1976; Cook and Hunter, 1978), but insufficient evidence is as yet available for this pathway. Chorionic gonadotrophin from the early embryo is well known to be involved in luteal maintenance in women, but the evidence for a chorionic (trophoblastic) gonadotrophin production in this context in farm species other than the horse is as yet slender (review by Cook and Hunter, 1978), although the production of an LH-like gonadotrophin by 10-day sheep blastocysts is under investigation (Wintenberger-Torrès, 1977, 1978). The embryo may also produce prolactin-like hormones (see Martal and Djiane, 1977), there being evidence of placental lactogens in cattle, sheep and goats, but it is uncertain if they are available early enough to influence the destiny of the cyclic corpus luteum. Nonetheless, the idea of involvement of hormones—steroids, gonadotrophins or lactogens—from the conceptus in the mechanism of maintenance of the corpora lutea remains attractive, and an anti-luteolytic protein, trophoblastin, in Day 14–16 sheep embryos has been reported (Martal *et al.*, 1979). In the equine situation, it is of course PMSG of endometrial cup origin that leads to the formation of luteinised follicles or accessory corpora lutea and thereby contributes to the maintenance of pregnancy (see Chapter III).

Luteotrophins versus luteolysins

The balance between luteotrophic and luteolytic factors that determines the establishment of pregnancy is well illustrated in the domestic pig, where the presence of less than four viable embryos in the uterus on Day 14 or 15 is incompatible with prolongation of the lifespan of the corpora lutea (Polge *et al.*, 1966). Although the studies of Bazer's group suggest that uterine $PGF_{2\alpha}$ is not secreted extensively into the venous drainage at this stage but rather by an exocrine route into the uterine lumen under the influence of oestrogens from the conceptus (Bazer and Thatcher, 1977; Frank *et al.*, 1977, 1978), there is evidence for some elevation in uterine venous $PGF_{2\alpha}$ on Days 14–16 in mated animals compared with on previous days. Hormonal establishment of pregnancy in this species would therefore seem to depend on a delicate balance between uterine $PGF_{2\alpha}$ (as a luteolysin) and embryonic oestrogens (as a source of luteotrophin or more correctly an anti-luteolysin). The requirement for a minimum number of embryos should be interpreted in this context as secretion of sufficient "embryonic luteotrophins" to overcome the luteolytic influence of non-gravid portions of the uterus (Hunter, 1977b). But the embryo may act in other species also to suppress active secretion of $PGF_{2\alpha}$ into the uterine veins at the end of the luteal phase, as apparently occurs in sheep (Barcikowski *et al.*, 1974).

Even after this critical phase is passed, and the balance between luteotrophic and luteolytic factors has been tilted in favour of pregnancy, maintenance of the corpora lutea during gestation may still require support from anterior pituitary hormones. Once again, these have been noted to act as a luteotrophic complex in some species (Table VI.4), and in cattle, sheep and pigs, LH and prolactin either singly or together serve as the pituitary luteotrophin. Pregnancy maintenance in the pig requires continuing pituitary endocrine support of the ovaries, hypophysectomy or ovariectomy leading to failure of the pregnancy (Du Mesnil du Buisson and Dauzier, 1957; Du Mesnil du Buisson, 1966). But in species where the placenta can synthesize a sufficient level of progesterone for maintenance of pregnancy, as occurs in the sheep after approximately Day 50 of gestation when the ovaries can be removed without terminating the pregnancy (Denamur and Martinet, 1955), hypophysectomy may not lead to termination. In other words, once the corpus luteum is no longer essential for the maintenance of pregnancy, luteotrophic factors from the pituitary may no longer be mandatory since the placenta seemingly regulates its own production of progesterone.

Components of the placenta

Although the placenta has been referred to already in this chapter, its specific membranous components that function to permit metabolic exchange between foetal and maternal vascular systems have not been mentioned systematically. The objective here is simply to portray their general form, essentially diagrammatically, and not to dwell on their embryological derivation or details of their rates of growth under experimental conditions. Likewise, growth of the foetus, regarded here as an embryo whose organ systems have already differentiated but have yet to undergo their principal phase of prenatal growth, is considered beyond the scope and aims of this book.

The major placental membranes are the chorion, the allantois, the yolk sac and the amnion, and their disposition around the foetus is portrayed in Fig. VI.12. Endometrial cups associated with the region of attachment of the yolk sac are a specific feature in the horse and other Equidae, being transient structures concerned with production of PMSG but they begin to be sloughed off after about the 20th week of

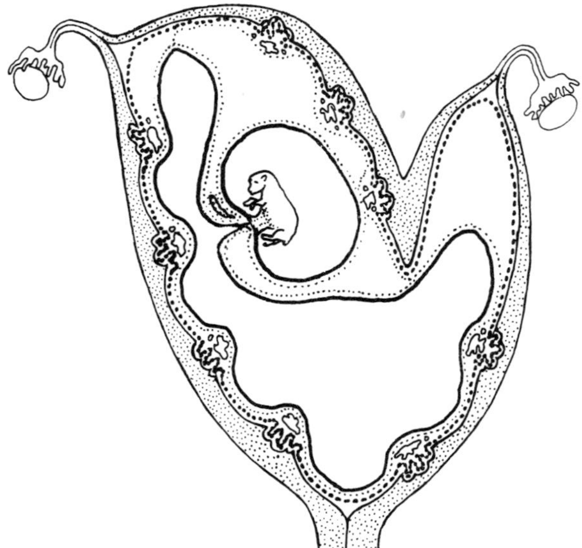

Fig. VI.12. Dissection of a gravid sheep uterus to illustrate the arrangement of placental membranes and the specialised sites of attachment to the uterine wall.

gestation, at which time the secretion of PMSG diminishes quite abruptly (Fig. III.1). Although conspicuous at an early stage of gestation and contributing in the horse in this specialised manner, the yolk sac is destined to become vestigial in the developing placenta of most higher mammals, whereas the amnion and allantois remain functional and fluid-filled units encompassed by the chorion. Fusion of the allantoic surface with the overlying chorionic membrane may occur, forming a well-vascularised allanto-chorion as in the farm species, and fusion may also occur between a portion of the amnion and the chorion. Detailed descriptions of the morphology of placentation in the domestic animals can be found in the reviews of Patten (1948), Amoroso (1952), Boyd and Hamilton (1952) and Marrable (1971).

On the maternal side of the placenta, the caruncular specialisations of the ruminant uterus that permit interdigitation of chorionic villi and compose the unit of the placentome have already been mentioned. These have a characteristic form in sheep and cattle, being concave in the former but convex in the latter (Fig. VI.8), and represent the regions of functional contact in contrast to the extensive epitheliochorial placentation of the pig. In the latter species especially, it is clear that the foetal membranes constitute the bulk of the placenta, although growth of this organ may be limited by the presence of neighbouring membranes which tend not to overlap but rather to give rise to necrotic tips (Marrable, 1968). Whilst many studies in laboratory and farm species have looked at the relationship between placental size and foetal growth at various stages of gestation, it is becoming increasingly apparent that placental efficiency—that is the dynamic aspects of placental transfer via its microcirculation—also exerts an important influence over development.

Because of the more complex eccentric or interstitial form of placentation found in rodents, carnivores and primates compared with the largely superficial form in the cow, sheep, pig and horse arising from their central mode of implantation, separation of the foetal from the maternal placental components at parturition is inevitably more traumatic in the former groups and may lead to substantial bleeding. However, as Nalbandov (1976) has pointed out, successful *post partum* breeding is paradoxically found in species such as rats, mice and guinea pigs in the absence of suckling, despite the placental tearing that must occur at birth, whereas in the pig with its simple epithelio-chorial placenta, conception is not found in the first week or two after parturition (see Chapter XI). Likewise in the cow, there is little evidence for ovulatory oestrus or fertile breeding in the first two to three weeks after delivery of the offspring and placental membranes.

Prenatal loss

Prenatal loss attains important levels in all the farm species: as many as 30–40% of the eggs ovulated at the time of mating or insemination fail to appear as viable foetuses at the end of gestation (Tables VI.5 and 6). Diverse studies have established that within this considerable prenatal loss, the bulk of it (i.e. approximately two-thirds) occurs during the early embroyonic stages that precede implantation; thus, in predominantly monotocous species such as the cow and sheep, it may not even have a detectable influence on the length of the oestrous cycle. Whilst many classical studies sought to ascribe this extensive prenatal loss to factors such as infection, nutritional constraints, or gross endocrine imbalances, it is now generally conceded that much of the mortality cannot be explained on such grounds. Indeed, it may in part be a predetermined loss and represent a form of prenatal biological selection (see Bishop, 1964), and to that extent the main value in the considerable literature on prenatal mortality must reside in its analysis of the distribution of the loss (e.g. Hammond, 1921; Brambell, 1948; Perry, 1954, 1960; Casida, 1956; Adams, 1960; Hanly, 1961). The same comment is essentially true of more recent observations on embryonic mortality in cattle in which estimations of the concentration of progesterone in milk have been used to diagnose the presence of a functional corpus luteum (Ball, 1978; Bulman and Lamming, 1978). Nonetheless, there are factors known to exacerbate the incidence of prenatal loss and which are susceptible to some degree of regulation. Such factors clearly require discussion, and their influence must be taken into account in animal breeding programmes.

Perhaps the most important of these factors is the damaging influence of ageing of the gametes in the female reproductive tract due to mating or, more probably, to artificial insemination performed at the wrong time in relation to ovulation. The advantages of precise control of the time of ovulation have already been mentioned (Chapter III) and the deleterious influence of postovulatory ageing of the eggs before sperm penetration has been discussed in some detail (Chapter V and Austin, 1970).

Concerning *in vivo* ageing of spermatozoa in the preovulatory inter-

Table VI.5. To illustrate the extent of embryonic loss in two groups of cows classified as of normal fertility or repeat breeders when slaughtered at different stages of gestation. (Modified from the data of Ayalon, 1978.)

	Normal cows			Repeat-breeders		
Time of slaughter after mating (days)	No. of animals	No. of embryos	Loss of embryos (%)	No. of animals	No. of embryos	Loss of embryos (%)
2–3[1]	12	10	17	17	12	29
11–13	18	16	11	18	9	50
14–16	20	16	20	20	10	50
17–19	21	12	43	21	9	57
35–42	13	9	31	24	8	65

[1] These 2–3 day values represent eggs fertilised in relation to eggs recovered.

Table VI.6. To illustrate the extent of embryonic loss in domestic pigs examined at different stages of gestation. (Adapted from Perry, 1954.)

Stage of gestation at autopsy (days)	Number of animals examined	Total number of			Overall percentage loss	Average number of surviving embryos
		corpora lutea	embryos	eggs[1] lost		
16–24	28	535	358	177	33·1	12·8
25–49	33	604	390	214	35·4	11·8
>50	37	646	386	260	40·3	10·4

[1] The number of corpora lutea not represented by embryos at autopsy.

val, the evidence for a deleterious influence on fertilisation is less clear. This may in part be due to the massive population of cells in the ejaculate amongst which a form of selection would be exerted during the events of sperm transport and penetration of the egg investments (see Chapter V). Certainly, there is some suggestion that in situations where *in vivo* ageing of sperm would have occurred, fertilisation in mammals (as distinct from in birds) is an all-or-none event, and if it takes place then there is little or no harmful influence on the developmental potential of the zygote (Soderwall and Young, 1940; Soderwall and Blandau, 1941; Tesh and Glover, 1966; Tesh, 1969). Nonetheless, evidence from inseminations made at the very beginning of oestrus and hence some 25–40 hours before ovulation in cattle, sheep and pigs strongly suggests that conception rates are diminished by such *in vivo* ageing (e.g. Trimberger, 1948; Hancock and Hovell, 1962).

Other factors relating to the male input to fertility are the differences attributable to the use of semen from different bulls or boars, as revealed in conception rates obtained in artificial insemination programmes where data are available from very large numbers of females. Such potential differences between males may be revealed by heterospermic insemination trials in cattle (Beatty *et al.*, 1969, 1976; Stewart *et al.*, 1974) or competitive mating experiments in pigs using two males sequentially (Dziuk, 1970; Martin and Dziuk, 1977). The excessive use of males (i.e. overworking) in mating as distinct from artificial insemination situations may also adversely influence fertility if the concentration of spermatozoa in the ejaculate becomes too low and when the epididymal reserves of mature spermatozoa are seriously depleted. Although there is some controversy as to whether the male would continue to mate freely in such situations of depletion, ejaculation of semen containing low sperm concentrations at the beginning of oestrus would undoubtedly depress conception rate.

Turning to *in vitro* factors, considerable care is required in the conditions of handling ejaculates for semen dilution, storage and use in artificial insemination under field conditions. This topic has been reviewed in depth by several experts (Foote, 1974; Gomes, 1977) and there is no reason to duplicate the evidence here. Sufficient to mention that conception rates accruing from large numbers of inseminations using poorly prepared or checked semen samples may be seriously depressed, and there is thus a strong argument for monitoring the pregnancy status of all cows inseminated with batches of semen found retrospectively to have given poor conception rates.

In the case of the female, factors that may exacerbate the extent of early embryonic loss over and above those directly concerned with the gametes are a variety of stresses in the preimplantation interval such as the influence of high ambient temperature (Elliott *et al.*, 1968; Bellve, 1972), metabolic stress as seen in gilts with marked fluctuations in dietary intake around the time of ovulation and mating (Bazer *et al.*, 1968; Frobish, 1970), and possibly even pheromonal stresses—if the large domestic species under conditions of intensive housing show responses to strange odours similar to those shown by laboratory rodents (see Chapter XII). There may also be endocrine disturbances arising, for example, in the case of sheep grazing pastures rich in oestrogenic plants or perhaps arising as a carry-over effect after synchronisation treatments with injected progestagens. Cattle in poor condition may be less able to establish pregnancies in winter, and there is circumstantial evidence that inadequate pituitary (luteotrophic) support of the corpus luteum may then be a limiting factor.

More closely concerned with the number of embryos in the uterus is the question of the balance between luteotrophic and luteolytic factors at the time of reprogramming the lifespan of the corpora lutea (see above), this being most clearly demonstrable in the pig with a small number of embryos in the uterus. The contrary situation of overcrowding in the uterus has been provoked by the experimental approaches of either induced superovulation or embryo transplantation, and there is evidence for increased mortality in such "overcrowded" uteri, especially in pigs. But the controversy as to whether excessive mortality precipitated in the above experimental situations occurs before or after, for example, Days 25–30 of gestation in pigs (see Dziuk, 1977) is largely academic and sheds little light on the causes of preimplantation mortality in untreated animals, where it is currently thought that the uterus develops to accommodate the viable products of conception.

As already pointed out, the bulk of the prenatal loss usually occurs in the preimplantation period. This early expression of inherent defects in the embryo or environmental deficiencies in the female tract could be interpreted as biologically advantageous in that it would avoid committing females, especially those of monotocous species, to extensive periods of non-fruitful gestation; this would become the situation if the bulk of the loss occurred after establishment of the corpus luteum of pregnancy. Once the latter critical stage is passed and the placenta is fully established, the principal problem for the embryo soon ceases to be one of differentiation of organ systems, but rather becomes one of placental efficiency and the growth of tissues. This is not to infer that further prenatal loss may not occur, but if and when it does, it is more likely to be an expression of constraints imposed by the uterine environment than the failure of foetal development due to intrinsic factors.

References

Adams, C. E. (1960). Studies on prenatal mortality in the rabbit, *Oryctolagus cuniculus*: the amount and distribution of loss before and after implantation. *J. Endocr.* **19**, 325–344.

Aitken, R. J. (1977). Embryonic diapause. *In* "Development in Mammals", Vol. 1 (M. H. Johnson, ed.), pp. 307–359. North Holland Publ. Co., Amsterdam.

Amoroso, E. C. (1952). Placentation. *In* "Marshall's Physiology of Reproduction", 3rd edition, Vol. 2 (A. S. Parkes, ed.), pp. 127–311. Longmans, Green and Co., London.

Anderson, L. L. (1973). Effects of hysterectomy and other factors on luteal function. *In* "Handbook of Physiology", Endocrinology II, Part 1 (R. O. Greep and E. B. Astwood, eds), pp. 69–86. American Physiological Society, Washington.

Anderson, L. L. (1978). Growth, protein content and distribution of early pig embryos *Anat. Rec.* **190**, 143–154.

Anderson, L. L., Dyck, G. W., Mori, H., Henricks, D. M. and Melampy, R. M. (1967). Ovarian function in pigs following hypophysial stalk transection or hypophysectomy. *Amer. J. Physiol* **212**, 1188–1194.

Anderson, L. L., Bland, K. P. and Melampy, R. M. (1969). Comparative aspects of uterine-luteal relationships. *Recent Progr. Horm. Res.* **25**, 57–99.

Asdell, S. A. (1964). "Patterns of Mammalian Reproduction". 2nd edition. Constable and Co. Ltd., London.

Austin, C. R. (1970). Ageing and reproduction: post-ovulatory deterioration of the egg. *J. Reprod. Fert., Suppl.* **12**, 39–53.

Ayalon, N. (1978). A review of embryonic mortality in cattle. *J. Reprod. Fert.* **54**, 483–493.

Ball, P. J. H. (1978). The relationship of age and stage of gestation to the incidence of embryo death in cattle. *Res. vet. Sci.* **25**, 120–122.

Barcikowski, B., Carlson, J. C., Wilson, L. and McCraken, J. A. (1974). The effect of endogenous and exogenous oestradiol-17β on the release of prostaglandin $F_{2\alpha}$ from the ovine uterus. *Endocrinology* **95**, 1340–1349.

Bazer, F. W. and Thatcher, W. W. (1977). Theory of maternal recognition of pregnancy in swine based on estrogen controlled endocrine versus exocrine secretion of prostaglandin $F_{2\alpha}$ by the uterine endometrium. *Prostaglandins* **14**, 397–401.

Bazer, F. W., Clawson, A. J., Robison, O. W., Vincent, C. K. and Ulberg, L. C. (1968). Explanation for embryo death in gilts fed a high energy intake. *J. Anim. Sci.* **27**, 1021–1026.

Beatty, R. A., Bennett, G. H., Hall, J. G., Hancock, J. L. and Stewart, D. L. (1969). An experiment with heterospermic insemination in cattle. *J. Reprod. Fert.* **19**, 491–502.

Beatty, R. A., Stewart, D. L., Spooner, R. L. and Hancock, J. L. (1976). Evaluation by the heterospermic insemination technique of the differential effect of freezing at −196°C on fertility of individual bull semen. *J. Reprod. Fert.* **47**, 377–379.

Bellvé, A. R. (1972). Viability and survival of mouse embryos following parental exposure to high temperature. *J. Reprod. Fert.* **30**, 71–81.

Betteridge, K. J. (1977). Embryo transfer in farm animals: a review. *Canad. Dept. Agric., Monograph* No. **16**, Ottawa.

Bishop, M. W. H. (1964). Paternal contribution to embryonic death. *J. Reprod. Fert.* **7**, 383–396.

Blandau, R. J. (1949). Observations on implantation of the guinea pig ovum. *Anat. Rec.* **103**, 19–47.

Blandau, R. J. and Rumery, R. E. (1957). The attachment cone of the guinea-pig blastocyst as observed under time-lapse cinematography. *Fert. Steril.* **8**, 570–585.

Boshier, D. P. (1969). A histological and histochemical examination of implantation and early placentome formation in sheep. *J. Reprod. Fert.* **19**, 51–61.

Boyd, J. D. and Hamilton, W. J. (1952). Cleavage, early development and implantation of the egg. *In* "Marshall's Physiology of Reproduction", 3rd edition, Vol. 2 (A. S. Parkes, ed.), pp. 1–126. Longmans, Green and Co., London.

Brambell, F. W. R. (1948). Prenatal mortality in mammals. *Biol. Rev.* **23**, 370–407.

Bulman, D. C. and Lamming, G. E. (1978). Milk progesterone levels in relation to conception, repeat breeding and factors influencing acyclicity in dairy cows. *J. Reprod. Fert.* **54**, 447–458.

Burdick, H. O. and Pincus, G. (1935). The effect of oestrin injections upon the developing ova of mice and rabbits. *Amer. J. Physiol.* **111**, 201–208.

Casida, L. E. (1956). Variations in the maternal environment affecting pre-natal survival and development. Proc. 3rd Int. Congr. Anim. Reprod. and A. I., Cambridge, pp. 19–25.

Chang, M. C. (1950). Development and fate of transferred rabbit ova or blastocyst in relation to the ovulation time of recipients. *J. exp. Zool.* **114**, 197–225.

Chang, M. C. (1952). Development of bovine blastocyst with a note on implantation. *Anat. Rec.* **113**, 143–162.

Chang, M. C. (1966). Effects of oral administration of medroxyprogesterone acetate and ethinyl estradiol on the transportation and development of rabbit eggs. *Endocrinology* **79**, 939–948.

Chang, M. C. and Rowson, L. E. A. (1965). Fertilisation and early development of Dorset Horn sheep in the spring and summer. *Anat. Rec.* **152**, 303–316.

Cook, B. and Hunter, R. H. F. (1978). Systemic and local hormone requirements for implantation in domestic animals. *J. Reprod. Fert.* **54**, 471–482.

Corner, G. W. (1921). Internal migration of the ovum. *Bull. Johns Hopkins Hosp.* **32**, 78–83.

Crombie, P. R. (1970). Ultrastructure of the foetal-maternal attachment in the pig. *J. Physiol., Lond.* **210**, 101P.

Daniel, J. C., Jr. (1972). Preliminary attempts to terminate pregnancy by immunological attack on uterine protein. *Experientia* **28**, 700–701.

Denamur, R. and Martinet, J. (1955). Effets de l'ovariectomie chez la Brebis pendant la gestation. *C.r. Séanc. Soc. Biol.* **149**, 2105–2107.

Denamur, R., Martinet, J. and Short, R. V. (1973). Pituitary control of the ovine corpus luteum. *J. Reprod. Fert.* **32**, 207–220.

Dhindsa, D. S. and Dziuk, P. J. (1968). Effect on pregnancy in the pig after killing embryos or fetuses in one uterine horn in early gestation. *J. Anim. Sci.* **27**, 122–126.

Dhindsa, D. S., Dziuk, P. J. and Norton, H. W. (1967). Time of transuterine migration and distribution of embryos in the pig. *Anat. Rec.* **159**, 325–330.

Dickmann, Z. (1969). Shedding of the zona pellucida. *Adv. Reprod. Physiol.* **4**, 187–206.

Dickmann, Z., Dey, S. K. and Sen Gupta, J. (1976). A new concept: control of early pregnancy by steroid hormones originating in the preimplantation embryo. *Vitams. Horm.* **34**, 215–242.

Du Mesnil du Buisson, F. (1966). Contribution à l'étude du maintien du corps jaune de la truie. Thesis, University of Paris.

Du Mesnil du Buisson, F. and Dauzier, L. (1957). Influence de l'ovariectomie chez la truie pendant la gestation. *C.r. Séanc. Soc. Biol.* **151**, 311–313.

Dziuk, P. J. (1970). Estimation of optimum time for insemination of gilts and ewes by double-mating at certain times relative to ovulation. *J. Reprod. Fert.* **22**, 277–282.

Dziuk, P. J. (1977). Reproduction in pigs. *In* "Reproduction in Domestic Animals", 3rd edition (H. H. Cole and P. T. Cupps, eds), pp. 455–474. Academic Press, New York.

Dziuk, P. J., Polge, C. and Rowson, L. E. (1964). Intra-uterine migration and mixing of embryos in swine following egg transfer. *J. Anim. Sci.* **23**, 37–42.

Elliott, D. S., Burfening, P. J. and Ulberg, L. C. (1968). Subsequent development during incubation of fertilized mouse ova stressed by high ambient temperatures. *J. exp. Zool.* **169**, 481–485.

Enders, A. C. (Ed.) (1963). "Delayed Implantation". University of Chicago Press, Chicago.

Evans, C. A. and Kennedy, T. G. (1978). The importance of prostaglandin synthesis for the initiation of blastocyst implantation in the hamster. *J. Reprod. Fert.* **54**, 255–261.

Ewart, J. C. (1897). "A Critical Period in the Development of the Horse". Adam and Charles Black, London.

Finn, C. A. (1971). The biology of decidual cells. *Adv. Reprod. Physiol.* **5**, 1–26.

Fléchon, J. E. and Renard, J.-P. (1978). A scanning electron microscope study of the hatching of bovine blastocysts *in vitro*. *J. Reprod. Fert.* **53**, 9–12.

Fléchon, J. E., Guillomot, M. and Wintenberger-Torrès, S. (1978). Étude ultrastructurale du blastocyste et de l'implantation chez la brebis. *In* "L'Implantation de l'Oeuf", Colloque S.N.E.F.S., pp. 107–119. Masson, Paris.

Foote, R. H. (1974). Artificial insemination. *In* "Reproduction in Farm Animals", 3rd edition (E. S. E. Hafez, ed.), pp. 409–431. Lea and Febiger, Philadelphia.

Frank, M., Bazer, F. W., Thatcher, W. W. and Wilcox, C. J. (1977). A study of prostaglandin $F_{2\alpha}$ as the luteolysin in swine. III. Effects of estradiol valerate on prostaglandin F, progestins, estrone and estradiol concentrations in the utero-ovarian vein of non-pregnant gilts. *Prostaglandins* **14**, 1183–1196.

Frank, M., Bazer, F. W., Thatcher, W. W. and Wilcox, C. J. (1978). A study of prostaglandin $F_{2\alpha}$ as the luteolysin in swine: IV. An explanation for the luteotrophic effect of estradiol. *Prostaglandins* **15**, 151–160.

Frobish, L. T. (1970). Effect of energy intake on reproductive performance and estrous synchronisation of gilts. *J. Anim. Sci.* **31**, 486–490.

Gadsby, J. E., Burton, R. D., Heap, R. B. and Perry, J. S. (1976). Steroid metabolism and synthesis in early embryonic tissue of pig, sheep and cow. *J. Endocr.* **71**, 45P–46P.

Gardner, M. L., First, N. L. and Casida, L. E. (1963). Effect of exogenous estrogens on corpus luteum maintenance in gilts. *J. Anim. Sci.* **22**, 132–134.

Goding, J. R. (1974). The demonstration that $PGF_{2\alpha}$ is the uterine luteolysin in the ewe. *J. Reprod. Fert.* **38**, 261–271.

Goding, J. R., McCracken, J. A. and Baird, D. T. (1967). A study of ovarian function in the ewe by means of a vascular autotransplantation technique. *J. Endocr.* **39**, 37–52.

Gomes, W. R. (1977). Artificial insemination. *In* "Reproduction in Domestic Animals", 3rd edition (H. H. Cole and P. T. Cupps, eds), pp. 257–284. Academic Press, New York.

Green, W. W. and Winters, L. M. (1945). Prenatal development of the sheep. *Univ. of Minn. Agric. Exp. Statn Tech. Bull.* No. **169**.

Green, W. W. and Winters, L. M. (1946). Cleavage and attachment stages of the pig. *J. Morph.* **78**, 305–315.

Greenwald, G. S. (1959). The comparative effectiveness of oestrogens in interrupting pregnancy in the rabbit. *Fert. Steril.* **10**, 155–161.

Hammond, J. (1921). Further observations on the factors controlling fertility and foetal atrophy. *J. Agric. Sci.* **11**, 337–366.

Hancock, J. L. and Hovell, G. J. R. (1962). Insemination before and after the onset of heat in sows. *Anim. Prod.* **4**, 91–96.

Hanly, S. (1961). Prenatal mortality in farm animals. *J. Reprod. Fert.* **2**, 182–194.

Hansel, W. (1975). Luteal regression in domestic animals. *Annls Biol. anim. Biochim. Biophys.* **15**, 147–160.

Hansel, W. and Fortune, J. E. (1978). The applications of ovulation control.

In "Control of Ovulation" (D. B. Crighton, G. R. Foxcroft, N. B. Haynes and G. E. Lamming, eds), pp. 237–263. Butterworths, London and Boston.

Hansel, W., Concannon, P. W. and Lukaszewska, J. H. (1973). Corpora lutea of the large domestic animals. *Biol. Reprod.* **8**, 222–245.

Heap, R. B. (1972). Role of hormones in pregnancy. *In* "Reproduction in Mammals", Vol. 3 (C. R. Austin and R. V. Short, eds), pp. 73–105. University Press, Cambridge.

Heap, R. B., Perry, J. S. and Challis, J. R. G. (1973). Hormonal maintenance of pregnancy. *In* "Handbook of Physiology", Endocrinology II, Part 1 (R. O. Greep and E. B. Astwood, eds), pp. 217–260. American Physiological Society, Washington.

Heuser, C. H. and Streeter, G. L. (1929). Early stages in the development of pig embryos, from the period of initial cleavage to the time of appearance of limb-buds. *Contrib. Embryol. Carneg. Instn* **20**, 1–30.

Hunter, J. (1787). An experiment to determine the effect of extirpating one ovarium upon the number of young produced. *Phil. Trans. R. Soc.* **77**, 233–239.

Hunter, R. H. F. (1974). Chronological and cytological details of fertilisation and early embryonic development in the domestic pig, *Sus scrofa*. *Anat. Rec.* **178**, 169–186.

Hunter, R. H. F. (1977a). Function and malfunction of the Fallopian tubes in relation to gametes, embryos and hormones. *Europ. J. Obstet. Gynec. Reprod. Biol.* **7**, 267–283.

Hunter, R. H. F. (1977b). Physiological factors influencing ovulation, fertilisation, early embryonic development and establishment of pregnancy in pigs. *Brit. vet. J.* **133**, 461–470.

Kaltenbach, C. C., Graber, J. W., Niswender, G. D. and Nalbandov, A. V. (1968). Effect of hypophysectomy on the formation and maintenance of corpora lutea in the ewe. *Endocrinology* **82**, 753–759.

Knight, J. W., Bazer, F. W. and Wallace, H. D. (1973). Hormonal regulation of porcine uterine protein secretion. *J. Anim. Sci.* **36**, 546–553.

Leiser, R. (1975). Kontaktaufnahme zwischen Trophoblast und Uterusepithel während der frühen Implantation beim Rind. *Anat. Histol. Embryol.* **4**, 63–86.

Longenecker, D. E. and Day, B. N. (1972). Maintenance of corpora lutea and pregnancy in unilaterally pregnant gilts by intrauterine infusion of embryonic tissue. *J. Reprod. Fert.* **31**, 171–177.

McCracken, J. A. (1971). Prostaglandin $F_{2\alpha}$ and corpus luteum regression. *Ann. N.Y. Acad. Sci.* **180**, 456–472.

McCracken, J. A., Baird, D. T. and Goding, J. R. (1971). Factors affecting the secretion of steroids from the transplanted ovary in the sheep. *Recent Progr. Horm. Res.* **27**, 537–582.

McLaren, A. (1970). The fate of the zona pellucida in mice. *J. Embryol. exp. Morph.* **23**, 1–19.

Marrable, A. W. (1968). The ischaemic extremities of the allanto-chorion of the pig and their relation to the endometrium. *Res. vet. Sci.* **9**, 578–582.

Marrable, A. W. (1971). "The Embryonic Pig: a Chronological Account". Pitman and Sons Ltd., London.

Martal, J. and Djiane J. (1977). The production of chorionic somatomammotrophin in sheep. *J. Reprod. Fert.* **49**, 285–289.

Martal, J., Lacroix, M.-C., Loudes, C., Saunier, M. and Wintenberger-Torrès, S. (1979) Trophoblastin, an antiluteolytic protein present in early pregnancy in sheep. *J. Reprod. Fert.* **56**, 63–73.

Martin, P. A. and Dziuk, P. J. (1977). Assessment of relative fertility of males (cockerels and boars) by competitive mating. *J. Reprod. Fert.* **49**, 323–329.

Menke, T. M. and McLaren, A. (1970). Mouse blastocysts grown *in vivo* and *in vitro*: carbon dioxide production and trophoblast outgrowth. *J. Reprod. Fert.* **23**, 117–127.

Moor, R. M. (1968). Effect of embryo on corpus luteum function. *J. Anim. Sci.* **27**, Suppl. 1, 97–118.

Moor, R. M. and Rowson, L. E. A. (1964). Influence of the embryo and uterus on luteal function in the sheep. *Nature, London* **201**, 522–523.

Moor, R. M. and Rowson, L. E. A. (1966). The corpus luteum of the sheep: functional relationship between the embryo and the corpus luteum. *J. Endocr.* **34**, 233–239.

Moor, R. M., Allen, W. R. and Hamilton, D. W. (1975). Origin and histogenesis of equine endometrial cups. *J. Reprod. Fert., Suppl.* **23**, 391–396.

Moore, N. W., Adams, C. E. and Rowson, L. E. A. (1968). Developmental potential of single blastomeres of the rabbit egg. *J. Reprod. Fert.* **17**, 527–531.

Murray, F. A., Bazer, F. W., Wallace, H. D. and Warnick, A. C. (1972). Quantitative and qualitative variation in the secretion of protein by the porcine uterus during the estrous cycle. *Biol. Reprod.* **7**, 314–320.

Nalbandov, A. V. (1976). "Reproductive Physiology of Mammals and Birds", 3rd edition. W. H. Freeman and Co., San Francisco.

Nishikawa, Y. and Waide, Y. (1958). Effect of stilboestrol injected at the corpus luteum stage in the sow, with special reference to the long maintenance of corpus luteum. *Bull. Nat. Inst. agric. Sci. (Chiba), Ser. G,* **(14)**, 1–13.

Noyes, R. W. and Dickmann, Z. (1960). Relationship of ovular age to endometrial development. *J. Reprod. Fert.* **1**, 186–196.

Patek, C. E. and Watson, J. (1976). Prostaglandin F and progesterone secretion by porcine endometrium and corpus luteum *in vitro*. *Prostaglandins* **12**, 97–111.

Patten, B. M. (1948). "Embryology of the Pig", 3rd edition. McGraw-Hill, New York, Toronto and London.

Perry, J. S. (1954). Fecundity and embryonic mortality in pigs. *J. Embryol. exp. Morph.* **2**, 308–322.

Perry, J. S. (1960). The incidence of embryonic mortality as a characteristic of the individual sow. *J. Reprod. Fert.* **1**, 71–83.

Perry, J. S. and Rowlands, I. W. (1962). Early pregnancy in the pig. *J. Reprod. Fert.* **4**, 175–188.

Perry, J. S., Heap, R. B. and Amoroso, E. C. (1973). Steroid hormone production by pig blastocysts. *Nature, London* **245**, 45–47.

Perry, J. S., Heap, R. B., Burton, R. D. and Gadsby, J. E. (1976). Endocrinology of the blastocyst and its role in the establishment of pregnancy. *J. Reprod. Fert., Suppl.* **25**, 85–104.

Pharriss, B. B. and Wyngarden, L. J. (1969). The effect of prostaglandin $F_{2\alpha}$ on the progestagen content of ovaries from pseudopregnant rats. *Proc. Soc. exp. Biol. Med.* **130**, 92–94.

Piper, P. J., Vane, J. R. and Wyllie, J. H. (1970). Inactivation of prostaglandins by the lungs. *Nature, London* **225**, 600–604.

Polge, C. and Dziuk, P. J. (1970). Time of cessation of intrauterine migration of pig embryos. *J. Anim. Sci.* **31**, 565–566.

Polge, C., Rowson, L. E. A. and Chang, M. C. (1966). The effect of reducing the number of embryos during early stages of gestation on the maintenance of pregnancy in the pig. *J. Reprod. Fert.* **12**, 395–397.

Psychoyos, A. (1960). La réaction déciduale est précédée des modifications précoces de la perméabilité capillaire de l'utérus. *C.r. Séanc. Soc. Biol.* **154**, 1384–1387.

Psychoyos, A. (1973). Endocrine control of egg implantation. *In* "Handbook of Physiology", Endocrinology II, Part 1 (R. O. Greep and E. B. Astwood, eds), pp. 187–215. American Physiological Society, Washington.

Reimers, T. J., Dziuk, P. J., Bahr, J., Sprecher, D. J., Webel, S. K. and Harmon, B. G. (1973). Transuterine embryonal migration in sheep, anteroposterior orientation of pig and sheep fetuses, and presentation of piglets at birth. *J. Anim. Sci.* **37**, 1212–1217.

Robertson, H. A. and King, G. J. (1974). Plasma concentrations of progesterone, oestrone, oestradiol-17β and of oestrone sulphate in the pig at implantation, during pregnancy and at parturition. *J. Reprod. Fert.* **40**, 133–141.

Rowson, L. E. A. (1971). The role of reproductive research in animal production. *J. Reprod. Fert.* **26**, 113–126.

Rowson, L. E. A. and Moor, R. M. (1966a). Embryo transfer in the sheep: the significance of synchronising oestrus in the donor and recipient animal. *J. Reprod. Fert.* **11**, 207–212.

Rowson, L. E. A. and Moor, R. M. (1966b). Development of the sheep conceptus during the first fourteen days. *J. Anat.* **100**, 777–785.

Rowson, L. E. A. and Moor, R. M. (1967). The influence of embryonic tissue homogenate infused into the uterus, on the lifespan of the corpus luteum of the sheep. *J. Reprod. Fert.* **13**, 511–516.

Rowson, L. E. A., Moor, R. M. and Lawson, R. A. S. (1969). Fertility following egg transfer in the cow: effect of method, medium and synchronisation of oestrus. *J. Reprod. Fert.* **18**, 517–523.

Rowson, L. E. A., Lawson, R. A. S., Moor, R. M. and Baker, A. A. (1972). Egg transfer in the cow: synchronization requirements. *J. Reprod. Fert.* **28**, 427–431.

Scanlon, P. F. (1972). Frequency of transuterine migration of embryos in ewes and cows. *J. Anim. Sci.* **34**, 791–794.

Scaramuzzi, R. J. and Baird, D. T. (1976). The oestrous cycle of the ewe after active immunization against prostaglandin $F_{2\alpha}$. *J. Reprod. Fert.* **46**, 39–47.

Soderwall, A. L. and Blandau, R. J. (1941). The duration of the fertilizing capacity of spermatozoa in the female genital tract of the rat. *J. exp. Zool.* **88**, 55–64.

Soderwall, A. L. and Young, W. C. (1940). The effect of ageing in the female genital tract on the fertilizing capacity of guinea pig spermatozoa. *Anat. Rec.* **78**, 19–29.

Stewart, D. L., Spooner, R. L., Bennett, G. H., Beatty, R. A. and Hancock, J. L. (1974). A second experiment with heterospermic insemination in cattle. *J. Reprod. Fert.* **36**, 107–116.

Tesh, J. M. (1969). Effects of the ageing of rabbit spermatozoa *in utero* on fertilization and prenatal development. *J. Reprod. Fert.* **20**, 299–306.

Tesh, J. M. and Glover, T. D. (1966). The influence of ageing of rabbit spermatozoa on fertilization and prenatal development. *J. Reprod. Fert.* **12**, 414–415.

Trimberger, G. W. (1948). Breeding efficiency in dairy cattle from artificial insemination at various intervals before and after ovulation. *Univ. Nebraska Agric. Exp. Sta. Res. Bull.* No. **153**.

Van Niekerk, C. H. and Allen, W. R. (1975). Early embryonic development in the horse. *J. Reprod. Fert., Suppl.* **23**, 495–498.

Warwick, B. L. (1926). Intra-uterine migration of ova in the sow. *Anat. Rec.* **33**, 29–33.

Wimsatt, W. A. (1975). Some comparative aspects of implantation. *Biol. Reprod.* **12**, 1–40.

Wintenberger-Torrès, S. (1977). Signal embryonnaire et maintien du corps jaune. *Ann. Méd. Vet.* **121**, 383–398.

Wintenberger-Torrès, S. (1978). Rôle actif de l'embryon avant l'implantation. *In* "L'Implantation de l'Oeuf", Colloque S.N.E.F.S., pp. 181–192. Masson, Paris.

Wintenberger-Torrès, S. and Fléchon, J. E. (1974). Ultrastructural evolution of the trophoblast cells of the pre-implantation sheep blastocyst from day 8 to day 18. *J. Anat.* **118**, 143–153.

Winters, L. M., Green, W. W. and Comstock, R. E. (1942). Prenatal development of the bovine. *Univ. Minn. Agric. Exp. Sta. Tech. Bull.* No. **151**.

Chapter VII

Techniques for *in vitro* Maturation of Gametes, *in vitro* Fertilisation, Embryo Culture and Long Term Storage

Introduction

This chapter is concerned with studies which suggest that a range of fairly sophisticated techniques, initially developed using the gametes and embryos of mice, hamsters and rabbits, could have a contribution to make in the field of animal production. The reader may immediately suspect that there is some incompatibility between sophisticated laboratory techniques and their eventual application to large farm animals. That this need not be the case is well illustrated by the extensive use of deep frozen semen in artificial insemination schemes, and the more recent application of the technique of embryo transplantation in export and upgrading programmes (see Chapter VIII). Having therefore stressed the relevance of modern techniques, the material to be considered in the following pages is primarily concerned with achieving a more flexible use of the reproductive potential of domestic animals, both in terms of exploiting the germ cell population of the ovaries and in discussing the means whereby gametes and preimplantation embryos may be manipulated in the laboratory. If knowledge gleaned from the ever proliferating fundamental studies into reproductive processes is to have a significant impact on animal productivity, it seems important to set some of the findings in a meaningful sequence; this is attempted in the following pages.

Maturation of gametes in vitro

At the outset, it should be recalled that neither the complement of oocytes formed in the ovaries by the time of birth nor the male gametes still within the testes are fully functional cells. They must undergo further processes of maturation and of course liberation before becoming competent to participate in the events of fertilisation. The spermatozoa require a post testicular phase of maturation which is completed during slow passage along the highly convoluted epididymal duct, and a post ejaculatory phase of maturation—capacitation—which occurs in the female tract (see Chapters IV and V). In some contrast, the phases of maturation of the primary oocytes that will concern us here occur within the developing follicle rather than in the genital ducts, although there may be a phase of membranous maturation of the female gamete after release from the ovary into the Fallopian tube (see Chapter V). The techniques to be described deal with treatment of oocytes initially present in Graafian follicles, and with the seemingly less demanding requirements for *in vitro* maturation of epididymal or ejaculated spermatozoa.

Culture of ovarian oocytes

Even though the importance of the follicular environment in the physiology of oocyte maturation has just been indicated, many experiments have attempted to obtain viable oocytes by treating them after artificial release from the ovaries (Pincus and Enzmann, 1935; Noyes, 1952; Chang, 1955). Before discussing the technical details, it is worth mentioning the practical justification on which this approach to oocyte maturation is based.

The population of germ cells in the ovaries of young mammals, and in the present context especially in cattle, sheep and pigs of known genetic merit, represents a source of reproductive potential that is possibly amenable to exploitation in the sense of a much enhanced use compared with in the normal ovulatory situation. In these species, as in other mammals, the number of oocytes ovulated spontaneously at the time of oestrus is only a minute proportion of the thousands of oocytes in the ovaries at birth. The remainder degenerate within the ovary, this condition generally being referred to as atresia. A fundamental question to be asked, therefore, is whether the great majority of oocytes within antral follicles must be abandoned to the atretic process, which may imply qualitative deficiencies in the germ cells or their associated somatic cells, or whether an increased proportion of oocytes could be liberated by appropriate treatments, in due course to undergo fertilisation in the Fallopian tubes or *in vitro*. A satisfactory answer to this question is not available at present, but considerable numbers of oocytes have been obtained directly from large Graafian follicles in the laboratory shortly after surgery or slaughter; these oocytes have achieved the full diameter of an ovulated egg and have therefore already completed an important aspect of cytoplasmic maturation.

A high proportion of oocytes liberated by aspirating, puncturing or scraping the follicular contents will resume meiotic maturation *in vitro* from the dictyate nucleus stage through to metaphase I or II (e.g. Table VII.1), and the time course of this maturation usually corresponds quite closely with that observed *in vivo* after the preovulatory surge of gonadotrophin. This approach to oocyte maturation has been successful using as a culture medium a relatively simple salt solution plus a protein source such as serum albumin. But in the light of this general observation, further questions that must be asked are:

Table VII.1. Timing of the response of the oocyte nucleus after the preovulatory gonadotrophin surge or following a systemic injection of a luteinising hormone preparation. (All times are in hours.)

Species	Latent period[1] after stimulus (hours)	Stage of meiotic maturation: First metaphase	First anaphase	First telophase	Second metaphase
Cow	10–12	14–21	22	23	24
Sheep	10–11	12–20	21	22	24
Pig	17–18	26–34	35	36	37

[1] The oocyte nucleus remains as a germinal vesicle during this latent period

(a) Does the nuclear maturation observed in these oocytes also imply functional (i.e. physiological) maturation?

(b) If cultured oocytes do become functionally mature, has further use been made of them in studies of fertilisation and embryonic development, and if so by what means?

These questions are answered with respect to studies in cattle, sheep and pigs, and with some reference also to experiments in rabbits.

In cattle, quantitative aspects of the germ cell population at various foetal and postnatal stages have been presented in detail (Erickson, 1966a, b; Fig. VII.1), and also in terms of the complement of oocytes within Graafian follicles (Rajakoski, 1960; Marion *et al.*, 1968). The number of antral follicles remains fairly constant from two months of postnatal age until eight to ten years, after which it slowly declines

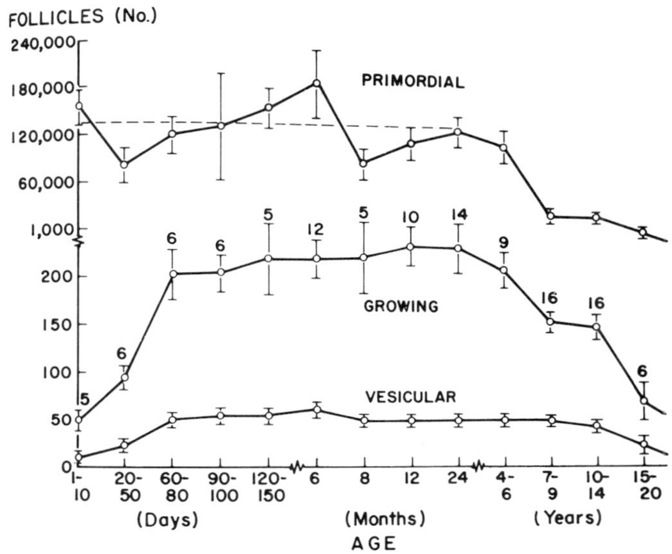

Fig. VII.1. Analysis of the population of primordial, growing and vesicular follicles in bovine ovaries after birth. The number of pairs of ovaries studied is indicated. (Reproduced from Erickson, 1966a. Courtesy of the Editor, *Journal of Animal Science*.)

(Erickson, 1966a). Data on ovarian morphology at earlier stages are less precise, but Desjardins and Hafs (1969) could not detect ovarian follicles macroscopically at birth whereas Black and Hansel (1972) observed both follicular growth and atresia in calves as early as one month of age. Thus, notwithstanding the finding of extreme variation in the postnatal population of germ cells and follicles, even the prepuberal animal should lend itself to treatments aimed at enhancing the utilisation of oocytes (Hunter, 1976).

Liberation of cow or calf oocytes can be accomplished mechanically by aspirating the contents of the follicular antrum (Edwards, 1965; Sreenan, 1970; Hunter *et al.*, 1972), and those still within a healthy investment of cumulus cells will then progress to metaphase I or II in culture, as for example in microdrops of medium under oil in plastic Petri dishes at 37°C (Table VII.2). Moreover, cow oocytes that have at least partially matured *in vitro* and have then been transplanted into the Fallopian tubes of inseminated recipent heifers have been shown to be capable of normal fertilisation (Hunter *et al.*, 1972; Fig. VII.2). Embryonic development and the birth of calves have also been recorded in two instances following the above approach with the additional step

Table VII.2. Maturation of bovine ovarian oocytes when cultured in a modified Tyrode's solution containing homologous follicular fluid. (From Hunter *et al.*, 1972.)

Period of culture (hours.minutes)	Number of oocytes examined	Number of oocytes maturing to[1]				
		GV	PM	M_I	A_I	Degenerating
25·00	9	–	1	8	–	–
28·25	3	–	2	1	–	–
28·25	6	–	–	6	–	–
29·10	14	–	–	14	–	–
36·00	11	–	–	10	–	1
44·00	10	–	–	7	2	1
Total	53	–	3	46	2	2

[1] Stages of the first meiotic division: GV, germinal vesicle; PM, prometaphase; M_I, first metaphase; A_I, first anaphase.

of blastocyst transfer (Newcomb *et al.*, 1978), but the proportion of oocytes demonstrated to be viable after use of these extra-follicular techniques is remarkably low. Very few oocytes after transplantation to inseminated recipients progress even to the stage of blastocyst formation (Newcomb *et al.*, 1978; Table VII. 3). By contrast, if oocytes are recovered from follicles some 24 hours after *in vivo* stimulation with HCG, and then transplanted to inseminated recipients, normal blastocyst formation and foetal development can be obtained (Trounson *et al.*, 1977).

Culture of Graafian follicles

Because *in vitro* maturation of primary oocytes removed from Graafian follicles is not ultimately a fruitful technique, it is possible in many oocytes that the ooplasm is deficient from a biochemical point of view, even though the process of meiotic maturation may appear morphologically normal. Such has been shown to be the case for cultured rabbit oocytes, in which the liberated cell is unable to support development of a normal male pronucleus (Thibault and Gérard, 1970, 1973). Accordingly, a more physiological approach to oocyte maturation may be to dissect whole follicles from the ovary, taking the oocyte through the

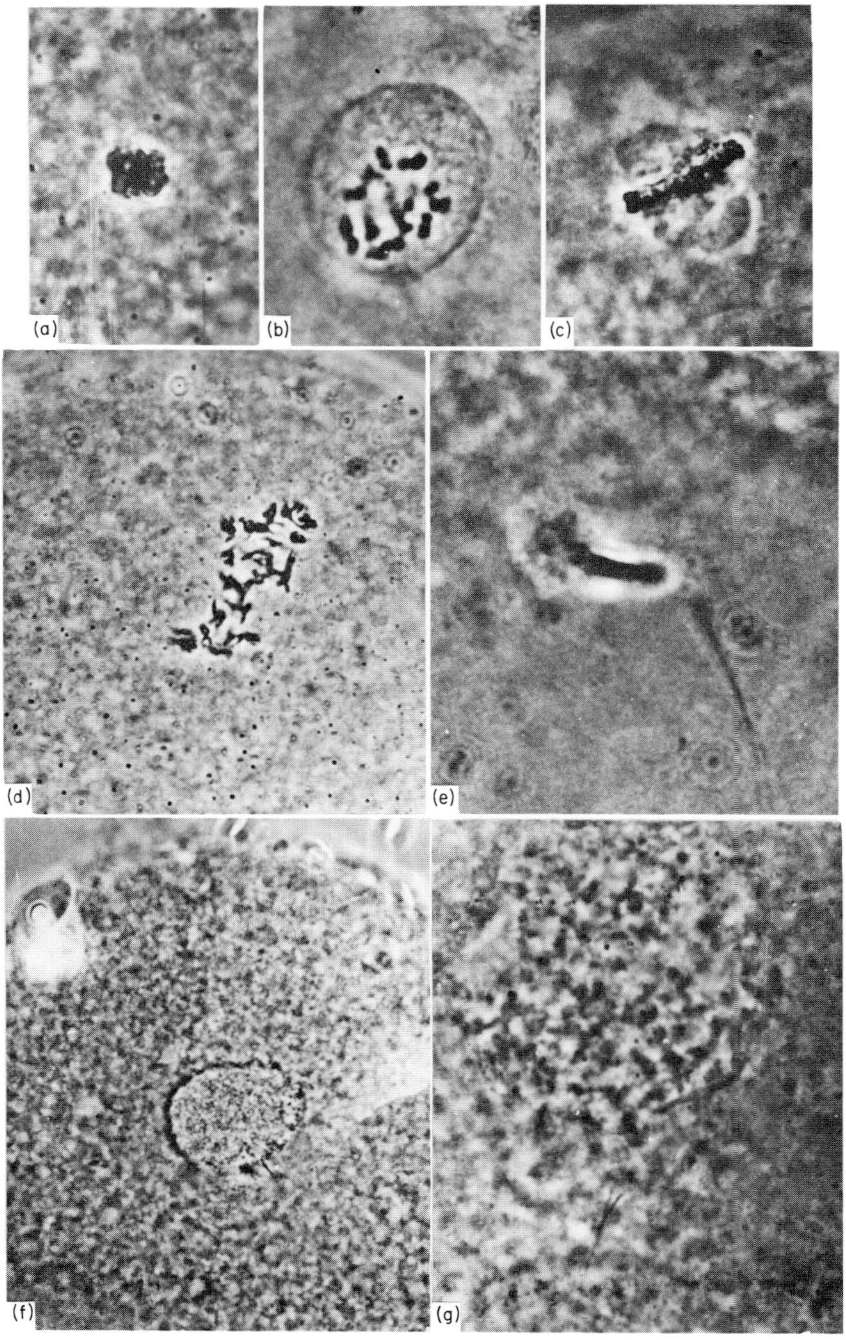

Fig. VII.2. *In vitro* matured cow oocytes showing nuclear changes and aspects of fertilisation obtained *in vivo* after transplantation. The figures show (a) condensation of the chromatin, (b, c) metaphase arrangements of the chromosomes, (d) dispersal of chromosomes at first meiotic metaphase, (e) the fertilising spermatozoon in an oocyte 22 hours after transplantation, and (f, g) fertilised pronuclear eggs recovered from the oviducts 21 hours after transplantation. (From Hunter *et al.*, 1972.)

Table VII.3. The stages of development attained by *in vitro* matured bovine follicular oocytes seven days after their transfer to the Fallopian tubes of three inseminated heifers. (Modified from Newcomb *et al.*, 1978.)

Number of oocytes		Apparent stage of development						Percentage of recovered eggs >8-cells
transferred	recovered	1-cell	Difficult to classify	2–4 cells	8–16 cells	Late morula	Blastocyst	
45	5	5	–	–	–	–	–	0
50	23	6	7	4	1	3	2	26
40	32	19	–	5	5	3	–	25
135	60	30	7	9	6	6	2	23

resumed first meiotic division under conditions of organ culture (Baker and Neal, 1969, for the mouse; Foote and Thibault, 1969, for cow and pig) still encompassed in the intact Graafian follicle. The advantage of selection of suitable vesicular follicles from gonadotrophin stimulated ovarian tissue is also inherent in this system, not least since it is known that the physiological competence of the oocyte is related to the size of the cultured follicle in comparison with its mature diameter.

The first studies of this kind demonstrated that maturation of rabbit oocytes within cultured follicles followed by transplantation of the secondary oocytes to the Fallopian tubes of mated recipients could lead to the birth of normal young (Thibault *et al.*, 1975). There is every reason to suppose that the technique would yield some success in cattle, and studies on cow and calf follicles in culture are in progress (Thibault *et al.*, 1975, 1976). Transplantation of oocytes from calf follicles matured in this manner is an experiment whose outcome is urgently awaited, since it avoids problems associated with *in vitro* fertilisation (Menezo *et al.*, 1976) or those of obtaining embryos from gonadotrophin treated calves. Nonetheless, it should be noted that transplantation of secondary oocytes to the Fallopian tubes will only be accomplished satisfactorily by a surgical as distinct from an intracervical approach, and therefore will not be suitable for field-scale development.

In the case of sheep oocytes, the possibility of extra-follicular nuclear maturation was shown by Edwards (1965), and the timing of the meiotic stages after release of the oocytes into culture was in line with the figures of Berry and Savery (1958) and of Dziuk (1965). None of these early studies attempted fertilisation, and the more recent work with sheep material has usually emphasised initial culture of whole follicles before testing the physiological state of the oocyte (Moor and Trounson, 1977). At the fundamental level, Warnes *et al.* (1977) have been able to demonstrate distinct phases of cytoplasmic maturation involving the appearance of specific protein components in sheep oocytes, and some modification of the oocyte membrane during the culture period has also been suggested (Moor, 1978). At the practical level, oocytes removed from 3–5 mm diameter follicles cultured *in vitro* for 24 hours with steroid and gonadotrophin support, and then transplanted into the Fallopian tubes of inseminated recipients, have been shown to yield normal lambs (Moor and Trounson, 1977). Although the success rate in this work was approximately 60% in terms of the blastocysts (selected from the matured oocytes) that ultimately produced lambs, the extent to which the techniques will have an application in sheep production, as distinct from in basic research, is not easy to imagine.

In the pig, once again the study of Edwards (1965) showed that oocytes would resume meiosis when liberated from the follicle into culture, and that the time course of the meiotic stages was in line with observations from *in vivo* material (Spalding *et al.*, 1955; Hunter and Polge, 1966). But McGaughey and Polge (1971) were able to demonstrate certain abnormalities in the chromosomes of oocytes subjected to extra-follicular maturation, such as heteroploidy, fragments, chromosomal bridges, tripolar spindles and lagging chromosomes. However, the emphasis in culture studies has once more moved to attempted intra-follicular maturation (Foote and Thibault, 1969; Baker *et al.*, 1975; Baker and Hunter, 1978), not least since the rôle of the follicular environment in bestowing competence on the oocytes for normal maturation and normal fertilisation had been clearly shown by *in vivo* work (Leman and Dziuk, 1971); and the oestrogenic milieu within the preovulatory follicle seems to facilitate functional maturation of the pig oocyte (Hunter *et al.*, 1976). But despite using a selected preovulatory population of 6–10 mm diameter Graafian follicles in culture, and testing a range of concentrations of steroid and gonadotrophic hormones, the proportion of pig oocytes undergoing maturation to metaphase II in individual experiments rarely exceeded 50% (Baker and Hunter, 1978), and viable embryos have yet to be obtained after transplantation of oocytes to mated recipients.

Summarising the argument so far, female calves, lambs and pigs are known to have a complement of oocytes far in excess of that found in the mature animal, and there is evidence to suggest that an increased proportion of these cells can be liberated from suitable follicles by mechanical (or hormonal) techniques to undergo subsequent development *in vivo*. Matured oocytes would be utilised by means of transplantation into recipients. In this manner, the extremely small proportion of ovarian oocytes that is ever expressed as calves, lambs or piglets in the biological situation might be enhanced, this proposition having considerable attraction particularly in cows or calves of superior genetic merit. Even so, the manner of maturation of mammalian follicles and oocytes remains complex (Thibault, 1977), and attempts to exploit the germ cell complement of the ovaries are limited by knowledge and by currently available techniques. It seems reasonable to conclude that a clearer understanding of the dynamics of follicular growth from birth to puberty must underlie approaches to exploiting the ovarian potential of domestic animals.

Final maturation of spermatozoa

In contrast to the condition of the egg at ovulation, meiosis has already been completed in the spermatozoon before it enters the epididymis. The final phase of sperm maturation after liberation into the female tract at coitus involves changes in the nature of the swimming activity and a modification in the disposition of membranes on the sperm head. The proportion of spermatozoa in the ejaculate competent at any one time to undergo these functional changes may be limited, but the changes are almost certainly mandatory before penetration of the egg membranes can occur. In large measure, they may reflect removal of a surface coating from the spermatozoa, probably acquired mainly in the rete testis and epididymis, and the provision of appropriate metabolic substrates to the previously immotile cell in order to support its swimming activity. In fact, therefore, the final phase of maturation of spermatozoa that one attempts to obtain *in vitro* corresponds essentially to the process of capacitation (see Chapter V).

Techniques have been applied to both epididymal and ejaculated spermatozoa, and involve suspending the sperm cells at suitable concentrations in a physiological medium, usually and perhaps surprisingly containing a large molecular weight compound such as serum albumin, followed by mild centrifugation and resuspension in a fresh aliquot of medium. The objective here is to remove elements of the seminal plasma from the sperm surface, and in due course to permit modifications in the sperm head membranes leading to the vesiculation that constitutes the acrosome reaction (Chapter V). Under appropriate conditions of incubation, both an increase in sperm motility and development of the acrosome reaction can be obtained *in vitro*: these changes may be promoted by the addition of eggs in cumulus investments or by addition of follicular fluid to the culture. This general technique for obtaining functional maturation of spermatozoa has been fruitful in many of these species where *in vitro* fertilisation has been demonstrated (see below), but the time course of the reaction is certainly no faster than in the *in vivo* situation. However, a means for reliably obtaining capacitation of bull, ram and boar spermatozoa *in vitro* has so far not been documented. Whether this reflects a failure in adequately modifying the sperm surface or whether achieving and sustaining appropriate motility is the principal snag remains to be clarified.

Fertilisation in vitro

The fertilisation of mammalian eggs outside the body, that is penetration of the egg by a spermatozoon in some form of cell culture system, has long been a research target for physiologists and scientists in related disciplines. Because mammalian fertilisation normally occurs within the confines of the Fallopian tubes, one much repeated argument has been that a system of *in vitro* fertilisation would be a valuable analytical tool for examining biochemical and physiological factors that contribute to the successful union of gametes. Furthermore, in the context of a technology that might one day be applied in the field of animal breeding, *in vitro* fertilisation has also been suggested as a possible sequel to *in vitro* maturation of large numbers of oocytes and as a means of fertilising eggs obtained from immature animals or after excessive responses to superovulatory treatments. And, of course, in the case of human medicine, a technique of *in vitro* fertilisation has already been used to overcome problems of infertility due to blocked Fallopian tubes (Steptoe and Edwards, 1978), the embryo being instilled into the uterus of the mother at an early stage of cleavage (8-cells), so there is no question here of the emotive test-tube baby. The subject of *in vitro* fertilisation has been reviewed on numerous occasions, but reference to the works of Austin (1961a), Chang (1968), Thibault (1969), and Gwatkin (1977) may be found helpful at the outset.

Because *in vitro* fertilisation must first depend on obtaining suitable preparations of eggs and spermatozoa, it should be clear that fertilisation itself will represent only one step in a sequence of manipulations, and that in order for an embryo formed *in vitro* to develop to full term, the technique of transplantation will here again have to be applied. The paragraphs that follow will not attempt to treat the subject of *in vitro* fertilisation exhaustively, but will present some of the key details involved in bringing male and female gametes together under appropriate conditions of culture.

The history of attempted *in vitro* fertilisation in mammals is a lengthy one, dating back at least to the work of Shenk (1878) with rabbit and guinea-pig eggs; details of many of the well documented attempts are provided by Austin (1961b) and Thibault (1969). It is generally believed, however, that before the requirement for capacitation of spermatozoa was appreciated, the supposed fertilisation of eggs *in vitro* with samples of ejaculated spermatozoa or those prepared from the male tract was either a chance occurrence or, more probably, an incorrect diagnosis. In the latter case, the most common pitfall was to regard eggs containing two pronuclei or other nuclear structures as penetrated and fertilised, when such a situation could have arisen by parthenogenetic activation, especially under less than perfect conditions of culture. The essential criteria for establishing fertilisation in a recently penetrated egg should include the presence of first and second polar bodies, male and female pronuclear elements, and detection of some remnant of the sperm mid-piece or tail. Other misleading observations have been those on 2- and 4- cell eggs that were in fact fragmenting, and the appearance of unswollen sperm heads seemingly within the cytoplasm of sectioned eggs when it is now clear that they must have been displaced by the microtome. Yet a further source of error, albeit leading to eventual fertilisation, has been the transplantation of eggs soon after attempted fertilisation *in vitro* to the Fallopian tubes of recipients in order to follow their development. The obvious danger here is of transferring spermatozoa in association with unfertilised eggs, which then undergo successful penetration and cleavage in the recipient.

Having stressed the more frequent sources of error in interpreting putative claims for *in vitro* fertilisation, and inferred the need for caution even in evaluating contemporary attempts, let us now consider the first authentic successes and the state of the art today. Capacitation was recognised as a specific phenomenon by Austin (1951) and Chang (1951) (see Chapter V), and the first systematic studies of *in vitro* fertilisation using spermatozoa recovered from the female tract (i.e. spermatozoa presumed to have been capacitated *in vivo*) were almost certainly by Dauzier *et al*. (1954) in the rabbit. In addition to the use of spermatozoa flushed from the tubes and uterus some 12 hours after mating, these authors employed an incubator system with rotating glass tubes in which the recently-ovulated eggs were agitated for 30–120 minutes in a small volume of Locke's solution before introducing the sperm suspension. This washing treatment was held to improve the proportion of fertilised eggs due to removal of a fertilisin-like substance from the egg or its investments (Thibault and Dauzier, 1960, 1961). Although Thibault and his colleagues presented impeccable histological evidence of *in vitro* fertilisation as shown in pronucleate and cleaved eggs, it was Chang (1959) who consummated the experimental procedure of the French workers by transplanting rabbit eggs after fertilisation and cleavage *in vitro* to suitable recipients, in due course obtaining viable young. But these early successes with the rabbit egg failed to point the way to a fruitful *in vitro* technique with the gametes of large domestic animals, even though Thibault and Dauzier (1961) recorded incidental successes for the sheep and pig.

Whilst these first successful studies and much later work obtained sperm samples from the tract of mated females, more recent experiments, especially in rodents but also in humans, have obtained capacitation of epididymal or ejaculated spermatozoa *in vitro*. The forerunners here were Yanagimachi and Chang (1963, 1964) who obtained *in vitro* fertilisation of golden hamster eggs with suspensions of epididymal spermatozoa, this step forward being followed by fertilisation *in vitro* of hamster ovarian oocytes using preparations of epididymal spermatozoa (Barros and Austin, 1967). Thus capacitation of hamster spermatozoa together with any final maturation of the oocyte must have been obtained in the culture system, neither gamete being exposed to the fluid environment of the female tract. As far as spermatozoa of the golden hamster are concerned, Bavister (1969) has put forward the view that capacitation in this species may simply require provision of a suitable, albeit critically defined, fluid milieu in which this intrinsically regulated process will occur, but as yet there is no evidence for such a suggestion in farm species. Use of oocytes aspirated from follicles shortly before or at the moment of ovulation has proved valuable in the rabbit (Seitz *et al.*, 1970) and human (Edwards *et al.*, 1970), and possibly gave impetus to subsequent experiments on *in vitro* fertilisation of *in vitro* matured oocytes.

Species in which *in vitro* fertilisation has been obtained in at least several trials are listed in Table VII.4; cow and sheep are not shown as positive although there have been sporadic but poorly documented claims of success. In the pig, Harms and Smidt (1970) recorded some apparent fertilisations using ejaculated boar spermatozoa, and more recently Iritani, Niwa and Imai (1978) have presented photographic evidence with which to support their own observations. An impressive fact in reviewing much of the data on *in vitro* fertilisation is the enormous variability in the proportion of eggs fertilised between experiments in any given species, the only exceptions to date being the golden hamster and possibly the mouse which may regularly yield an incidence of fertilisation of 80–90% or more. Why should this variability exist?

Table VII.4. Species in which *in vitro* fertilisation has been recorded. In the three farm species, a subjective judgement has been offered as to whether authentic instances of fertilisation have occurred that would give rise to a viable embryo.

Species	Repeatable instances of *in vitro* fertilisation[1]	Foetuses and/or viable offspring produced after transplantation
Rabbit	+	+
Rat	+	+
Mouse	+	+
Golden hamster	+	−
Chinese hamster	+	−
Guinea pig	+	−
Cow	?	−
Sheep	?	−
Pig	?	−
Man	+	+

+ indicates positive evidence; − indicates not determined

[1] Sperm penetration of oocytes alone has not been considered sufficient evidence of fertilisation, as this condition may be found in immature or atretic eggs. The criteria insisted upon include activation of the second meiotic division with extrusion of the second polar body and formation of a female pronucleus together with development of a male pronucleus, or stages subsequent to these.

Apart from differences in skills between individuals, and the possible advantage of working with certain strains of animals whose gametes may be particularly amenable to manipulations *in vitro*, the actual conditions in the culture system play a vital rôle.

Factors influencing in vitro *fertilisation*

Factors that must be considered include the following:

(1) The source and condition of spermatozoa. Suspensions of spermatozoa recovered from the female tract several hours after mating featured in many experiments during the 1960s, but washed ejaculated or preincubated epididymal spermatozoa are extensively used nowadays in primate and rodent studies. A washing treatment usually implies mild centrifugation to remove most of the free seminal plasma followed by a resuspension of the sperm cells in the culture medium.

(2) The presence or absence of follicular cells and fluid around the eggs at the commencement of incubation. The majority of studies, especially in rodents, have used recently-ovulated eggs recovered still within a plug of cumulus cells by perfusion or rupture of the Fallopian tubes.

(3) The choice of a medium, physiological salt solutions such as Locke's, Tyrode's or Kreb's-Ringer being at first favoured. Recent studies have used more complex media such as TC 199 or Ham's F 10. An osmolarity of 290–310 m. osmols and a pH of 7·4–7·6 are common, the latter frequently being maintained with a bicarbonate buffer and a gas phase of 5% CO_2 in humidified air. In addition to follicular or oviducal fluid constituents, the medium usually contains a large molecular weight substance such as bovine serum albumin (concentration of 1–5 gm $litre^{-1}$) or a synthetic substance such as polyvinyl pyrrolidone. In studies during the last decade, the presence of lactate and calcium ions in the culture medium has been considered desirable or essential.

(4) The volume of the microdrops used for culture and the concentration of the sperm suspension therein, respective figures often being of the order of 40–100 μl and 1–2 $\times$ 10^6 cells $ml.^{-1}$. Although such sperm numbers would greatly exceed those reaching the ampulla at the time of fertilisation, it is generally accepted that only a small proportion of the spermatozoa would be competent to penetrate the eggs (see Chapter IV).

(5) The time relationship between placing the separate gamete preparations in culture, and then mixing the two. A period of preincubation of the sperm suspension, perhaps in a medium rich in ovarian follicular fluid, may be found beneficial.

(6) Motility of the sperm suspension. A high degree of progressive motility seems to be essential to penetration of the egg membranes *in vitro*.

(7) The absolute minimisation of temperature fluctuations during the *in vitro* manipulations; the avoidance of contamination of the cell suspensions with blood which may rapidly cause agglutination of the spermatozoa; and the strict observation of aseptic precautions. The culture system is normally held in darkness, except during periodic examination under the dissecting microscope.

As to the actual system of culture, microdrops of medium in Falcon plastic dishes have been most frequently used, with the advantage that the cell preparations can be inspected without disturbance. But as already mentioned, rotating glass tubes containing the medium between droplets of paraffin oil have been used by Thibault and Dauzier (1961), and others have used glass or plastic micro test-tubes, watch glasses or cavity slides. To avoid problems of evaporation, a common practice with the Falcon dishes has been to place the microdrops of medium beneath a thin layer of paraffin or mineral oil, the oil having been previously equilibrated against a small volume of the culture medium to avoid subsequent osmotic disturbances in the microdrops. A laboratory sequence for obtaining *in vitro* fertilisation of hamster eggs is illustrated in outline in Fig. VII.3.

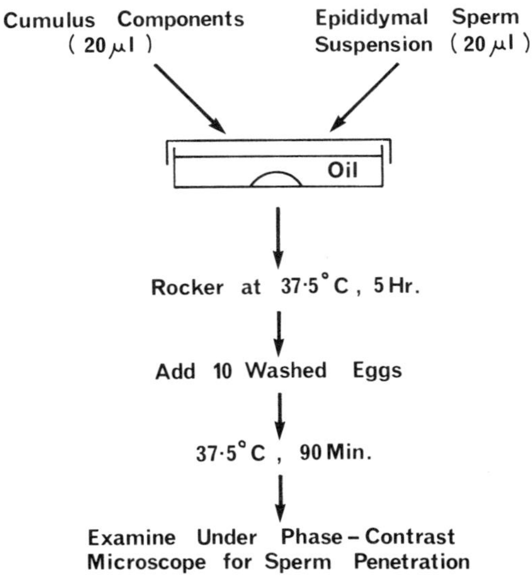

Fig. VII.3. Diagrammatic representation of a simplified sequence for obtaining *in vitro* fertilisation of golden hamster oocytes. The first stages in the sequence permit capacitation of epididymal spermatozoa. (Redrawn from Gwatkin, 1977.)

As far as the large domestic animals are concerned, it is perplexing that attempts at *in vitro* fertilisation should prove somewhat refractory when compared with successes in laboratory species and primates. In the case of cattle and sheep, relatively little experimentation has so far been undertaken, but some progress is being made with pig eggs (Iritani *et al.*, 1978). Obtaining capacitation and vigorous motility of boar spermatozoa may be a major problem, since these cells are notoriously prone to diminished activity *in vitro* without some provision for substantial aeration and agitation. A method taking the latter factors into account, and perhaps also using corpus epididymal spermatozoa (Hunter *et al.*, 1978), will doubtless soon be developed to enable at least a reasonable proportion of pig eggs to be fertilised *in vitro*. Nonetheless, the necessary proof of viable offspring arising from these manipulations will require transplantation of genetically marked zygotes and identification of the offspring at birth.

Embryo culture and storage: general remarks

Apart from the desirability of a period of *in vitro* culture before transplanting eggs presumed to have been fertilised *in vitro* (see reasons presented in the previous section of this chapter), a means of storing embryos *in vitro* or of culturing them through various cleavage stages could be valuable in practical and experimental situations. For example, when *in vitro* fertilisation of cow eggs becomes a reliable technique, a period of culture through to the morula stage would almost certainly be necessary before a reasonable conception rate could be obtained from transplantation into the uterus, the latter site being essential for a non-surgical method of transfer through the cervix. And, as is argued in Chapter VIII, a bank of embryos seems an important element in any large scale transplant programme in cattle, and a temporary form of storage is certainly required during shipment of embryos in export programmes. At the experimental level, a system of culture is clearly desirable or necessary in studies examining the metabolic requirements of the developing embryo, its response to potentially damaging drugs, or where microsurgical modifications are performed such as removal of cells for sex determination (see Chapter XII). But it is worth stressing that only preimplantation stages of the embryo are under consideration here, even though techniques for the culture of differentiated mammalian embryos, such as those of the rat at the stage of the yolk sac placenta, have received much attention (New, 1966).

Culture of rabbit and rodent embryos

Early studies with rabbit embryos showed that culture from the 1-cell to the blastocyst stage was possible if the medium contained a large proportion (≥ 50%) of homologous or heat-treated heterologous serum (Chang, 1949), but further development was not observed *in vitro*. More recent experiments have attempted to define culture media appropriate to individual preimplantation stages (Onuma *et al.*, 1968), and have examined the amino acid requirements for rabbit embryos rather than adding whole serum to the medium (Daniel and Olson, 1968; Kane and Foote, 1970).

The single-cell zygote from many laboratory rodents has often proved difficult or refractory to culture, whereas embryos obtained at 4–8 cells will usually develop *in vitro* to the blastocyst stage, especially if appropriate protein fractions are included in the culture medium (see Whitten, 1956). However, 2-cell mouse embryos have proved amenable to culture, both fully *in vitro* (Whitten, 1957; Brinster, 1963) and within cultured explants of the Fallopian tubes (Whittingham and Biggers, 1967), and have given rise to viable young upon transplantation (McLaren and Biggers, 1958). In fact, culture of the mouse embryo has been extensively employed as a model system for studying the metabolic activities and requirements of preimplantation embryos in chemically defined media (Brinster, 1965, 1973), and it is largely on the basis of this work that the need for pyruvate, lactate and some source of nitrogen in cultures of the early stages was established. Single cell unfertilised mouse eggs as well as this stage of zygote can metabolise only pyruvate and oxaloacetate, but if follicular cells are still present to effect conversion, lactate, phosphoenolpyruvate and glucose can be utilised to support cleavage (Biggers *et al.*, 1967). Because of the extensive use of mouse gametes and zygotes in this kind of study, it is appropriate to mention here that mouse oocytes have now been frozen and thawed, fertilised *in vitro*, and cultured to the blastocyst stage as part of the same experimental sequence (Whittingham, 1977). Subsequent transfer of embryos into the uterus of appropriate recipients in due course led to the birth of viable young. However, the incidence of fertilisation *in vitro* of frozen-thawed oocytes was significantly lower than that of freshly collected oocytes.

Culture of embryos from farm species

Brief reference has already been made to culture of cow or calf eggs in the section on maturation of primary oocytes. The conditions of culture for developing embryos must clearly aim to duplicate the essential components of the tubal and uterine fluids during the preimplantation period, and some encouraging successes have now been reported. In these studies, embryos have been recovered from cattle principally at slaughter, but experiments have also been conducted with material flushed from the donor tract at laparotomy with a medium such as TC 199 (Rowson *et al.*, 1971) or flushed non-surgically with modified phosphate-buffered saline. Much of the early literature was summarised by Foote and Onuma (1970), whilst Seidel (1977) has presented a detailed and tabulated review of work in the following six years.

Apparently normal cleavage to the 24-blastomere stage was obtained with 4–8 cell cow eggs cultured in bovine follicular fluid (Thibault, 1966). Division proceeded for up to 31 hours in the medium, but between 44 and 54 hours only a proportion of the eggs continued to divide normally and beyond that cleavage ceased entirely. No embryos developed in culture beyond the 24-cell stage, at which time passage from the Fallopian tube to the uterus would have already occurred; these culture findings were essentially endorsed by Sreenan *et al.* (1968). However, Tervit *et al.* (1972) reported a system of culture that was successful in taking small numbers of 1-cell and 8-cell cow eggs through to the morula and blastocyst stage and, by means of transplantation, two blastocysts arising from 8-cell embryos were reported to be viable foetuses on Day 35 of gestation. An important modification in the culture system of Tervit *et al.* appeared to be the use of a synthetic oviduct fluid and a gas phase of 5% CO_2, 5% O_2 and 90% N_2, but even so the overall results indicated that less than 25% of cultured 8-cell embryos continued development after transfer to recipients (Tervit, 1973).

The work of Wright *et al.* (1976) showed that culture in Ham's F-10 medium supplemented with heat-treated foetal calf serum permitted about 20% of 8-cell embryos to develop to the expanded blastocyst stage, but the number of embryos involved in this study was small and the subsequent viability of the embryos was not tested. By contrast, in the experiments of Renard and his colleagues using medium Menezo B2, about 80% of cow blastocysts collected on Day 7 and cultured for three days were able to emerge from the zona pellucida. Whether this modification closely simulated *in vivo* hatching is not easy to resolve, but following surgical transplantation into recipients on Day 10, such embryos developed normally until at least Days 20–25 (Renard and Du Mesnil du Buisson, 1976). This observation is in line with a conclusion of Rowson's group that bovine blastocysts are more tolerant of *in vitro* manipulations than are younger embryos (Willadsen *et al.*, 1976), even though Trounson *et al.* (1976) themselves obtained a 50% survival of morulae cultured for two days and then transplanted to recipient heifers.

Judging from the approaches in this work so far, and taking account of the limited number of bovine embryos usually available, it is easy to imagine culture studies continuing for many years whilst searching for an ideal technique. Perhaps fundamental studies into the nature of genital tract fluids and of embryo metabolism in the cow should receive greater emphasis, whilst the empirical approach is nevertheless continued. In the search for ideal conditions of culture and in the enthusiastic pursuit of high success rates, it must not be overlooked, however, that even under the *in vivo* conditions obtaining in the mother, the proportion of embryos succumbing to early mortality may reach 30–40%.

Whilst the arguments for developing *in vitro* culture systems for sheep embryos may be less compelling than those for cattle, basic studies have been undertaken and the results have also contributed to the success of transplantation programmes. In preliminary work, sheep and goat zygotes cultured from the 1-cell stage rarely cleaved beyond 8-cells, but the 16-cell stage could be cultured through to a blastocyst (Wintenberger *et al.*, 1953). These findings have been largely confirmed in subsequent work using a variety of media and protein sources, and have suggested that development of the sheep embryo through the 8-cell stage may require rather special conditions of culture. Nonetheless, this step has been achieved by Tervit *et al.* (1972) with apparent emphasis again being given to the use of a synthetic oviduct fluid medium and a low oxygen tension in the gas phase. Culture of sheep embryos for three days from the morula stage has been successful, and led to the development of viable foetuses upon transplantation to suitable recipients (Moor and Cragle, 1971). Culture of earlier cleavage stages for up to six days gave rise to healthy lambs after transfer (Table VII.5), even though the process of cleavage *in vitro* was somewhat retarded (Tervit and Rowson, 1974).

Turning to pig embryos, Thibault (1969) reported failure of cleavage when these were cultured in either homologous serum or a mixture of Locke's solution and serum, but Polge and Frederick (1968) did obtain limited cleavage of 1–4 cell embryos although usually only one division occurred in culture. But using Brinster's medium supplemented with 0·1% glucose as an energy source, Pope and Day (1970) found that

Table VII. 5. The development and viability of sheep embryos cultured in a synthetic oviduct fluid medium. (Modified from Tervit and Rowson, 1974.)

Stage of development at start of culture	Number of embryos cultured	Length of culture (days)	Stage of development at end of culture (no. of embryos/stage)						No. of lambs born to recipient animals
			12-cells	16-cells	Early morula (approx. 20 cells)	Morula (approx. 32 cells)	Late morula (approx. 64 cells)	Early blastocyst	
1-cell	6	3	5	1	–	–	–	–	1
1-cell	17	6	–	1	7	8	1	–	0
2-4 cells	6	3	–	2	–	4	–	–	1
2-4 cells	27	5	–	–	11	14	2	–	10
2-4 cells	4	6	–	–	–	–	–	4	1
8 cells	18	3	–	–	4	10	4	–	10
8 cells	4	4	–	–	–	–	1	3	2

approximately 90% of 1- and 2-cell pig embryos would divide *in vitro*. Moreover, using Whitten's medium containing 1·5% (w/v) bovine serum albumin, Wright (1977) grew 8–12 cell embryos and morulae to the blastocyst stage, although hatching from the zona pellucida was rarely observed. Further work by Lindner and Wright (1978) has shown that the overall proportion of 1- and 2-cell pig embryos that can be predictably cultured through to blastocysts is low (Table VII.6), and the question arises as to whether a longer initial period of exposure to the fluid environment of the tract confers some metabolic or other fundamental advantage on embryos subsequently placed *in vitro*. In any event, the success of these relatively short term cultures has usually been assessed on morphological grounds. When the proportion of viable foetuses has been examined after retransfer of embryos into a recipient animal, the results have been disappointing. Thus, after 48 hours of culture of 1–4 cell pig embryos and then surgical transplantation, Pope and Day (1977) obtained pregnancies in only 2 of 13 gilts. A study by Polge (1977) showed that after culture of 8-cell embryos for 48 hours to the blastocyst stage and then transfer into 19 recipient pigs, 10 animals were pregnant at autopsy 21 days later, but only 51 of the 229 embryos were represented by living foetuses. None the less, Davis and Day (1978) reported considerable success in culturing 4-cell pig embryos through to blastocysts and, upon transplantation to recipient gilts, survival at three to four weeks of gestation was seen to be best after 24 hours of culture (60 of 100 embryos surviving) compared with 72 hours (19 of 100 embryos surviving).

There is little published information on the specific culture of horse embryos *in vitro*, that is for periods of one to three days or more, although embryos have of course been held *in vitro* on a short term basis during the manipulations between recovery and transplantation.

In vitro *and* in vivo *techniques of culture*

Whilst there are many variations and, at present, individual preferences in the actual techniques of culture, most studies have used either Falcon plastic petri dishes containing the embryos in microdrops of medium under oil or glass tubes (mini test tubes) with silicone rubber or glass stoppers. Others have employed glass cavity slides, embryological watch glasses, or even slowly rotating glass tubes. An exceptional technique for short-term storage (10 hours) has been the use of a length

Table VII.6. Development of pig embryos cultured in Whitten's medium supplemented with 1·5% bovine serum albumin and incubated in an atmosphere of 90% N_2, 5% CO_2 and 5% O_2. (From Lindner and Wright, 1978.)

Stage of development at start of culture		Number of embryos							
		3–4 cells	5–8 cells	9–16 cells	Morula	Blastocysts	Expanded blastocysts	Partially hatched blastocysts[1]	Hatched blastocysts
1–2 cells	23	6	1	1	1	1	0	0	0
3–4 cells	74	–	53	47	40	33	20	7	2
5–8 cells	40	–	–	30	21	14	9	4	0
9–16 cells	5	–	–	–	5	2	2	0	0
Morula	5	–	–	–	–	3	2	2	0
Blastocyst	3	–	–	–	–	–	2	1	0
Expanded blastocyst	4	–	–	–	–	–	–	1	1

[1] Denotes embryos that commenced the process of hatching in the culture system.

of silicone rubber capillary tubing (dimensions: 50 cm × 2·1 mm OD) by Baker and Dziuk (1970) to transport 2–8 cell pig embryos by air from Urbana, Illinois to Montreal before successful transplantation into the uterus of a synchronised recipient. But it is not clear from the report if a further division occurred in this *in vitro* situation in which the tubing was immersed in medium at 24°C. General comments concerning media for culture, attention to pH and osmolarity, the use of equilibrated oil and other such details are, in principle, in line with those mentioned above under techniques for *in vitro* fertilisation trials. But use of a phosphate rather than bicarbonate based buffer has sometimes been favoured to avoid the necessity of the CO_2-enriched gas phase for maintaining pH. When blood serum is included as a component of a culture medium, it is essential that it be heat-treated (i.e. 56°C for 30 min) to denature an ovicidal factor first described by Chang (1949).

An alternative means of culturing embryos outside the tract of the mother that is still much in favour is within the ligated Fallopian tubes of a pseudopregnant, or less commonly oestrous, rabbit. Using this system of *in vivo* culture, cattle, sheep and pig embryos have been supported for up to three days, frequently during long distance transfer experiments such as those involving intercontinental flights. Examples of the use of this technique include the pioneer studies of Averill *et al.* (1955) who reported that sheep embryos could survive for at least five days and develop to the early blastocyst stage in the genital tract of rabbits; and Hunter *et al.* (1962) who used this approach to transport 4–8 cell sheep embryos from Cambridge, UK, to Natal, South Africa. Sreenan *et al.* (1968) also used the pseudopregnant rabbit successfully as a culture system, having first obtained only the limited cleavage *in vitro* of cow eggs mentioned above: they recorded a maximum cleavage stage of 64 blastomeres following storage of an 8-cell egg for four days. Lawson *et al.* (1972) have retransferred cow embryos after periods of storage in the rabbit oviduct of two to four days, and have recorded the birth of a number of viable calves. As far as pig embryos are concerned, Polge *et al.* (1972) have cultured the 1–4 cell stages for two days in rabbit oviducts, and horse embryos have similarly undergone short term culture in the rabbit, in due course yielding a viable foal upon transfer to an appropriate recipient mare (Allen *et al.*, 1976).

Insulated glass tubes have been substituted for the rabbit oviduct in some further studies involving export of embryos, but it is certainly arguable that the conditions of culture are more sensitively controlled in the rabbit than in an *in vitro* system, and may therefore justify the extra costs involved. On the other hand, the possibility of losing a small proportion of the embryos may be increased by use of the rabbit system.

As a footnote, it should be mentioned that human eggs fertilised *in vitro* seem to be able to continue development in culture at least to the morula or early blastocyst stage (Edwards *et al.*, 1970), and some may develop further after retransfer to the uterus (Steptoe and Edwards, 1976) or indeed give rise to a healthy baby (Steptoe and Edwards, 1978). Thus, the notion of there being at least one sensitive stage in the early cleavage of mammalian embryos, as discussed above for certain farm species, may not have a parallel in the primate.

Deep-freezing technology

Techniques for preserving gametes or embryos at very low temperature are discussed briefly at this stage since, in many situations, they might follow logically from manipulations already described such as maturation of oocytes *in vitro*, fertilisation *in vitro*, or culture of embryos to a stage of development compatible with non-surgical transplantation to the uterus. Likewise, in terms of long distance transport or international movement of sperm samples or embryos, packaging in the deep-frozen condition may well be the method of choice. However, as far as export to developing countries is concerned, it must be stressed that either the packaged product needs to be in a form convenient for putting to the desired use, as is the case with deep-frozen ampoules or plastic straws of diluted semen, or appropriate laboratory equipment must be available with which to achieve the necessary conditions of thawing and final preparation of the gametes or embryos. It is also assumed that adequate consideration has been given to the possible risk of transmission of diseases, even with deep-frozen material, and that animal health regulations have been fully complied with in the countries concerned. Detailed reviews on the early methods of deep-freezing mammalian spermatozoa include those of Polge (1953a) and Polge and Rowson (1952a). Less comprehensive information is available from studies on oocytes and embryos, although successes have been achieved after freezing of both these forms in a small number of species (Whittingham, 1976).

Perhaps the dominating principle in studies involving freezing and thawing of cells, be they somatic cells such as red blood cells or gametes, is at all costs to avoid damaging the integrity of the cell during the quite separate steps of cooling and warming. This stricture focuses not only on the cell membrane but also on the cell organelles, and is achieved by inclusion of some form of protective agent in the medium in which the cells are suspended and which penetrates the cells during a period of equilibration. The protective agent is included not only to stabilise the cellular contents, but to prevent the highly damaging effects that would result from intracellular formation of ice. The original work on successful freezing of gametes appears to be that of Rostand (1946, 1949) using amphibian material, but the study of Polge *et al.* (1949) on fowl spermatozoa is more commonly cited. The latter authors noted that inclusion of glycerol in the suspending medium conferred considerable protection on the sperm cells during freezing and thawing, and the subsequent successful use of glycerol in essentially similar procedures for the preservation of bull spermatozoa underlay the enormous expansion of artificial insemination in cattle breeding (Polge, 1953b; Polge and Rowson, 1952a, b, c). Other protective agents have since been used for spermatozoa and in the more recent studies on oocytes and embryos; these will be mentioned where appropriate.

Preservation of spermatozoa

The technology for diluting and deep-freezing bull semen has been described repeatedly (Salisbury and Van Demark, 1961; Maule, 1962; Perry, 1968; Foote, 1974; Gomes, 1977), and does not warrant further detail here as frozen semen is already used extensively in the cattle breeding industry of many countries. Nonetheless, it is important to note that for the freeze-preservation and ultimate use of gametes (and embryos), a number of factors critical to the success of the procedures must be taken into account. These factors include:

(1) Medium of suspension (i.e. diluent) and concentration of cells.
(2) A suitable cryoprotective agent.
(3) Concentration of protective agent in medium.
(4) Time and temperature of equilibration.
(5) Nature of cooling curve (i.e. shape and rate).
(6) Nature of thawing curve.
(7) Use of a specific thawing medium.
(8) Means of removal of protective agent (e.g. dilution or dialysis).

Whilst the temperature for storage of frozen semen used to be that of solid carbon dioxide (−79°C), the advent of convenient supplies of liquid nitrogen together with suitable containers for holding this material has meant that storage at −196°C, either actually in the liquid nitrogen or its vapour, is now the more common practice in developed countries. Even though there may be a slight and slow deterioration in the fertilising ability of spermatozoa held at the latter temperature, deterioration is certainly very significantly reduced compared with that of samples stored at −79°C.

Having chosen a suitable semen diluent or extender (see Chapter IV), a cryoprotective agent such as glycerol can be added to the sperm suspension at concentrations of 4–10% in procedures for freezing bull spermatozoa; 7% glycerol is a frequently selected concentration for yolk-citrate based diluents and 10% with milk based diluents (Gomes, 1977). Such concentrations of glycerol are known to be damaging to spermatozoa of some species, whereas 1–2% glycerol may prove suitable for rams and boars. But spermatozoa of this latter species are peculiarly sensitive to glycerol at ambient temperatures, for glycerol and several other protective agents may in fact damage the acrosomes of boar spermatozoa during the freezing and thawing procedures (Wilmut and Polge, 1977). These remarks notwithstanding, there does seem to be some difference in sensitivity between individual boars to the effects of glycerol treatments.

The time and temperature during which equilibration of the sperm suspension with the added glycerol takes place varies according to the species and laboratory in question, as does the shape of the cooling curve (see details in Foote, 1974). The actual packaging of sperm suspensions to be frozen is usually in the form of:

(1) Glass ampoules containing approximately 1·0 ml of sperm suspension,
(2) Polyvinyl chloride (PVC) straws, containing 0·25–0·5 ml, or
(3) Pellets containing approximately 0·1 ml.

Use of glass ampoules developed as the conventional method for storing bull semen, but in recent years the various forms of storage in straws have won increasing favour, not least for their ease of use at the time of insemination. Pelletting has been employed for freezing of boar spermatozoa, appropriate suspensions of cells with protective agent being added dropwise to rounded cavities drilled in a block of solid carbon dioxide before final storage of the pellets at the temperature of liquid nitrogen (Polge *et al.*, 1970). Thawing of pellets is less convenient than for ampoules or straws, but in the case of boar spermatozoa this step may be achieved by adding the pellets to a biological fluid such as sperm-free seminal plasma—the protein or phospholipid components being thought to confer some protection during this process. Straws are generally thawed in air or at body temperature and ampoules in water at 5°C or at 40°C (Gomes, 1977). A fact of considerable practical importance is that frozen-thawed spermatozoa deteriorate rapidly outside the female genital tract; hence thawing should be undertaken only shortly before the time of insemination.

Preservation of oocytes

Since it is now feasible to store the spermatozoa of a number of species by deep-freezing, an analogous methodology for storing oocytes should permit particular genetic combinations to be assembled in the future, either by an *in vivo* or perhaps *in vitro* approach. Successful treatments for the preservation of mouse oocytes have been described by Whittingham (1977), the viability of the frozen-thawed gametes being evaluated by both *in vivo* and *in vitro* studies of fertilisation. It remains for these techniques to be extended to the oocytes of farm livestock, and thus provide the basis for establishing banks of suitable female gametes.

In brief, the methodology used by Whittingham (1977) involved the following steps. Oocytes were obtained from the oviducts of gonadotrophin treated mice shortly after ovulation and used either within the investment of cumulus cells or denuded. They were transferred to glass test tubes or ampoules in a small volume of physiological medium, placed in an ice bath at 0°C, and equilibrated with 1·5 M dimethyl sulphoxide (the cryoprotective agent) 15 minutes later. The temperature was taken down to −80°C, and the samples then transferred directly to liquid nitrogen for storage over a range of times lasting up to three months. Full details of the cooling and warming rates are given in the original paper, the dimethyl sulphoxide being diluted out during the recovery phase at 0°C.

Examination of the morphology of the oocytes after these procedures revealed that 66–72% were apparently normal. Tests of their competence involved exposure to *in vitro* and *in vivo* fertilisation, and showed that there were strain differences in the susceptibility of the gametes to the treatments, and that oocytes achieved a higher incidence of fertilisation when transferred to the oviducts of mated recipients rather than when exposed to sperm suspensions *in vitro*. But the highest rates of foetal survival (45%), which clearly involved a selection of material prior to transfer, were obtained after *in vitro* fertilisation and subsequent transplantation at the 2-cell or blastocyst stage to pseudopregnant recipients (Table VII.7).

As yet, the numbers of oocytes involved in this kind of study are relatively small and definitive conclusions as to the most suitable procedures cannot be drawn. But the success of these experiments in mice suggests that techniques may be developed for the ovulated (i.e. secondary) oocytes of farm species, even though there may well prove to be significant differences in the sensitivity of gametes to freezing and thawing procedures between breeds and strains of animal. Perhaps the principal problem, though, will be to obtain sufficient numbers of oocytes for such studies, and in this situation techniques of preliminary maturation *in vitro* of ovarian oocytes will almost certainly need to be called upon, especially for bovine material.

Preservation of embryos

From many points of view, deep-freeze preservation of embryos rather than gametes might be considered to offer the greater attraction to animal production systems of the future; and if reliable freezing and thawing procedures could be developed, the only remaining manipulation of the embryo would be its transplantation into the recipient. The latter statement is perhaps an oversimplification since it may be found that very young embryos (i.e. the early cleavage stages) of some species are more amenable to freezing procedures than later ones, in which case a post-thawing phase of culture would probably become necessary before transplantation to the uterus. Moreover, continued development of the embryo during a relatively brief post-thawing phase could be introduced as a useful assessment of viability before committing the material to transplantation. Quite apart from successes obtained after freezing and thawing of mouse, rabbit and rat embryos (Table VII.8), cow embryos have also been shown to survive the necessary procedures and to give rise to normal calves following uterine transplantation (Wilmut and Rowson, 1973; Betteridge, 1977; Willadsen *et al.*, 1978). Sheep embryos have also been successfully frozen (Willadsen *et al.*, 1974), but studies so far suggest that pig embryos are refractory to low temperature manipulation and with currently available methods, 8-cell

Table VII.7. The development of frozen-thawed and control mouse oocytes fertilised *in vitro* and transferred to pseudopregnant recipients. (From Whittingham, 1977.)

		Embryos transferred		Number of			
Source and treatment of oocytes	Stage of development at transfer	Total no. (No. of recipients)	To females becoming pregnant (no.)	Implantation sites	Foetuses at 14 days of gestation	Live young at birth	% foetuses and live young
MF1							
Frozen	2-cell	104 (10)	63 (5)	7	6	–	15·7
			20 (2)	–	–	7	
Control	2-cell	35 (3)	23 (2)	8	8	–	34·3
			12 (1)	–	–	4	
F_1 (C57BL × CBA)							
Frozen	2-cell	87 (7)	49 (4)	29	17	–	38·4
			24 (2)	–	–	11	
	Early blastocyst	33 (3)	21 (2)	16	5	–	36·4
			12 (1)	–	–	7	
Control	2-cell	20 (2)	20 (2)	9	9	–	45·0
	Early blastocyst	12 (1)	12 (1)	7	3	–	25·0

Table VII. 8. Species in which successful storage of embryos has been achieved at the temperature of liquid nitrogen (−196°C.)

Species	Stages developing to birth after freezing, thawing and transplantation
Mouse	1-cell
	2- cell
	8-cell
	Blastocyst
Rat	2-cell
	4-cell
	8-cell
Rabbit	4-cell
	8-cell
	Morula
Sheep	Morula
	Blastocyst
Cow	Blastocyst

pig embryos appear not to survive cooling to temperatures below 15°C (Polge *et al*., 1974).

Details of the specific techniques for freezing the embryos of farm animals are given in the papers just cited. But general descriptions of the freezing procedure for mammalian embryos have been outlined by Whittingham (1976), Willadsen *et al*. (1976) and Leibo and Mazur (1978), stress being given to the fact that embryos are particularly sensitive to the rates of freezing and thawing: relatively slow rates of cooling (0·2–2·0°C per minute) and of warming (4–25°C per minute) are usually prescribed. A medium that has been extensively used prior to freezing is a modified Dulbecco's phosphate buffered solution whereas a modified Krebs-Ringer bicarbonate medium has been used for culturing the embryos after recovery from freezing and thawing (Whittingham, 1976). A procedure for deep-freezing and subsequent thawing of mouse embryos includes the following principal steps, and again employs the low molecular weight protective compound, dimethyl sulphoxide.

(1) The embryos are pipetted in a small volume of the modified Dulbecco's medium into glass ampoules, and placed in an ice bath at 0°C for 10 minutes.

(2) A further aliquot of the same medium is added, but also containing 3·0 M dimethyl sulphoxide, giving a final concen-

tration of 1·5 M. The ampoules are left for another 10 minutes for equilibration of the contents.

(3) The ampoules are next transferred to a "seeding" bath at −5°C where each is seeded with an ice crystal 2 minutes later to prevent supercooling. The ampoules are then sealed with an oxygen/butane flame, returned to the seeding bath for 5–10 minutes before transfer to the cooling bath at −6°C. Cooling proceeds to −80°C at approximately 0·5°C min^{-1}, after which the ampoules are immersed in liquid nitrogen and stored in a similarly based refrigerator (−196°C).

(4) Thawing is achieved by warming the ampoules at 6 to 8°C min^{-1} to 0°C. The ampoule seal is broken, and three separate volumes of medium are added at 1 min intervals to dilute out the hyperosmotic protective solution. The contents of the ampoule are withdrawn with plastic pipettes and added to embryological watch glasses, where the embryos are further rinsed with the modified Dulbecco to remove the dimethyl sulphoxide before culture in the modified Krebs-Ringer. Embryos selected on the basis of their morphology and development in culture are transplanted to suitable recipients.

So far, only in the mouse have all the preimplantation stages been shown to survive freezing and thawing (Whittingham, 1976). Owing to the limited supply of embryos available from the domestic species, experiments have not yet ranged widely enough for specific comment on the optimum developmental stage at which freezing could be imposed. But the procedure recommended by the Cambridge group for sheep and cow embryos is similar in principle to that used for the mouse work. Sheep morulae aged five days and cow blastocysts aged seven days have been demonstrated to tolerate these freezing and thawing procedures (Polge, 1977b; Willadsen, 1977; Polge and Willadsen, 1978; Willadsen *et al.*, 1978), and to yield fully viable offspring upon transfer to suitable recipients.

References

Allen, W. R., Stewart, F., Trounson, A. O., Tischner, M. and Bielanski, W. (1976). Viability of horse embryos after storage and long-distance transport in the rabbit. *J. Reprod. Fert.* **47**, 387–390.

Austin, C. R. (1951). Observations on the penetration of the sperm into the mammalian egg. *Aust. J. Sci. Res. (B)* **4**, 581–596.

Austin, C. R. (1961a). "The Mammalian Egg". Blackwell Scientific Publications, Oxford.

Austin, C. R. (1961b). Fertilisation of mammalian eggs *in vitro*. *Int. Rev. Cytol.* **12**, 337–359.

Averill, R. L. W., Adams, C. E. and Rowson, L. E. A. (1955). Transfer of mammalian ova between species. *Nature, London* **176**, 167.

Baker, R. D. and Dziuk, P. J. (1970). Aerial transport of fertilised pig ova. *Canad. J. Anim. Sci.* **50**, 215–216.

Baker, T. G. and Hunter, R. H. F. (1978). Interrelationships between the oocyte and somatic cells within the Graafian follicle of mammals. *Annls Biol. anim. Biochim. Biophys.* **18**, 419–426.

Baker, T. G. and Neal, P. (1969). The effects of X-irradiation on mammalian oocytes in organ culture. *Biophysika* **6**, 39–45.

Baker, T. G., Hunter, R. H. F. and Neal, P. (1975). Studies on the maintenance of porcine Graafian follicles in organ culture. *Experientia* **31**, 133–135.

Barros, C. and Austin, C. R. (1967). *In vitro* fertilisation and the sperm acrosome reaction in the hamster. *J. exp. Zool.* **166**, 317–324.

Bavister, B. D. (1969). Environmental factors important for *in vitro* fertilisation in the hamster. *J. Reprod. Fert.* **18**, 544–545.

Berry, R. O. and Savery, H. P. (1958). A cytological study of the maturation process of the ovum of the ewe during normal and induced ovulation. *In* "Proc. 3rd Symposium on Reproduction and Fertility", Colorado State University (F. X. Gassner, ed.), pp. 75–82. Pergamon Press, London and New York.

Betteridge, K. J. (1977). Editor. Embryo transfer in farm animals. *Canada Department of Agriculture Monograph* **16**.

Biggers, J. D., Whittingham, D. G. and Donahue, R. P. (1967). The pattern of energy metabolism in the mouse oocyte and zygote. *Proc. Nat. Acad. Sci., Washington* **58**, 560–567.

Black, D. L. and Hansel, W. (1972). Endocrine factors affecting reproduction in the bovine female. *Univ. of Mass. Agric. Exp. Sta. Res. Bull.* No. **596**.

Brinster, R. L. (1963). A method for *in vitro* cultivation of mouse ova from two-cell to blastocyst. *Expl Cell Res.* **32**, 205–208.

Brinster, R. L. (1965). Studies of the development of mouse embryos *in vitro*: energy metabolism. *In* "Ciba Foundation Symposium on Preimplantation Stages of Pregnancy" (G. E. W. Wolstenholme and M. O'Connor, eds), pp. 60–74. Churchill, London.

Brinster, R. L. (1973). Nutrition and metabolism of the ovum, zygote, and blastocyst. *In* "Handbook of Physiology", Section 7, Endocrinology II, (R. O. Greep and E. B. Astwood, eds), pp. 165–185. American Physiological Society, Washington.

Chang, M. C. (1949). Effects of heterologous sera on fertilised rabbit ova. *J. Gen. Physiol.* **32**, 291–300.

Chang, M. C. (1951). Fertilising capacity of spermatozoa deposited into the Fallopian tubes. *Nature, London* **168**, 697–698.

Chang, M. C. (1955). Development of fertilising capacity of rabbit spermatozoa in the uterus. *Nature, London* **175**, 1036–1037.

Chang, M. C. (1959). Fertilisation of rabbit ova *in vitro*. *Nature, London* **184**, 466–467.

Chang, M. C. (1968). *In vitro* fertilisation of mammalian eggs. *J, Anim. Sci.* **27**, Suppl. 1, 15–22.

Daniel, J. C. and Olson, J. D. (1968). Amino acid requirements for cleavage of the rabbit ovum. *J. Reprod. Fert.* **15**, 453–455.

Dauzier, L., Thibault, C. and Wintenberger, S. (1954). La fécondation *in vitro* de l'oeuf de la lapine. *C.r. Acad. Sci., Paris* **238**, 844–845.

Davis, D. L. and Day, B. N. (1978). Cleavage and blastocyst formation by pig eggs *in vitro*. *J. Anim. Sci.* **46**, 1043–1053.

Desjardins, C. and Hafs, H. D. (1969). Maturation of bovine female genitalia from birth through puberty. *J. Anim. Sci.* **28**, 502–507.

Dziuk, P. J. (1965). Timing of maturation and fertilisation of the sheep egg. *Anat. Rec.* **153**, 211–224.

Edwards, R. G. (1965). Maturation *in vitro* of mouse, sheep, cow, pig, rhesus monkey and human ovarian oocytes. *Nature, London* **208**, 349–351.

Edwards, R. G., Bavister, B. D. and Steptoe, P. C. (1969). Early stages of fertilisation *in vitro* of human oocytes matured *in vitro*. *Nature, London* **221**, 632–635.

Edwards, R. G., Steptoe, P. C. and Purdy, J. M. (1970). Fertilisation and cleavage *in vitro* of preovulatory human oocytes. *Nature, London* **227**, 1307–1309.

Erickson, B. H. (1966a). Development and senescence of the postnatal bovine ovary. *J. Anim. Sci.* **25**, 800–805.

Erickson, B. H. (1966b). Development and radio-response of the prenatal bovine ovary. *J. Reprod. Fert.* **11**, 97–105.

Foote, R. H. (1974). Artificial insemination. *In* "Reproduction in Farm Animals", 3rd edition, (E. S. E. Hafez, ed.), pp. 409–431. Lea and Febiger, Philadelphia.

Foote, R. H. and Onuma, H. (1970). Superovulation, ovum collection, culture and transfer. A review. *J. Dairy Sci.* **53**, 1681–1692.

Foote, W. D. and Thibault, C. (1969). Recherches expérimentales sur la maturation *in vitro* des ovocytes de truie et de veau. *Annls Biol. anim. Biochim. Biophys.* **9**, 329–349.

Gomes, W. R. (1977). Artificial insemination. *In* "Reproduction in Domestic Animals", 3rd edition, (H. H. Cole and P. T. Cupps, eds), pp. 257–284. Academic Press, New York and London.

Gwatkin, R. B. L. (1977). "Fertilization Mechanisms in Man and Mammals". 161pp. Plenum Press, New York and London.

Harms, von E. and Smidt, D. (1970). *In vitro* fertilisation of follicular and tubal eggs of swine. *Tieraerztl. Wochenschr.* **83**, 269–275.

Hunter, G. L., Bishop, G. P., Adams, C. E. and Rowson, L. E. (1962). Successful long-distance aerial transport of fertilized sheep ova. *J. Reprod. Fert.* **3**, 33–40.

Hunter, R. H. F. (1976). Reproductive physiology in the bovine female: a review. *In* "Beef Cattle Production in Developing Countries" (A. J. Smith, ed.), pp. 79–104. University of Edinburgh.

Hunter, R. H. F. and Polge, C. (1966). Maturation of follicular oocytes in the pig after injection of human chorionic gonadotrophin. *J. Reprod. Fert.* **12**, 525–531.

Hunter, R. H. F., Cook, B. and Baker, T. G. (1976). Dissociation of response to injected gonadotrophin between the Graafian follicle and oocyte in pigs. *Nature, London* **260**, 156–158.

Hunter, R. H. F., Holtz, W. and Herrmann, H. (1978). Stabilising rôle of epididymal plasma in relation to the capacitation time of boar spermatozoa. *Anim. Reprod. Sci.* **1**, 161–166.

Hunter, R. H. F., Lawson, R. A. S. and Rowson, L. E. A. (1972). Maturation, transplantation and fertilisation of ovarian oocytes in cattle. *J. Reprod. Fert.* **30**, 325–328.

Iritani, A., Niwa, K. and Imai, H. (1978). Sperm penetration *in vitro* of pig follicular oocytes matured in culture. *J. Reprod. Fert.* **54**, 379–383.

Kane, M. T. and Foote, R. H. (1970). Fractionated serum dialysate and synthetic media for culturing 2- and 4-cell rabbit embryos. *Biol. Reprod.* **2**, 356–362.

Lawson, R. A. S., Rowson, L. E. A. and Adams, C. E. (1972). The development of cow eggs in the rabbit oviduct and their viability after re-transfer to heifers. *J. Reprod. Fert.* **28**, 313–315.

Leibo, S. P. and Mazur, P. (1978). Methods for the preservation of mammalian embryos by freezing. *In* "Methods in Mammalian Reproduction" (J. C. Daniel, Jr., ed.), pp. 179–201. Academic Press, New York and London.

Leman, A. D. and Dziuk, P. J. (1971). Fertilisation and development of pig follicular oocytes. *J. Reprod. Fert.* **26**, 387–389.

Lindner, G. M. and Wright, R. W. (1978). Morphological and quantitative aspects of the development of swine embryos *in vitro*. *J. Anim. Sci.* **46**, 711–718.

McGaughey, R. W. and Polge, C. (1971). Cytogenetic analysis of pig oocytes matured *in vitro*. *J. exp. Zool.* **176**, 383–396.

McLaren, A. and Biggers, J. D. (1958). Successful development and birth of mice cultivated *in vitro* as early embryos. *Nature, London* **182**, 877–878.

Marion, G. B., Gier, H. T. and Choudary, J. B. (1968). Micromorphology of the bovine ovarian follicular system. *J. Anim. Sci.* **27**, 451–465.

Maule, J. P. (1962). The semen of animals and artificial insemination. *Commonwealth Bureau of Animal Breeding and Genetics, Edinburgh. Technical Communication* No. **15**.

Menezo, Y., Gérard, M. and Thibault, C. (1976). Attempts at *in vitro* fertilisation in calf. *In* "Egg Transfer in Cattle" (L. E. A. Rowson, ed.), pp. 189–196. Proceedings of an EEC Agricultural Research Seminar.

Moor, R. M. (1978). Role of steroids in the maturation of ovine oocytes. *Annls Biol. anim. Biochim. Biophys.* **18**, 477–482.

Moor, R. M. and Cragle, R. G. (1971). The sheep egg: enzymatic removal of the zona pellucida and culture of eggs *in vitro*. *J. Reprod. Fert.* **27**, 401–409.

Moor, R. M. and Trounson, A. O. (1977). Hormonal and follicular factors affecting maturation of sheep oocytes *in vitro* and their subsequent developmental capacity. *J. Reprod. Fert.* **49**, 101–109.

New, D. A. T. (1966). "The Culture of Vertebrate Embryos". Logos Press, London.

Newcomb, R., Christie, W. B. and Rowson, L. E. A. (1978). Birth of calves after *in vivo* fertilisation of oocytes removed from follicles and matured *in vitro*. *Vet. Rec.* **102**, 461–462.

Noyes, R. W. (1952). Fertilisation of follicular ova. *Fert. Steril.* **3**, 1–12.

Onuma, H., Maurer, R. R. and Foote, R. H. (1968). *In vitro* culture of rabbit ova from early cleavage stages to the blastocyst stage. *J. Reprod. Fert.* **16**, 491–493.

Perry, E. J. (1968). Editor. "The Artificial Insemination of Farm Animals", 4th edition. Rutgers University Press, New Jersey, U.S.A.

Pincus, G. and Enzmann, E. V. (1935). The comparative behaviour of mammalian eggs *in vivo* and *in vitro*. I. The activation of ovarian eggs. *J. exp. Med.* **62**, 665–675.

Polge, C. (1953a). The preservation of spermatozoa at low temperatures. *In* "Ciba Foundation Symposium on Mammalian Germ Cells" (G. E. W. Wolstenholme, ed.), pp. 108–118. J. and A. Churchill Ltd., London.

Polge, C. (1953b). The storage of bull semen at low temperatures. *Vet. Rec.* **65**, 557–559.

Polge, C. (1957). Low-temperature storage of mammalian spermatozoa. *Proc. Roy. Soc. B*, **147**, 498–508.

Polge, C. (1977a). Short-term maintenance and culture of embryos. *In* "Embryo Transfer in Farm Animals" (K. J. Betteridge, ed.), p. 44. Canada Department of Agriculture Monograph No. 16.

Polge, C. (1977b). The freezing of mammalian embryos: perspectives and possibilities. *In* "The Freezing of Mammalian Embryos", pp. 3–13. Ciba Foundation Symposium 52 (new series). Elsevier, Excerpta Medica, North Holland, Amsterdam and Oxford.

Polge, C. and Willadsen, S. M. (1978). Freezing eggs and embryos of farm animals. *Cryobiology* **15**, 370–373.

Polge, C. and Frederick, G. L. (1968). Culture and storage of fertilised pig eggs. *Proc. 6th Int. Congr. Anim. Reprod. Artif. Insem., Paris*, **1**, (Abstract).

Polge, C. and Rowson, L. E. A. (1952a). Long-term storage of bull semen frozen at very low temperatures (−79°C). *Proc. 2nd Int. Congr. Anim. Reprod. Artif. Insem., Copenhagen*, **3**, 90–98.

Polge, C. and Rowson, L. E. A. (1952b). Fertilising capacity of bull spermatozoa after freezing at −79°C. *Nature, London* **169**, 626–627.

Polge, C. and Rowson, L. E. A. (1952c). Results with bull semen stored at −79°C. *Vet. Rec.* **64**, 851.

Polge, C., Adams, C. E. and Baker, R. D. (1972). Development and survival of pig embryos in the rabbit oviduct. *Proc. 7th Int. Congr. Anim. Reprod. Artif. Insem., Munich*, **1**, 513–517.

Polge, C., Salamon, S. and Wilmut, I. (1970). Fertilising capacity of frozen boar semen following surgical insemination. *Vet. Rec.* **87**, 424–428.

Polge, C., Smith, A. U. and Parkes, A. S. (1949). Revival of spermatozoa after vitrification and dehydration at low temperatures. *Nature, London* **164**, 666.

Polge, C., Wilmut, I. and Rowson, L. E. A. (1974). The low temperature preservation of cow, sheep and pig embryos. *Cryobiology* **11**, 560 (Abstract).

Pope, C. E. and Day, B. N. (1970). Cleavage and survival of swine ova cultured *in vitro*. *J. Anim. Sci.* **31**, 1035 (Abstract).

Pope, C. E. and Day, B. N. (1977). Transfer of preimplantation pig embryos following *in vitro* culture for 24 or 48 hours. *J. Anim. Sci.* **44**, 1036–1040.

Rajakoski, E. (1960). The ovarian follicular system in sexually mature heifers with special reference to seasonal, cyclical and left-right variations. *Acta Endocr., Suppl.* **52**, 1–68.

Renard, J. P. and Du Mesnil du Buisson, F. (1976). Culture and storage of cow embryos. *Proc. 8th Int. Congr. Anim. Reprod. Artif. Insem., Krakow*, **3**, 309–312.

Rostand, J. (1946). Glycérine et résistance du sperme aux basses températures. *C.r. Acad. Sci.* **222**, 1524–1525.

Rostand, J. (1949). Conservation à basse température des ovules coelomiques de grenouille rousse (*Rana temporaria*). *C.r. Soc. Biol.* **143**, 17–18.

Rowson, L. E. A., Lawson, R. A. S. and Moor, R. M. (1971). Production of twins in cattle by egg transfer. *J. Reprod. Fert.* **25**, 261–268.

Salisbury, G. W. and Van Demark, N. L. (1961). "Physiology of Reproduction and Artificial Insemination of Cattle". W. H. Freeman and Co., San Francisco and London.

Seidel, G. E. (1977). Short-term maintenance and culture of embryos. *In* "Embryo Transfer in Farm Animals" (K. J. Betteridge, ed.), pp. 20–24. Canada Department of Agriculture Monograph No. 16.

Seitz, H. M., Brackett, B. G. and Mastroianni, L. (1970). *In vitro* fertilisation of ovulated rabbit ova recovered from the ovary. *Biol. Reprod.* **2**, 262–267.

Shenk, S. L. (1878). Das Säugethierei künstlich befruchtet ausserhalb der Mutterthieres. *Mitt. Embryol. Inst. Univ. Wien*, **1**, 107–118.

Spalding, J. F., Berry, R. O. and Moffit, J. G. (1955). The maturation process of the ovum of swine during normal and induced ovulations. *J. Anim. Sci.* **14**, 609–620.

Sreenan, J. (1970). *In vitro* maturation and attempted fertilisation of cattle follicular oocytes. *J. agric. Sci.* **75**, 393–396.

Sreenan, J., Scanlon, P. and Gordon, I. (1968). Culture of fertilized cattle eggs. *J. agric. Sci.* **70**, 183–185.

Steptoe, P. C. and Edwards, R. G. (1976). Reimplantation of a human embryo with subsequent tubal pregnancy. *Lancet* **i**, 880–882.

Steptoe, P. C. and Edwards, R. G. (1978). Birth after the reimplantation of a human embryo. *Lancet* **ii**, 366.

Tervit. H. R. (1973). Culture and transfer of sheep and cattle ova. Ph.D. thesis, University of Cambridge.

Tervit, H. R. and Rowson, L. E. A. (1974). Birth of lambs after culture of sheep ova *in vitro* for up to 6 days. *J. Reprod. Fert.* **38**, 177–179.

Tervit, H. R., Whittingham, D. G. and Rowson, L. E. A. (1972). Successful culture *in vitro* of sheep and cattle ova. *J. Reprod. Fert.* **30**, 493–497.

Thibault, C. (1966). La culture *in vitro* de l'oeuf de vache. *Annls Biol. anim. Biochim. Biophys.* **6**, 159–164.

Thibault, C. (1969). *In vitro* fertilisation of the mammalian egg. *In* "Fertilisation. Comparative Morphology, Biochemistry and Immunology" (C. B. Metz and A. Monroy, eds), Vol. 2, pp. 405–435. Academic Press, London and New York.

Thibault, C. (1977). Are follicular maturation and oocyte maturation independent processes? *J. Reprod. Fert.* **51**, 1–15.

Thibault, C. and Dauzier, L. (1960). 'Fertilisines' et fécondation *in vitro* de l'oeuf de lapine. *C.r. Acad. Sci., Paris* **250**, 1358–1359.

Thibault, C. and Dauzier, L. (1961). Analyse des conditions de la fécondation *in vitro* de l'oeuf de la lapine. *Annls Biol. anim. Biochim. Biophys.* **1**, 277–294.

Thibault, C. and Gérard, M. (1970). Facteur cytoplasmique nécessaire à la formation du pronucleus mâle dans l'ovocyte de lapine. *C.r. Acad. Sci.* **270**, 2025–2026.

Thibault, C. and Gérard, M. (1973). Cytoplasmic and nuclear maturation of rabbit oocytes *in vitro*. *Annls Biol. anim. Biochim. Biophys.* **13**, (hors–série), 145–155.

Thibault, C., Gérard, M. and Menezo, Y. (1975). Acquisition par l'ovocyte de lapine et de veau du facteur de décondensation du noyeau du spermatozoide fécondant (MPGF). *Annls Biol. anim. Biochim. Biophys.* **15**, 705–714.

Thibault, C., Gérard, M. and Menezo, Y. (1976). Attempts to mature calf oocytes in culture. *In* "Egg Transfer in Cattle" (L. E. A. Rowson, ed.), pp. 175–184. Proceedings of an EEC Agricultural Research Seminar.

Trounson, A. O., Willadsen, S. M. and Rowson, L. E. A. (1976). The influence of in-vitro culture and cooling on the survival and development of cow embryos. *J. Reprod. Fert.* **47**, 367–370.

Trounson, A. O., Willadsen, S. M. and Rowson, L. E. A. (1977). Fertilisation and development capability of bovine follicular oocytes matured *in vitro* and *in vivo* and transferred to the oviducts of rabbits and cows. *J. Reprod. Fert.* **51**, 321–327.

Warnes, G. M., Moor, R. M. and Johnson, M. H. (1977). Changes in protein synthesis during maturation of sheep oocytes *in vivo* and *in vitro*. *J. Reprod. Fert.* **49**, 331–335.

Whitten, W. K. (1956). Culture of tubal mouse ova. *Nature, London* **177**, 96.

Whitten, W. K. (1957). Culture of tubal ova. *Nature, London*, **179**, 1081–1082.

Whittingham, D. G. (1976). General aspects of egg culture and preservation. *In* "Egg Transfer in Cattle" (L. E. A. Rowson, ed.), pp. 101–113. Proceedings of an EEC Agricultural Research Seminar.

Whittingham, D. G. (1977). Fertilisation *in vitro* and development to term of unfertilised mouse oocytes previously stored at $-196°C$. *J. Reprod. Fert.* **49**, 89–94.

Whittingham, D. G. and Biggers, J. D. (1967). Fallopian tube and early cleavage in the mouse. *Nature, London* **213**, 942–943.

Willadsen, S. M. (1977). Factors affecting the survival of sheep embryos during deep-freezing and thawing. *In* "The Freezing of Mammalian Embryos", pp. 175–189. Ciba Foundation Symposium 52, (new series). Elsevier, Excerpta Medica, North Holland, Amsterdam and Oxford.

Willadsen, S. M., Polge, C., Rowson, L. E. A. and Moor, R. M. (1974). Preservation of sheep embryos in liquid nitrogen. *Cryobiology* **11**, 560 (Abstract).

Willadsen, S., Trounson, A. O., Polge, C., Rowson, L. E. A. and Newcomb, R. (1976). Low temperature preservation of cow eggs. *In* "Egg Transfer in Cattle" (L. E. A. Rowson, ed.), pp. 117–124. Proceedings of an EEC Research Seminar.

Willadsen, S., Polge, C. and Rowson, L. E. A. (1978). The viability of deep-frozen cow embryos. *J. Reprod. Fert.* **52**, 391–393.

Wilmut, I. and Polge, C. (1977). The low temperature preservation of boar spermatozoa. 2. *Cryobiology* **14**, 479–482.

Wilmut, I. and Rowson, L. E. A. (1973). Experiments on the low-temperature preservation of cow embryos. *Vet. Rec.* **92**, 686–690.

Wintenberger, S., Dauzier, L. and Thibault, C. (1953). Le développement *in vitro* de l'oeuf de la brebis et de celui de la chèvre. *C.r. Séanc. Soc. Biol.* **147**, 1971–1973.

Wright, R. W. (1977). Successful culture *in vitro* of swine embryos to the blastocyst stage. *J. Anim. Sci.* **44**, 854–858.

Wright, R. W., Anderson, G. B., Cupps, P. T. and Drost, M. (1976). Successful culture *in vitro* of bovine embryos to the blastocyst stage. *Biol. Reprod.* **14**, 157–162.

Yanagimachi, R. and Chang, M. C. (1963). Fertilisation of hamster eggs *in vitro*. *Nature, London* **200**, 281–282.

Yanagimachi, R. and Chang, M. C. (1964). *In vitro* fertilisation of golden hamster ova. *J. exp. Zool.* **156**, 361–376.

Chapter VIII

Transplantation of Embryos

Introduction

Although the transfer of preimplantation embryos may feature in the news headlines these days largely as a sequel to procedures of *in vitro* fertilisation of human eggs, or as a critical step in manipulations that may collectively be termed genetic engineering, the technique has a long and respectable scientific pedigree. Moreover, its use has contributed substantially to our knowledge of basic reproductive processes, and it will undoubtedly continue to be employed as a research tool in the foreseeable future.

In essence, the technique involves recovery of preimplantation embryos from the reproductive tract of a donor animal by perfusion with a suitable physiological medium and their deposition into the tract of a recipient with some form of glass pipette (Fig. VIII.1). Embryos have been removed at early cleavage stages from the Fallopian tubes and transferred to a similar site in the recipient or to the lumen of the uterus. More advanced stages of development such as blastocysts or even elongating conceptuses have also been transplanted successfully from one uterus to another.

The first recorded transfer of mammalian eggs was in the late nineteenth century, when Heape performed the manipulation between previously mated rabbits and noted that two of the offspring resembled the genetic mother, or embryo donor, rather than the foster mother or embryo recipient (Heape, 1891). Since that pioneering work, the technique appears to have remained dormant until its use by Pincus in the 1930s and its more extensive application by Hammond's research group in the late 1940s. Studies involving the technique of egg or embryo transplantation increased in number very considerably during the subsequent 25 years (see Adams and Abbott, 1971), and were used to examine or evaluate fundamental aspects of reproductive physiology such as the success of *in vitro* fertilisation procedures, as well as to contribute to animal breeding and selection schemes. The establish-

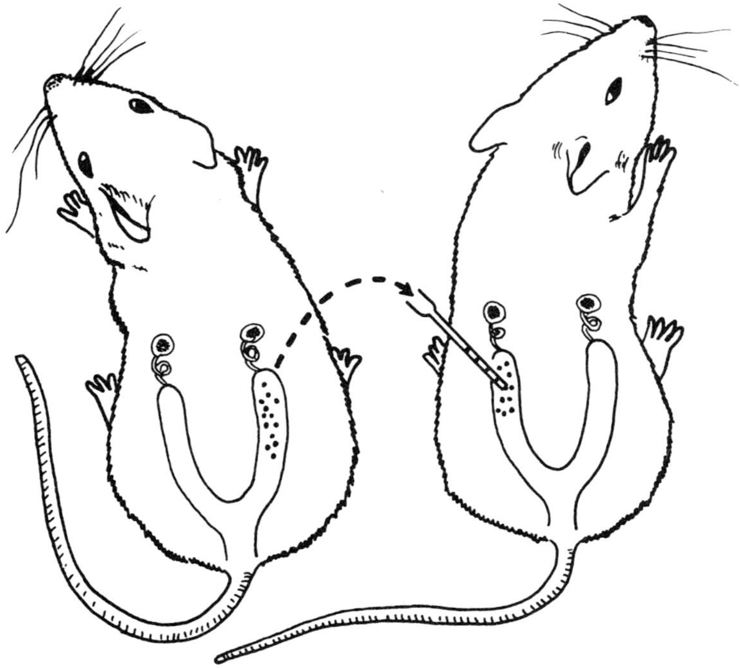

Fig. VIII.1. To illustrate some principal features underlying transplantation of preimplantation embryos from the reproductive tract of a donor animal to that of a recipient. Whilst a uterine site of recovery and transfer is common in the cow, transfer from one Fallopian tube to another has been practised extensively in sheep and pigs. (Redrawn after McLaren and Kerr, 1970.)

ment of a number of commercial embryo transfer units has been a notable feature in this field during the last ten years.

Underlying biological flexibility

In contrast to the more conventional situation of limited tolerance after tissue transplantation in mammals, transfer of embryos between donor and recipient animals takes advantage of two very significant physiological freedoms. First, the lumen of the Fallopian tube and uterus represents an immunologically privileged site to the extent that embryonic and placental tissues bearing foreign antigens are not normally rejected within a species. This, of course, is scarcely surprising in view of the paternal contribution of "foreign" antigens to an embryo, but it is the wider implication of the situation that is exploited in the technique of transplantation: both paternal and maternal components of the embryo are foreign to the recipient or foster mother and yet they are usually tolerated. Second, the mammalian embryo exists for a variable period in a free-living state in the lumen of the uterus before the process of implantation or attachment is initiated. In farm animals and other ungulates, this preimplantation period is relatively long (see Chapter VI), enabling blastocysts at quite advanced stages of development (e.g. as late as 12 or 13 days in the sheep, Moor and Rowson, 1964; and up to 16 days in cattle, Betteridge, 1977) to be recovered and transferred to the uterus of the recipient. This application of blastocyst transfer confers the added advantage that a morphological selection of embryos can be made at the time of transfer: those that appear retarded or show signs of degeneration are rejected, this procedure thereby raising the chances of a successful establishment of pregnancy.

As already inferred from Heape's study, there should be no conspicuous influence of the foster mother on the genetic composition of the foetus. But there may be a considerable influence of the foster mother on the tolerance to disease in the offspring, and an influence on the rate of prenatal growth and birth weight has been clearly documented (Venge, 1950; Hunter, 1956). Another non-genetic effect of the foster mother has been a change in the number of lumbar vertebrae in mouse foetuses developing in the uterus after transfer (McLaren and Michie, 1958).

Physiological and logistical requirements

Embryo transplantation has been accomplished successfully in most of the common species of laboratory animal (see Austin, 1961; Chang and Pickworth, 1969), in primates (Kraemer *et al.*, 1976; Marston *et al.*, 1977), and among the larger farm species (Willett, 1953; Dziuk, 1969; Foote and Onuma, 1970). Although embryo transfer is currently being applied quite extensively to the major farm species, as well as to certain species of wild animal, the largest agricultural application is almost certainly in the cattle breeding industry. Accordingly, the emphasis in this chapter will be placed on the physiological and logistical requirements for successful transfer between donor and recipient cattle, although the underlying principles apply to most mammals examined so far. A detailed review of the subject has been produced under the editorship of Betteridge (1977) and concise statements by Dziuk (1975) and Newcomb (1976). An economic evaluation of embryo transfer in cattle breeding in Britain has been published by Wilmut and Hume (1978).

An adequate supply of embryos

Bearing in mind that the technique of embryo transplantation is applied as a means of exploiting particular genotypes, a first objective should clearly be to exceed the spontaneous rate of reproduction of the selected dam. In other words, instead of a yield of one calf every nine months or a year from a given mating, the objective should be to obtain a minimum of two or more calves per year from the chosen donor (Fig. VIII.2). The number of embryos available is usually increased by the procedure of superovulation, this requiring an injection of approximately 2000–3000 iu PMSG given four days or so before breeding the dam (see Chapter III). The PMSG may be given at the beginning of the follicular phase during a spontaneous oestrous cycle or shortly before artificial regression of the corpus luteum by means of a prostaglandin injection. The procedure seeks to obtain a response of the order of 6–8 ovulations, but in most laboratories the response has been found to be unpredictable for individual animals, even when body weight of the animal and characteristics of the batch of PMSG are taken into consideration in formulation of the dose to be injected. When the superovulatory response has been good, the proportion of eggs recovered and/or fertilised has also been variable (Rowson, 1951), but double insemination or insemination of enhanced numbers of spermatozoa shortly before ovulation may diminish the latter problem.

Embryos are best recovered from the donor animal once they have entered the uterus, i.e. at least three days after ovulation, and this stricture becomes essential if a non-surgical means of recovery is employed. However, although non-surgical methods of flushing embryos from the uterine lumen (Rowson and Dowling, 1949) are being developed or applied in a number of laboratories (see below), a surgical means of recovery involving mid-ventral or flank laparotomy is still favoured and is more reliable in terms of the proportion of embryos recovered. But the surgical approach is limited by the small number of times the donor animal can be used due to the problems of adhesions in

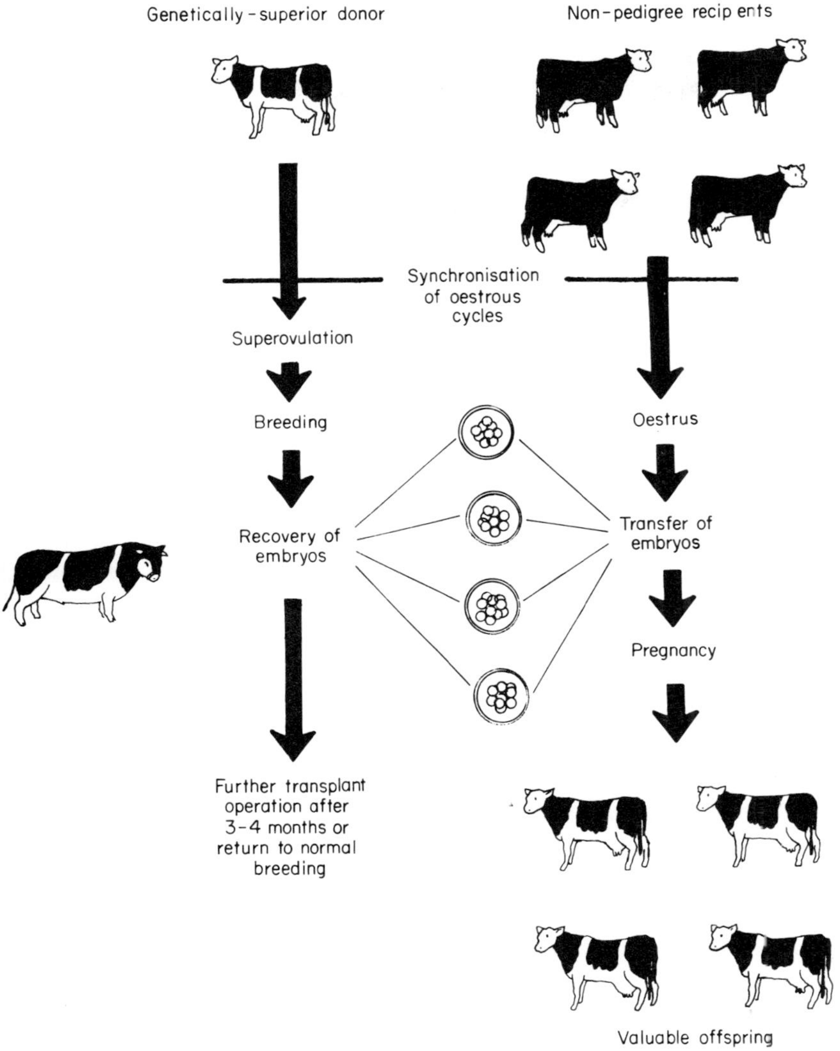

Fig. VIII.2. To illustrate the potential advantages of embryo transplantation from a genetically-superior dairy cow that had been bred with a nominated sire. Embryos are recovered from the superovulated donor and transferred to the non-pedigree recipients whose oestrous cycles have been synchronised with that of the donor.

the reproductive tract, weakening of the body wall, and the growth of scar tissue. Laparotomy is performed under general anaesthesia, the reproductive tract exposed, and the embryos collected into a glass dish after flushing the uterus with a physiological salt solution, usually containing a protein or other macromolecular component. This medium is introduced into the lumen with a needle and syringe, gently squeezed along the uterine horn by digital pressure, and collected through a suitably positioned rubber or glass cannula (Fig. VIII.3). This methodology and the preoperative procedures have been described in detail by Betteridge (1977) and Murray (1978). Donor animals are generally not used more than three times, after which they are returned to the herd to breed naturally.

The non-surgical approach (reviewed by Brand and Drost, 1977) aims to recover embryos from the upper portion of the uterine horns, commonly on Days 6–8 after ovulation. Most techniques now employ a simple means of introducing and recovering the flushing medium through the cervix, with the cow fully conscious in the standing position although preferably with the front feet somewhat lower than the hind feet. Administration of an epidural anaesthetic is customary to prevent defaecation and straining. The flushing apparatus usually consists of some form of 3-way Foley catheter, an inflatable balloon cuff being used to seal the caudal extremity of the uterine horn before flushing in the manner depicted in Figs VIII.4 and 5. Thus, one tube of the catheter leads to the inflatable cuff, a second permits the injected medium to escape from near the tip of the catheter, and the third permits return of the flushing medium to a glass collecting dish. At the time of insertion, a stiffening device is placed in the catheter for negotiation of the cervical canal and to aid its passage along the chosen uterine horn. Recovery of embryos is improved by using quite large volumes of medium for washing out the uterine contents (approximately 500 ml of medium at 37°C), and also by massaging the fluid-distended uterus *per rectum* during this procedure. In experienced hands and with repeated flushing, this approach to embryo recovery

---

Fig. VIII.3. Key stages in the process of surgical recovery and transfer of embryos in the cow using the approach of mid-ventral laparotomy. (Courtesy of Mr L. E. A. Rowson.) (a) Recovery of embryos by flushing the anterior portion of a uterine horn back through the oviduct, into the upper portion of which a catheter has been inserted. (b) Flushing a uterine horn by means of a Foley catheter with inflatable balloon cuff which is used to prevent escape of fluid into the body of the uterus. (c) Preparing to transplant an embryo into the recipient uterus by means of a Pasteur pipette and a puncture wound made in the wall of the uterus with a needle. (d) Introduction of embryo with a small volume of physiological medium into the uterine lumen.

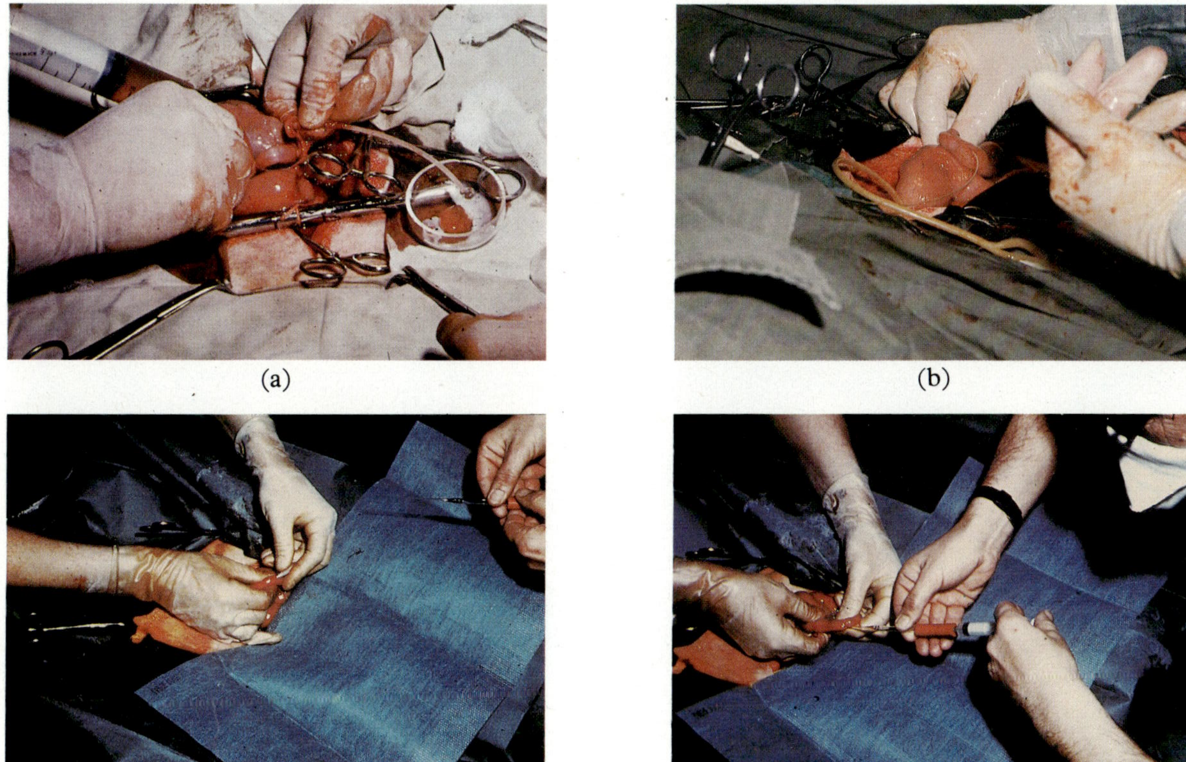

(a) (b) (c) (d)

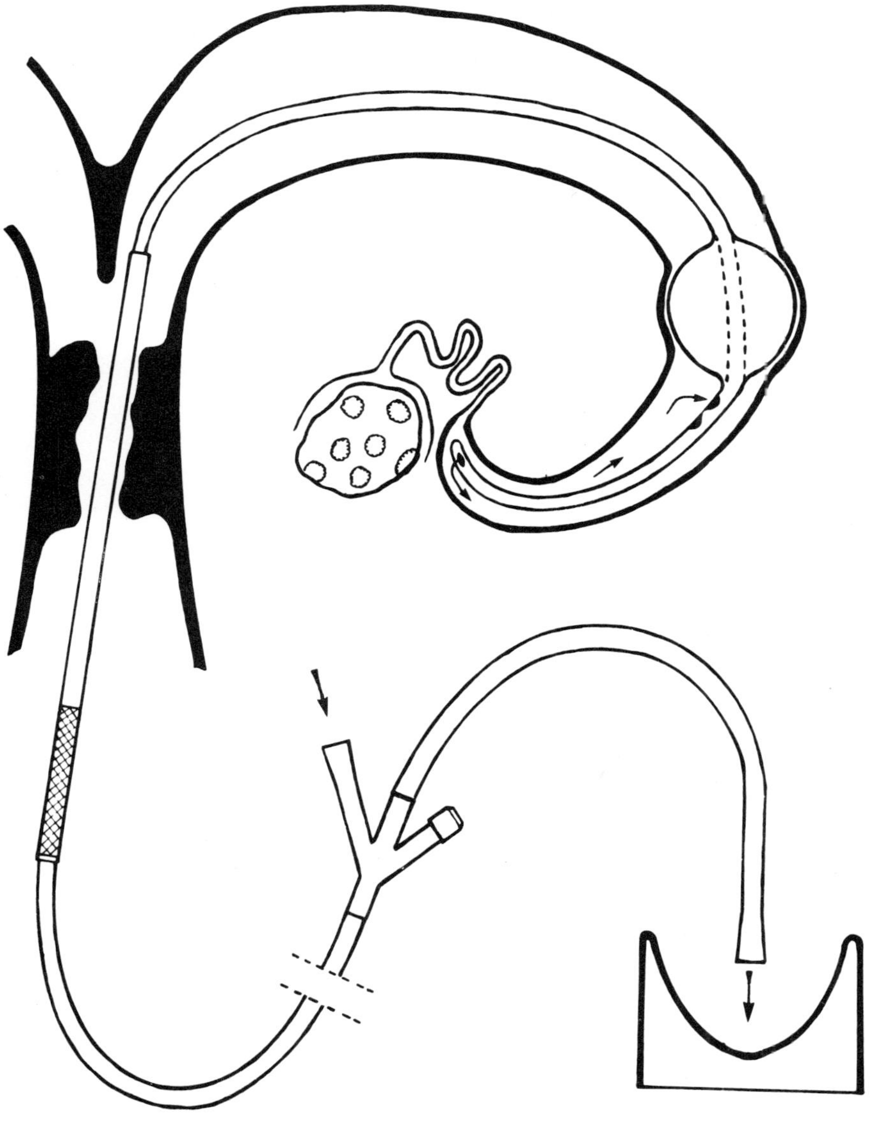

Fig. VIII.4. The procedure of non-surgical flushing of the uterine horn in a parous cow for recovery of preimplantation embryos. A 3-way Foley catheter with inflatable balloon cuff is introduced through the cervix, the cuff is distended, and the contents of the lumen flushed with physiological medium from the anterior end back into a collecting dish. (Redrawn from Newcomb *et al*., 1978).

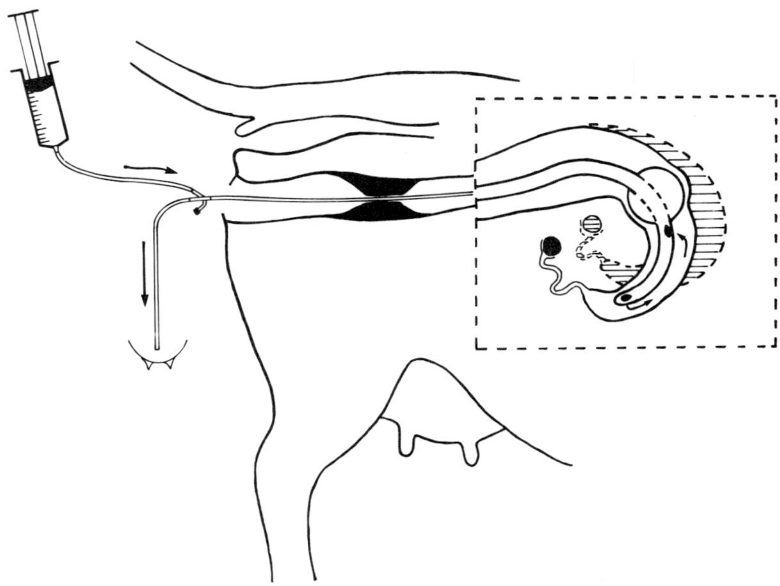

Fig. VIII.5. A further illustration of the non-surgical flushing procedure, with the uterine horn magnified to show the positioning of the Foley catheter.

can be quite successful (Newcomb *et al.*, 1978; Table VIII.1), and can be completed with little apparent trauma to the donor. As far as the embryos are concerned, searching for them amongst the endometrial débris at the bottom of the recovery dish can be time-consuming, and care must be taken to minimise the risks of chilling and bacterial contamination during these procedures.

Embryos or degenerating eggs are identified in the glass dish using a dissecting microscope, aseptic precautions being maintained throughout all manipulations. Embryos of good morphology that are selected for transfer can be held in an incubator (35–37°C) or at bench temperature (18–20°C) for several hours when in an appropriate culture medium, but care must be taken to prevent temperature shock, pH change and evaporation. If insufficient recipient animals are available to match the supply of embryos, longer term maintenance of embryos must resort either to temporary storage within the Fallopian tubes of a pseudopregnant rabbit (Hunter *et al.*, 1962; Adams *et al.*, 1968; Sreenan *et al.*, 1968; Lawson *et al.*, 1972) or prolonged storage at very low temperatures (Wilmut and Rowson, 1973a, b; Whittingham and Whitten, 1974; see Chapter VII). Use of pseudopregnant rabbits pro-

Table VIII.1. To illustrate the efficiency of non-surgical compared with surgical recovery of Day 7 bovine embryos from the the uterus; the results are presented as means. (Simplified from the data of Newcomb *et al.*, 1978.)

Hormone treatment	Details	Non-surgical recovery (first superovulation)			Surgical recovery (second superovulation)		
		Left horn	Right horn	Total	Left horn	Right horn	Total
	No of animals			20			17
1,000 iu	No of ovulations	2·0	2·7	4·7	1·1	1·1	2·2
PMSG	No of embryos	1·3	1·5	2·8	0·9	0·9	1·8
	% recovery	68	56	63	72	91	79
	No of animals			21			17
2,000 iu	No of ovulations	7·9	6·4	14·3	4·9	6·2	11·1
PMSG	No of embryos	4·6	3·1	7·7	3·0	3·2	6·2
	% recovery	58	48	55	69	52	64

vides better control of environment with regard to pH, osmolarity and nutritional substrates than is usually the case with an artificial medium. On the other hand, low temperature storage avoids significant interaction between the embryo and its fluid environment once freezing is achieved, and maintains the embryo in an arrested state of development.

Synchronisation of donor and recipient

One of the principal tenets underlying successful embryo transplantation is that the oestrous cycles of donor and recipient animals should be closely synchronised. The reason for this requirement is that the developing embryo shows a sensitive interaction with the maternal environment, the latter being in a dynamic state and changing according to the prevailing balance of ovarian steroid hormones. Synchronisation is best understood by recalling that under normal conditions of fertilisation, formation of the embryo and formation of the corpus luteum commence at almost the same time after completion of ovulation. As described in Chapter V, the free-living embryo is internally (genetically) programmed for its nutrient requirements on the basis of the molecules that can penetrate its membranes and be employed as metabolic substrates, and it is externally (environmentally) programmed on the basis of the chemical constituents available in oviducal and uterine fluids. Accordingly, any disruption in the synchrony between the phase of embryonic development and that of "reproductive tract development" may compromise the future of the embryo. This was first demonstrated experimentally in rabbits by Chang (1950), and has since been shown in other laboratory animals (McLaren and Michie, 1956; Noyes and Dickmann, 1960) and various species of farm animal (Polge, 1966, 1969; Rowson and Moor, 1966a; Rowson, 1970, 1971a, b). It may not be wholly true for primates (Harper and Pauerstein, 1976; Hunter, 1977; Marston *et al.*, 1977), where precocious entry of the embryo into the uterus may not prevent pregnancy or, of course, when ectopic pregnancies arise.

In practice, synchronisation between donor and recipient is achieved either by having a pool of recipients sufficiently large in number that some animals at the desired stage of the cycle are available on the day chosen for transfer or, more conveniently, by regulating the distribu-

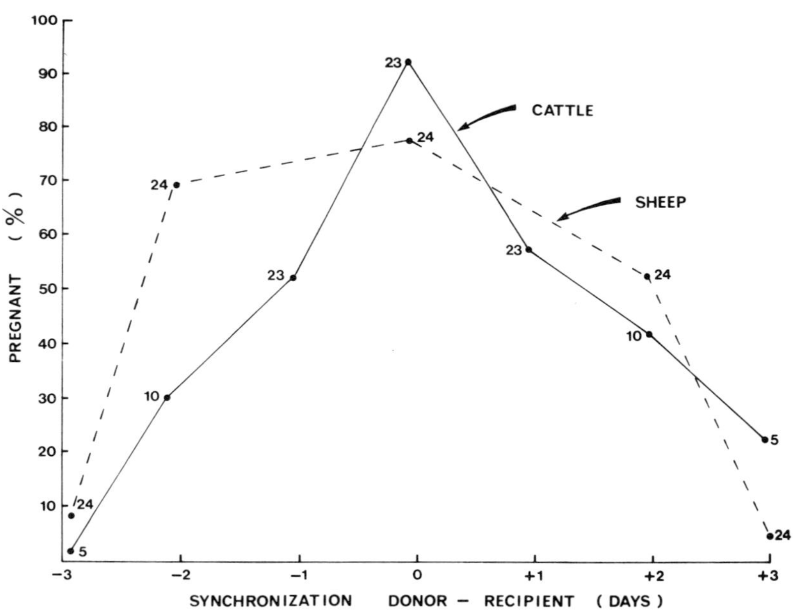

Fig. VIII.6. Graphical representation of the importance of synchronisation in the oestrous cycles of donor and recipient animals if embryo transplantation is to be followed by a high incidence of pregnancy. (Redrawn from Rowson *et al.*, 1972.)

tion of cycles by synchronisation of oestrus; treatment with prostaglandin analogues is currently favoured in this regard. Although optimal results in farm animals clearly require good synchronisation, results from Rowson's group and an Irish group (Sreenan and Beehan, 1974) indicate that there is some degree of tolerance in the precision of this synchronisation. In cattle, there is a tolerance of at least ± 1 day between the age of embryo and stage of uterus before conception rate drops significantly (Rowson *et al.*, 1969; Rowson *et al.*, 1972; Newcomb and Rowson, 1975), with the time of ovulation being taken as the baseline. In sheep, the tolerance may extend to ±2 days (Rowson and Moor, 1966a; Rowson, 1971a, b; see Fig. VIII.6). If the embryo is deposited somewhat precociously into the uterine environment, it appears better able to tolerate this situation than one in which the stage of endometrial proliferation is chronologically ahead of the development of the embryo, but this is not always the case (see Table VIII. 2).

A third approach to obtaining the desired synchronisation between the stages of embryonic and uterine development would be to have access to a supply of stored embryos. But this situation infers that the recipient animals would be determining the timetable of events, which may be neither convenient nor sensible in practice.

Table VIII.2. The incidence of pregnancy following synchronous and asynchronous transfer of embryos to the uterus of recipient cows by a surgical procedure. (Adapted from Rowson *et al.*, 1972.)

Variation from exact synchronisation (days)	No. of recipient animals	No. diagnosed as pregnant	Pregnant %
0	23	21	91
−1	23	12	52
+1	23	13	57
−2	10	3	30
+2	10	4	40

Method of embryo transfer

(a) *Surgical*

The more common and extensively used surgical approach has already been mentioned with reference to recovery of embryos. The procedure of transplantation usually involves a similar mid-ventral laparotomy in the recipient under general anaesthesia or less frequently a flank incision under local anaesthesia, exposure of the reproductive tract, and introduction of the embryo into the lumen through a very fine puncture wound made in the uterine wall (Fig. VIII.3). The manipulation employs a small volume of culture medium contained in the tip of a Pasteur pipette, into which the embryo is aspirated immediately before transfer by means of an attached hypodermic syringe, this stage being carefully monitored under a dissecting microscope. Although the surgical method of embryo transfer requires a skilled team (minimum of two surgeons, an anaesthetist, and animal attendants) and is relatively time consuming, the results obtained in terms of the proportion of recipient cows becoming pregnant after the transfer are usually good, ranging from 52% to >90% (Rowson *et al.*, 1969; Rowson *et al.*, 1972; Sreenan and Beehan, 1974; see Tables VIII.2 and 3). Recent results from several commercial organisations suggest that about 50% of all treated recipients become pregnant after embryo transfer. Transfer of two embryos, one to each horn of the uterus, can be used to induce twinning, and here a success rate close to 70% has been reported in one study (Rowson *et al.*, 1971), whereas transfer of two embryos to a single uterine horn frequently led to death through competition and a twinning rate of only 45%.

Table VIII.3. The influence of the degree of oestrous cycle synchronisation on the incidence of pregnancy in cattle following surgical transfer of embryos. (Adapted from Sreenan and Beehan, 1974.)

Degree of synchronisation (days)	No. of recipient animals	No. diagnosed as pregnant	Pregnant %
0	24	22	92
−1	7	5	71

(b) *Non-surgical*

In an effort to make the advantages of embryo transplantation less costly and more readily available to the commercial farmer, various techniques have been developed for depositing the embryo in the uterus without recourse to general anaesthesia and abdominal surgery. The most common of these has involved passage of a cannula or artificial insemination pipette through the cervical canal so that it protrudes beyond the internal cervical os, followed by instillation of the embryo into the uterine lumen in a small volume of medium. By analogy with the technique of artificial insemination, Sir John Hammond termed this procedure artificial *inovulation* and emphasized its value in any field scale programme of embryo transplantation. However, it is only in recent years that reasonable success has been obtained by a simple and straightforward approach in which the recipient uterus was distended at the time of transfer by infusion of gaseous carbon dioxide (Rowson and Moor, 1966b; Sreenan, 1975). This procedure may reduce myometrial contractions, as well as aiding deposition of the embryo in the lumen.

The underlying problems in much of the non-surgical work have been considered to be associated with dilatation of the cervix and invasion of the uterus by the transfer pipette at a stage during which the cervical canal would normally be closed. The specific cause of conception failure has been attributed to ejection of embryos back through the cervical canal into the vagina, probably due to enhanced contractions of the myometrium provoked by the pipette and manipulation of the cervix. Studies using radio-labelled artificial eggs (resin spheres) whose position in the reproductive tract could be monitored have indicated ejection of such eggs within a few hours of transfer in the very early luteal phase (Harper *et al.*, 1961; Rowson *et al.*, 1964). Enhanced secretion of oxytocin has been suggested *not* to be involved directly in this ejection response (Rowson *et al.*, 1964; Rowson *et al.*, 1972), although other smooth muscle stimulants such as prostaglandins may be released from the uterine wall. Infection of the uterus (pyometritis) at a stage of the oestrous cycle at which the uterine tissues cannot tolerate invasion of foreign bacteria may also be a cause of failure. The work of Rowson *et al.* (1953) indicated that the uterus was particularly susceptible to infection in the presence of an active corpus luteum.

Various modifications to the direct non-surgical approach have been attempted, such as the Japanese (Sugie, 1965) and French (Testart and Léglise, 1971) trans-vaginal approaches of entering the uterus via the peritoneal cavity (Figs VIII.7 and 8). These avoid breaching of the cervical canal, but the methods in practice have been time-consuming and required considerable skill and the results have not been good. The current emphasis in non-surgical transfers is once again on direct inovulation through the cervical canal, usually with embryos at the 6- or 7-day blastocyst stage of development. Several laboratories have

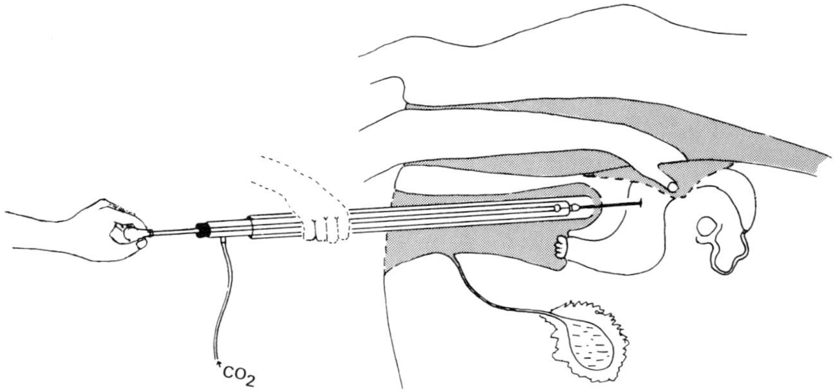

Fig. VIII.7. A non-surgical method of bovine embryo transplantation developed in Japan to avoid breaching the cervix. The transfer pipette enters the uterus via the vaginal wall and body cavity, and the uterine lumen is distended with carbon dioxide to aid this manoeuvre. (Redrawn from Sugie, 1965.)

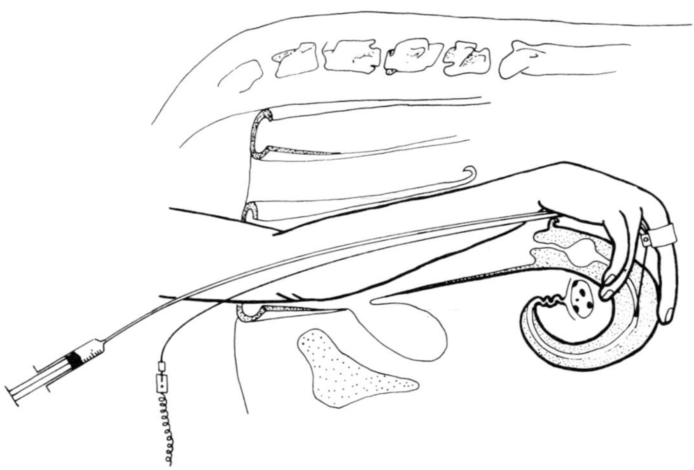

Fig. VIII.8. A trans-vaginal method of bovine embryo transplantation developed in France, again to avoid passage of the transfer pipette through the cervical canal. (Redrawn from Testart and Léglise, 1971.)

reported encouraging results (e.g. Boland *et al*., 1975, 1976; Lawson *et al*., 1975; Sreenan, 1975, 1978a; Heyman and Renard, 1978; Newcomb *et al*., 1978), which compare reasonably well with those obtained by surgical transfer (see Table VIII.4). A key factor in overcoming the

Table VIII.4. The results of a series of non-surgical transfers in cattle in which a blastocyst was introduced into each horn of the uterus via the cervix. (Adapted from Heyman and Renard, 1978.)

Stage of transfer	Days 7–9	Day 10	Day 13	Total
No. of recipients	18	16	13	47
No. of blastocysts transplanted	36	32	26	94
No. of recipients conceiving	9	11	7	27
Incidence of pregnancy %	50	68	54	57
No. of recipients with twins	3	8	3	14
Incidence of twin pregnancies %	33	73	43	52

ejection response may be to avoid manipulations in the early luteal phase (i.e. < day 5 of the cycle), thereby waiting until the extent of uterine contractions is diminished under the influence of elevated titres of progesterone from the maturing corpus luteum.

Stage and site of embryo transfer

As already indicated, a stage of development compatible with recovery from and transfer into the uterus is likely to prove most convenient: for a non-surgical means of transfer, this is essential. As the embryo spends approximately three days in the Fallopian tube in cattle, uterine transfers are unlikely to be performed before the embryo is at the morula or early blastocyst stage. The results of experiments in cattle suggest that transfer as 6–8 day blastocysts may give optimal results in terms of conception rate (Newcomb and Rowson, 1975); Fig. VIII.9), and recovery and examination of embryos at this stage rather than earlier permits selection of those blastocysts that appear most viable on morphological grounds. The site within the uterus at which the embryo is deposited should be in the horn adjacent to the ovary bearing the corpus luteum, often termed the ipsilateral horn (Rowson *et al.*, 1969), this situation underlying the local relationship that permits luteal maintenance and establishment of pregnancy (see Chapter VI). Within the ipsilateral horn, the actual position at which the blastocyst is deposited appears not to be critical for embryonic development (Sreenan and Beehan, 1976; Newcomb *et al.*, 1978) and, in any event, growth of the trophoblast soon ensures a more extensive relationship between embryo and uterus (see Fig. VI.9).

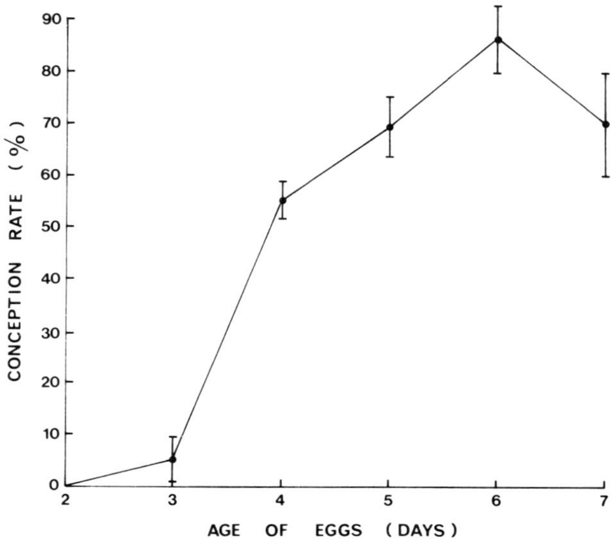

Fig. VIII.9. Graphical representation of the conception rates obtained after surgical transplantation of cow embryos to the uterus in the early luteal phase of the oestrous cycle. (Redrawn from Newcomb and Rowson, 1975.)

In sheep, embryos have been transplanted by the surgical approach at all stages up to that of the much elongated 12 or 13 day conceptus and, even at the latter stages, the surprisingly good conception rate of 40% was obtained (see Moor and Rowson, 1964). In pigs, by contrast, although 6–8 day blastocysts have been transferred with some success (Hunter *et al.*, 1967; Fig. VIII.10), the elongated form of trophoblast found after Day 9 of embryonic development has usually proved unsuitable for transfer followed by establishment of pregnancy (see Polge, 1972). Whether the conceptus at this stage is more susceptible to damage, or whether a failure reflects the activity of humoral factors preparing for regression of the corpora lutea is as yet unknown.

Non-surgical transfer of 8- to 16-cell embryos into the uterus has been shown to be feasible as a means of establishing pregnancy in the pig (Polge and Day, 1968) although the proportion of successes in this small study was extremely low; it may be significant that the procedure of transplantation was again associated with infusion of carbon dioxide into the uterus. But passage of a cannula through the lengthy cervical canal of pigs remains a difficult and skilful operation, with the attendant risk of puncturing the wall of the reproductive tract.

The results of limited embryo transplantation experiments in horses

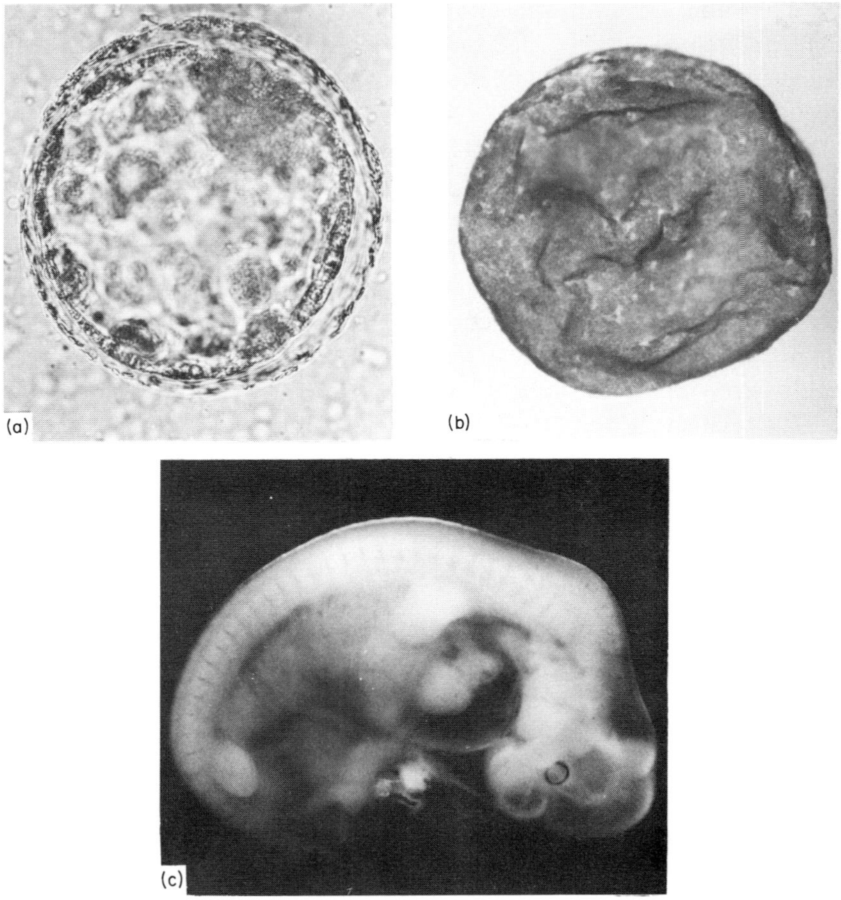

Fig. VIII.10. Blastocysts of the pig aged approximately six (a) and seven (b) days, at which stage of development surgical transplantation yielded some success as judged at autopsy on about Day 25 (c). (After Hunter *et al*., 1967).

are reviewed by Allen and Rowson (1975) and in the volume edited by Rowlands and Allen (1979).

Present and future uses of transplantation

The potential advantages of embryo transplantation in an agricultural context have been widely discussed in the last 30 years (Dowling, 1949; Hammond, 1950; Willett, 1953; Dziuk *et al*., 1958; Rowson, 1971a, b,

1976; Gordon 1975a, b; Betteridge, 1977; Du Mesnil du Buisson and Renard, 1978; Sreenan, 1978b). However, demonstration of large-scale biological and commercial feasibility is still awaited for some of the topics listed below.

(a) First and foremost, transplantation can be viewed as a counterpart to artificial insemination, permitting exploitation of superior female genotypes, giving more offspring from the same genetic donor than would arise under normal conditions of breeding (Fig. VIII.2). In conjunction with artificial insemination or selected mating, this would give complete control of the genotype of the embryo. Thus, a series of daughters could be produced from a high-yielding dairy cow or a rapidly-growing beef dam. In fact, with a non-surgical means of embryo recovery, the same donor animal might well be used three or four times in the same year and thus greatly exceed her spontaneous rate of reproduction.

(b) In recent years, transplantation has also been used for the rapid multiplication of rare, "exotic" or commercially desirable breeds. This has been particularly the case with some of the continental European breeds such as the Simmental, Limousin and Charolais, and has frequently been coupled with export of such breeds to North America, Japan or Australasia. Indeed, export of pregnant recipients may be a useful approach to meeting time-consuming quarantine regulations in the country of destination.

(c) Proliferation of a small pool of specific genotypes using superovulation and transplantation provides an attractive means of speeding up selection programmes, thereby facilitating an increased response to selection. Whilst the factors inherent in this argument have generally been appreciated by reproductive physiologists, it is only relatively recently that geneticists have published a mathematical assessment of the potential advantages (Land and Hill, 1975). In the mid- to long-term, selection programmes linked to a careful use of embryo transplantation could lead to establishment of élite herds, thereby upgrading the overall performance of the national herd.

(d) Increased reproductive output would also accrue from a non-surgical technique of inducing twin pregnancies, either by installing an embryo into each horn of a synchronous recipient or a single embryo into the non-gravid horn of a recently inseminated animal with only one corpus luteum (see Boland *et al.*, 1975; Gordon, 1975b; Renard *et al.*, 1977; Gordon and Boland, 1978; Heyman and Renard, 1978). This approach of "topping up" could overcome the disappointing results of gonadotrophin treatments aimed at producing twins (Gordon *et al.*, 1962), and could utilise embryos donated by beef heifers treated and inseminated shortly before slaughter. The embryos would be recovered aseptically at autopsy after evisceration of the carcase and removal of the reproductive tract. The production of beef calves from dairy cows or the induced twinning of dairy with beef calves by means of the transplant technique has been argued to permit greater commercial flexibility in the production system (Hammond, 1950; Rowson, 1971b). This would be especially so if the sex of the calves could be controlled, and the problem of freemartinism thereby avoided (see Chapter XI). Nonetheless, the overall financial advantages of induced twinning may be questionable except in very specific market circumstances (Wilmut and Hume, 1978).

(e) The objective of rapidly improving the performance of herds in developing countries could be well served by a programme of embryo transplantation, assuming satisfactory arrangements for temporary storage of the embryos were made. Export or import of embryos is clearly more convenient and cheaper than export or import of livestock. Moreover, by using dams of the indigenous breeds as recipients, valuable immunological advantages against local diseases may be conferred on the transplanted embryos (and subsequently via the colostrum), thus improving the ability of the young to tolerate the new environment.

(f) A major and costly component in the lifespan of a cow is the interval from birth until its first calf is born. Experimental shortening of this generation interval by induction of ovulation in prepuberal animals (Chapter III) coupled with transplantation of eggs or embryos to mature recipients has a number of attractive features, but the practice still awaits commercial testing and application.

(g) Long-term preservation of embryos in deep-frozen banks (i.e. at the temperature of liquid nitrogen) is reasonably certain to become the counterpart of deep-frozen semen reserves, enabling propagation of selected lines of livestock long after the donor is dead. Apart from the commercial desirability of such embryo banks, they could play a major rôle in the conservation of rare breeds and in the maintenance of genetic controls; the latter aspect will permit evaluation of progress in selection experiments after a number of years have elapsed and also allow estimation of the degree of genetic drift within a given breed (see Bailey, 1977). And, of course, the only means of realising the advantages of embryo storage remains the technique of transplantation.

(h) Finally, one should consider the involvement of egg or embryo transfer as a sequel to the fundamental research manipulations of *in vitro* maturation of oocytes, *in vitro* fertilisation, separation of blastomeres to give identical twins, and/or the fusion of blastomeres from different embryos to give chimaeras. Such manipulations are at present largely confined as research techniques to laboratory species, but they may well have contributions to make to the livestock industry of the future. Other areas of basic research that employ the technique of transplantation include examination of specific maternal effects on embryonic development; the timing of intra-uterine redistribution and "mixing" of embryos in a polytocous species such as the pig; the timing of embryonic influences on the lifespan and secretory activity of the corpora lutea; the potential uterine capacity when the number of embryos exceeds the number of ovulations; and the influence of the genotype of the foetus on the duration of pregnancy.

As a concluding remark, attention must be drawn away from basic research and back to the fact that whilst there are many attractive features in a programme of embryo transplantation, there may also be some notable disadvantages at the practical level. This is claimed to be the case in instances of induced twinning, which may give rise to problems of dystocia, to retained placentae, and to smaller calves. The extent to which the incidence of retained placentae and underweight calves can be alleviated by appropriate nutrition and husbandry is currently the subject of much research. Freemartinism may also be a problem with induced twinning in dairy herds, but this is clearly not the situation if the female calves are destined for meat production rather than breeding activity since the morphology of the genital tract would then be unimportant.

References

Adams, C. E. and Abbott, M. (1971). Bibliography on recovery and transfer of mammalian eggs and ovarian transplantation. *Bibliogr. Reprod.* No. **45**.

Adams, C. E., Moor, R. M. and Rowson, L. E. A. (1968). Survival of cow and sheep eggs in the rabbit oviduct. *Proc. 6th Int. Congr. Anim. Reprod. and Artif. Insem., Paris,* **1**, 573–574.

Allen, W. R. and Rowson, L. E. A. (1975). Surgical and non-surgical egg transfer in horses. *J. Reprod. Fert., Suppl.* **23**, 525–530.

Austin, C. R. (1961). "The Mammalian Egg". Blackwell Scientific Publications, Oxford.

Bailey, D. W. (1977). Genetic drift: the problem and its possible solution by frozen-embryo storage. *In* "The Freezing of Mammalian Embryos" (K. Elliott and J. Whelan, eds), pp. 291–299. Elsevier, Excerpta Medica, North-Holland, Amsterdam, Oxford and New York.

Betteridge, K. J. (1977) (Editor). Embryo transfer in farm animals: a review of techniques and applications. Canada Department of Agriculture, Monograph 16. Ottawa.

Boland, M. P., Crosby, T. F. and Gordon, I. (1975). Twin pregnancy in cattle established by non-surgical egg transfer. *Brit. vet. J.* **131**, 738–740.

Boland, M. P., Crosby, T. F. and Gordon, I. (1976). Induction of twin pregnancy in heifers using a simple non-surgical technique. *Proc. 8th Int. Congr. Anim. Reprod. and Artif. Insem., Krakow*, **3**, 241–244.

Brand, A. and Drost, M. (1977). Embryo collection by non-surgical methods. *In* "Embryo Transfer in Farm Animals" (K. J. Betteridge, ed.), pp. 16–19. Canada Department of Agriculture Monograph 16. Ottawa.

Chang, M. C. (1950). Development and fate of transferred rabbit ova or blastocyst in relation to the ovulation time of recipients. *J. exp. Zool.* **114**, 197–225.

Chang, M. C. and Pickworth, S. (1969). Egg transfer in the laboratory animal. *In* "The Mammalian Oviduct" (E. S. E. Hafez and R. J. Blandau, eds), pp. 389–405. University of Chicago Press.

Dowling, D. F. (1949). Problems of the transplantation of fertilised ova. *J. agric. Sci.* **39**, 374–396.

Du Mesnil du Buisson, F. and Renard, J. P. (1978). Transfert d'embryons chez les mammifères domestiques. *In* "L'Implantation de l'Oeuf" Colloque S.N.E.F.S., pp. 217–229. Masson, Paris.

Dziuk, P. J. (1969). Egg transfer in cattle, sheep and pigs. *In* "The Mammalian Oviduct" (E. S. E. Hafez and R. J. Blandau, eds), pp. 407–417. University of Chicago Press.

Dziuk, P. (1975). Embryo transfer: an experimental tool with practical applications. *BioScience* **25**, 102–104.

Dziuk, P. J., Donker, J. D., Nichols, J. R. and Petersen, W. E. (1958). Problems associated with the transfer of ova between cattle. *Univ. Minnesota Agric. Exp. Sta. Tech. Bull.* **222**.

Foote, R. H. and Onuma, H. (1970). Superovulation, ovum collection, culture and transfer. A review. *J. Dairy Sci.* **53**, 1681–1692.

Gordon, I. (1975a). Problems and prospects in cattle egg transfer. *Irish vet. J.* **29**, 21–30.

Gordon, I. (1975b). Problems and prospects in cattle egg transfer. *Irish vet. J.* **29**, 39–62.

Gordon, I. and Boland, M. P. (1978). Cattle twinning by non-surgical egg transfer. *In* "Control of Reproduction in the Cow" (J. M. Sreenan, ed.), pp. 336–355. Martinus Nijhoff, The Hague, Boston, London.

Gordon, I., Williams, G. and Edwards, J. (1962). The use of serum gonadotrophin (PMS) in the induction of twin pregnancy in the cow. *J. agric. Sci.* **59**, 143–198.

Hammond, J. (1950). Problems concerning the transplantation of fertilised ova or "artificial pregnancy". *Anales de la Facultad de Medicina, Montevideo*, **35**, 810–819.

Harper, M. J. K. and Pauerstein, C. J. (1976). "Ovum Transport and Fertility Regulation". Scriptor, Copenhagen.

Harper, M. J. K., Bennett, J. P. and Rowson, L. E. A. (1961). A possible explanation for the failure of non-surgical ovum transfers in the cow. *Nature (London)* **190**, 789–790.

Heape, W. (1891). Preliminary note on the transplantation and growth of mammalian ova within a uterine foster-mother. *Proc. R. Soc. B.* **48**, 457–458.

Heyman, Y. and Renard, J. P. (1978). Production de jumeaux par transplantation d'embryons chez les bovins de race à viande. *Ann. Méd. Vet.* **122**, 157–163.

Hunter, G. L. (1956). The maternal influence on size in sheep. *J. agric. Sci.* **48**, 36–60.

Hunter, G. L., Bishop, G. P., Adams, C. E. and Rowson, L. E. (1962). Successful long-distance aerial transport of fertilised sheep ova. *J. Reprod. Fert.* **3**, 33–40.

Hunter, R. H. F. (1976). Reproductive physiology in the bovine female: a review. *In* "Beef Cattle Production in Developing Countries" (A. J. Smith, ed.), pp. 79–104. University of Edinburgh.

Hunter, R. H. F. (1977). Function and malfunction of the Fallopian tubes in relation to gametes, embryos and hormones. *Europ. J. Obstet. Gynec. Reprod. Biol.* **7**, 267–283.

Hunter, R. H. F., Polge, C. and Rowson, L. E. A. (1967). The recovery, transfer and survival of blastocysts in pigs. *J. Reprod. Fert.* **14**, 501–502.

Kraemer, D. C., Moore, G. T. and Kramen, M. A. (1976). Baboon infant produced by embryo transfer. *Science (New York)* **192**, 1246–1247.

Land, R. B. and Hill, W. G. (1975). The possible use of superovulation and embryo transfer in cattle to increase response to selection. *Anim. Prod.* **21**, 1–12.

Lawson, R. A. S., Rowson, L. E. A. and Adams, C. E. (1972). The development of cow eggs in the rabbit oviduct and their viability after retransfer to heifers. *J. Reprod. Fert.* **28**, 313–315.

Lawson, R. A. S., Rowson, L. E. A., Moor, R. M. and Tervit, H. R. (1975). Experiments on egg transfer in the cow and ewe: dependence of conception rate on the transfer procedure and stage of the oestrous cycle. *J. Reprod. Fert.* **45**, 101–107.

McLaren, A. and Kerr, M. G. (1970). Egg transplantation. *Science J.* **6**, 51–58.

McLaren, A. and Michie, D. (1956). Studies on the transfer of fertilised mouse eggs to uterine foster-mothers. 1. Factors affecting the implantation and survival of native and transferred eggs. *J. exp. Biol.* **33**, 394–416.

McLaren, A. and Michie, D. (1958). An effect of the uterine environment upon skeletal morphology in the mouse. *Nature (London)* **181**, 1147–1148.

Marston, J. H., Penn, R. and Sivelle, P. C. (1977). Successful autotransfer of tubal eggs in the rhesus monkey (*Macaca mulatta*). *J. Reprod. Fert.* **49**, 175–176.

Moor, R. M. and Rowson, L. E. A. (1964). Influence of the embryo and uterus on luteal function in the sheep. *Nature (London)* **201**, 522–523.

Murray, F. A. (1978). Embryo transfer in large domestic mammals. *In* "Methods in Mammalian Reproduction" (J. C. Daniel, Jr., ed.), pp. 285–305. Academic Press, New York and London.

Newcomb, R. (1976). Fundamental aspects of ovum transfer in cattle. *Vet. Rec.* **99**, 40–44.

Newcomb, R. and Rowson, L. E. A. (1975). Conception rate after uterine transfer of cow eggs, in relation to synchronisation of oestrus and age of eggs. *J. Reprod. Fert.* **43**, 539–541.

Newcomb, R., Christie, W. R. and Rowson, L. E. A. (1978). The non-surgical recovery and transfer of bovine embryos. *In* "Control of Reproduction in the Cow" (J. M. Sreenan, ed.), pp. 292–304. Martinus Nijhoff, The Hague, Boston and London.

Noyes, R. W. and Dickmann, Z. (1960). Relationship of ovular age to endometrial development. *J. Reprod. Fert.* **1**, 186–196.

Polge, C. (1966). Egg transplantation in the pig. *World Rev. Anim. Prod.* **4**, 79–86.

Polge, C. (1969). Advances in reproductive physiology in pigs. *J. Aust. Inst. Agric. Sci.* **35**, 147–153.

Polge, C. (1972). Increasing reproductive potential in farm animals. *In* "Reproduction in Mammals" (C. R. Austin and R. V. Short, eds), Book 5, pp. 1–31. Cambridge University Press.

Polge, C. and Day, B. N. (1968). Pregnancy following non-surgical egg transfer in pigs. *Vet. Rec.* **82**, 712.

Renard, J. P., Heyman, Y. and Du Mesnil du Buisson, F. (1977). Unilateral and bilateral cervical transfer of bovine embryos at the blastocyst stage. *Theriogenology* **7**, 189–194.

Rowlands, I. W. and Allen, W. R. (1979). Equine Reproduction II. *J. Reprod. Fert., Suppl.* **27**.

Rowson, L. E. (1951). Methods of inducing multiple ovulation in cattle. *J. Endocr.* **7**, 260–270.

Rowson, L. E. A. (1970). Techniques of livestock improvement. *Outlook on Agriculture* **6**, 108–113.

Rowson, L. E. A. (1971a). Egg transfer in domestic animals. *Nature (London)* **233**, 379–381.

Rowson, L. E. A. (1971b). The rôle of reproductive research in animal production. *J. Reprod. Fert.* **26**, 113–126.

Rowson, L. E. A. (1976). (Editor). Egg transfer in cattle. *Proc. EEC Cambridge Seminar, Eur. 5491*, Commission of the European Communities, Luxembourg.

Rowson, L. E. A. and Dowling, D. F. (1949). An apparatus for the extraction of fertilised eggs from the living cow. *Vet. Rec.* **61**, 191.

Rowson, L. E. A. and Moor, R. M. (1966a). Embryo transfer in the sheep: the significance of synchronising oestrus in the donor and recipient animal. *J. Reprod. Fert.* **11**, 207–212.

Rowson, L. E. A. and Moor, R. M. (1966b). Non-surgical transfer of cow eggs. *J. Reprod. Fert.* **11**, 311–312.

Rowson, L. E. A., Bennet, J. P. and Harper, M. J. K. (1964). The problem of non-surgical egg transfer to the cow uterus. *Vet. Rec.* **76**, 21–23.

Rowson, L. E. A., Lamming, G. E. and Fry, R. M. (1953). The relationship between ovarian hormones and uterine infection. *Vet. Rec.* **65**, 335–340.

Rowson, L. E. A., Lawson, R. A. S. and Moor, R. M. (1971). Production of twins in cattle by egg transfer. *J. Reprod. Fert.* **25**, 261–268.

Rowson, L. E. A., Lawson, R. A. S., Moor, R. M. and Baker, A. A. (1972). Egg transfer in the cow: synchronisation requirements. *J. Reprod. Fert.* **28**, 427–431.

Rowson, L. E. A., McNeilly, A. S. and O'Brien, C. A. (1972). The effect of vaginal and cervical stimulation on oxytocin release during the luteal phase of the cow's oestrous cycle. *J. Reprod. Fert.* **30**, 287–288.

Rowson, L. E. A., Moor, R. M. and Lawson, R. A. S. (1969). Fertility following egg transfer in the cow; effect of method, medium and synchronisation of oestrus. *J. Reprod. Fert.* **18**, 517–523.

Sreenan, J. M. (1975). Successful non-surgical transfer of fertilised cow eggs. *Vet. Rec.* **96**, 490–491.

Sreenan, J. M. (1978a). Non-surgical egg recovery and transfer in the cow. *Vet. Rec.* **102**, 58–60.

Sreenan, J. M. (1978b) (Editor). "Control of Reproduction in the Cow". Martinus Nijhoff, The Hague, Boston and London.

Sreenan, J. M. and Beehan, D. (1974). Egg transfer in the cow: pregnancy rate and egg survival. *J. Reprod. Fert.* **41**, 497–499.

Sreenan, J. M. and Beehan, D. (1976). Effect of site of transfer on pregnancy and twinning rates following bilateral egg transfer in the cow. *J. Reprod. Fert.* **48**, 223–224.

Sreenan, J., Scanlon, P. and Gordon, I. (1968). Culture of fertilised cattle eggs. *J. agric. Sci.* **70**, 183–185.

Sugie, T. (1965). Successful transfer of a fertilised bovine egg by non-surgical techniques. *J. Reprod. Fert.* **10**, 197–201.

Testart, J. (1974). Obtention et transplantation d'oeufs fécondés chez les bovins. *Élevage et Insémination,* **140**, 3–18.

Testart, J. and Léglise, P. C. (1971). Transplantation d'oeufs divisés chez la vache, par voie transvaginale. *C.R. Séanc. Acad. Sci., Paris* **272**, 2591–2592.

Venge, O. (1950). Studies of the maternal influence on the birth weight of rabbits. *Acta Zool. (Stockh.)* **31**, 1–148.

Whittingham, D. G. and Whitten, W. K. (1974). Long-term storage and aerial transport of frozen mouse embryos. *J. Reprod. Fert.* **36**, 433–435.

Willett, E. L. (1953). Egg transfer and superovulation in farm animals. *Iowa State College J. Sci.* **28**, 83–100.

Wilmut, I. and Hume, A. (1978). The value of embryo transfer to cattle breeding in Britain. *Vet. Rec.* **103**, 107–110.

Wilmut, I. and Rowson, L. E. A. (1973a). The successful low-temperature preservation of mouse and cow embryos. *J. Reprod. Fert.* **33**, 352–353.
Wilmut, I. and Rowson, L. E. A. (1973b). Experiments on the low temperature preservation of cow embryos. *Vet. Rec.* **92**, 686–690.

Chapter IX

Diagnosis of Pregnancy

Introduction

Whilst there may exist a clear understanding of the optimum time for insemination relative to ovulation, and of the numbers of viable spermatozoa that need to be deposited in a particular region of the female tract, there is no guarantee that pregnancy will always follow mating or a carefully timed insemination. Of course, in systems of animal production where females have free access to a mature male, and at an appropriate ratio of males to females, failure of conception will frequently be remedied by mating at the following period of oestrus. But under systems employing artificial insemination, or other circumstances where a male may not be continuously present, a means of diagnosing pregnancy—or rather the failure to establish pregnancy—is of prime importance if reproductive output is to be improved and the duration of the breeding cycle kept to a minimum.

The duration of an oestrous cycle represents a significant proportion of gestation length, and whilst loss of one cycle due to unfruitful insemination may be economically damaging under an intensive system of production, loss of a second cycle from failure to detect and re-inseminate barren animals could prove economically disastrous. Accordingly, as intensification of livestock production continues, and as the pressure to optimise output from breeding females increases, so the desirability of confirming the successful establishment of pregnancy as soon as possible after breeding also increases. In the following pages, the physiological and tactical objectives of pregnancy diagnosis are considered before discussing methods that have been applied to cattle, sheep, pigs and horses.

Rationale for pregnancy diagnosis

The principal characteristics of an ideal method of pregnancy diagnosis are that it should be accurate, inexpensive, easily and quickly performed, and give an immediate result to avoid secondary handling of animals. Perhaps most important of all, it should be applicable very early in the presumed gestation. The rationale behind the last point is that the means of diagnosis should enable one to *predict* which animals are going to return to oestrus approximately one cycle length after service or insemination, so that they can be appropriately identified, surveyed and rebred at the optimum time. The inference here is that at least one of the methods of pregnancy diagnosis will be attempting to monitor ovarian activity, and specifically the divergence in luteal function between animals that have successfully conceived and those re-entering the follicular phase of the oestrous cycle after spontaneous luteolysis. Another approach would be to try more directly to detect some form of activity emanating from the developing embryo or its uterine environment, bearing in mind that in ungulates the embryo will have entered the phase of a much elongated conceptus whether or not the processes associated with attachment have commenced (Chapter VI). In either case, the test would almost certainly need to be applied, in terms of the oestrous cycle length of a species, as late as possible compatible with obtaining a reliable result and predicting which animals will return to oestrus. Laboratory procedures that need to be imposed after clinical surveillance of the animals must not take so long as to compromise the objective of being able to identify animals at least a day before expected oestrus.

Not to be overlooked, however, is the fact that a small proportion of cattle, sheep and pigs have prolonged oestrous cycles after breeding (i.e. a delayed return to oestrus) as a result of the early failure of embryonic development. In these circumstances, with a delayed regression of the corpus luteum or corpora lutea, too early a pregnancy test would probably give an erroneous result in terms of a false positive. But having argued the desirability of an early pregnancy diagnosis, let us examine to what extent the methods so far applied to domestic animals match up to this requirement.

General procedures and limitations

A survey of diagnostic tests quickly reveals that the approaches most widely used in cattle, sheep and pigs have usually depended upon physical detection of a developing foetus. Therefore, by definition, they have been applied after the phase of the developing embryo and after attachment or implantation has occurred. Two examples are examination of the uterine contents and ovarian status *per rectum* in cattle, and ultrasonic detection of the uterine contents in sheep and pigs. In some contrast stand the methods of vaginal biopsy as applied in pigs and those estimating the concentration of progesterone in body fluids such as blood or milk. Methods that depend on detecting an enlarged uterine vascular supply in pregnant animals probably fall between the two previous groups in terms of timing. Finally, there is the approach of attempting to detect specific embryonic antigens in blood plasma or urine, such as proteins of trophoblastic origin, since this might well be applicable to the preimplantation stage of blastocyst development. This, of course, is the basis of one of the earliest pregnancy tests as applied to women to detect the presence of human chorionic gonadotrophin in urine (Aschheim and Zondek, 1928).

Two other points deserve mention before discussing individual methods applicable to the farm species. First, not all animals diagnosed as pregnant early in gestation will produce viable offspring. This applies especially to monotocous species such as cattle and sheep, but on occasions also to pigs and other polytocous (litter-bearing) species. Apart from the method of diagnosis occasionally yielding false positives, the situation would also be brought about by foetal loss resulting for instance from some form of stress or constraint. Second, many of the more sophisticated methods of pregnancy diagnosis, such as estimation of the concentration of progesterone in body fluids, do not as yet enable accurate prediction of the number of embryos or foetuses, and this may well be seen as a desirable or even essential criterion in future methods for cattle and sheep. This might therefore suggest that research should focus upon an embryonic or trophoblastic (placental) component or secretion, the activity of which may well be more closely correlated with fecundity than the concentration of an ovarian hormone in the peripheral circulation.

Pregnancy diagnosis in cattle

The traditional, although only presumptive, method of determining the establishment of pregnancy in dairy cows is the failure of mated or inseminated animals to return to oestrus. Animals are closely surveyed at least twice a day at milking, and those in which a return to oestrus is imminent can usually be judged both from behavioural signs (i.e.

mounting and riding; undue restlessness) and clinical symptoms (raised tail head; abundant mucus on vulva). Because these indications are all readily apparent in the absence of a male, dairy herds that resort to breeding solely by artificial insemination should not be at a significant disadvantage in terms of detection of oestrus. Nonetheless, careful surveillance of inseminated animals should be made not only 18–20 days or so after insemination but continuously for two to three months, since a delayed return to oestrus is not infrequent in cattle and is presumed to be associated with early embryonic loss. This phenomenon of a delayed return is taken into account by insemination services when calculating conception rates, which are usually expressed as the proportion of animals *not* returning to oestrus 45–60 days and 60–90 days after insemination; the latter percentage is invariably lower. In contrast to dairy cattle, the majority of beef herds have an intact bull running with the cows, so that a remating should occur in those animals returning to oestrus, but the consequent spread in calving time associated with this approach can be disadvantageous in some production systems. On the other hand, beef herds that employ artificial insemination may have a raddled vasectomised bull constantly associated with the cows, so that a good indication is obtained of the proportion of cows conceiving to insemination. Animals that return can then be identified, inseminated once more, and possibly separated into sub-groups according to the expected time of calving.

Rectal palpation

Tangible evidence for the establishment of pregnancy has been obtained most commonly by palpation of the ovaries and uterus with the hand introduced *per rectum* (Fig. IX.1). This procedure is best undertaken by a veterinary surgeon, not least since undue handling of the uterus in the early stages of pregnancy may compromise development of the foetus and on occasions may precipitate foetal death and/or abortion (see Rowson and Dott, 1963). The objective of the palpation procedure is firstly to detect the swelling associated with a gravid uterus and, with extreme care, to note the presence of foetal contents together with the enlarged placental cotyledons and uterine arteries. Before the stage of significant enlargement of the cotyledons, the feeling of slipping membranes can be detected if a fold of the appropriate uterine horn is gently manipulated between finger and thumb. At the same time, palpation of the ovaries for the presence of a mature and functional corpus luteum enables further confidence to be placed in the diagnosis. A skilled veterinary surgeon can detect pregnancy from about 35 days of gestation onwards, although greater accuracy is obtained after 45–50 days. But if the precise time of breeding is known,

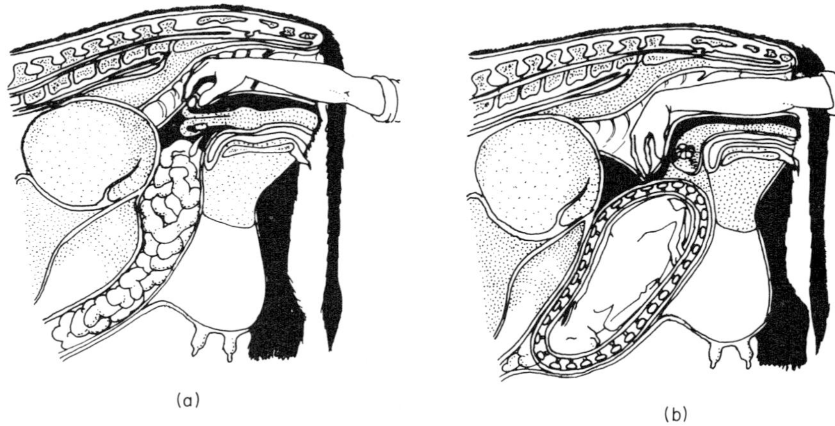

Fig. IX.1. The technique of rectal palpation in cattle for determining (a) the status of the ovaries with regard to a corpus luteum or mature follicle, and (b) the presence of a gravid uterus.

then verification of a fully developed corpus luteum 19–22 days later provides >85% evidence of conception, and Roche *et al.* (1978) reported 93% accuracy in a group of heifers when using this criterion some 21 days after service.

Progesterone estimation

Chemical evidence of successful conception can be sought at a comparatively early stage after breeding (Shemesh *et al.*, 1968), and depends on measuring the concentration of progesterone in peripheral blood or in milk where this method of sampling is appropriate. The basis of the diagnosis is the expected divergence in progesterone concentration between a successfully bred animal where the corpus luteum is maintained and one failing to conceive with a consequent regression of this structure commencing on about Day 17 or 18 (Fig. IX.2). Blood sampling is usually recommended on Day 20 or 21, and involves a simple jugular collection of approximately 10 ml made from the distended vein with a needle and glass vacuum tube. Plasma or serum is

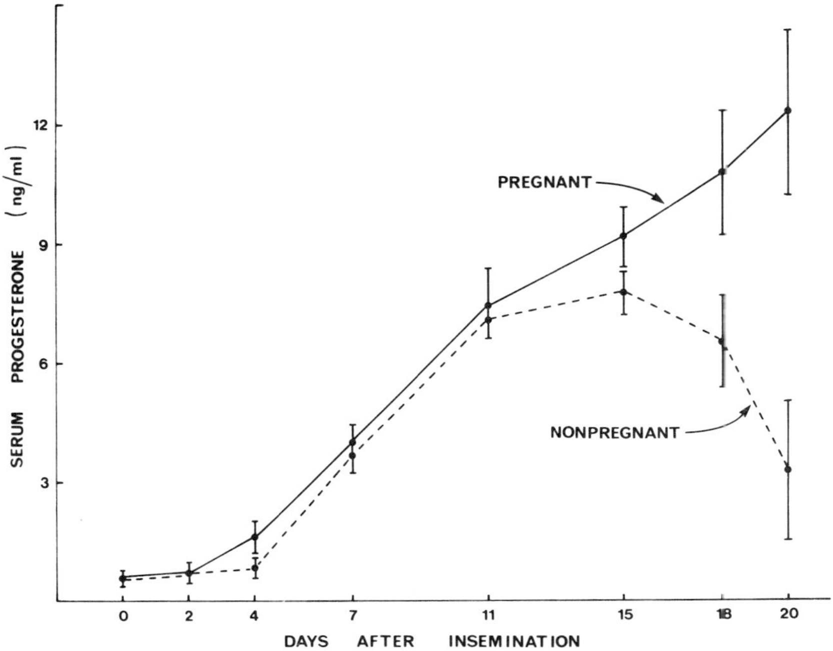

Fig. IX.2. Simplified curve of the progesterone concentration in peripheral blood serum to illustrate the divergence in values obtained between pregnant and non-pregnant cows some 17 or 18 days after oestrus. (Redrawn from Edgerton and Hafs, 1973.)

prepared according to the method of assay, which would normally be based on competitive protein binding (Robertson and Sarda, 1971) or radioimmunoassay (Edgerton and Hafs, 1973). Diagnosis based on 20–24 day values for progesterone gave an accuracy of 88–100% (Table IX.1), whereas forecasts based on plasma progesterone levels on the 19th day gave only a 74% accuracy (Robertson and Sarda, 1971). Clearly, a balance must be struck between the expected level of accuracy of the test and the desire to predict which animals will return to oestrus. Perhaps the most critical factor here will be the time required for the laboratory procedures.

Milk progesterone

An alternative, and in many ways more attractive, means of detecting pregnancy is to estimate the concentration of progesterone in milk

Table IX.1. The accuracy of pregnancy diagnosis in cows based on measuring the concentration of progesterone in peripheral blood. (Extracted from Robertson and Sarda, 1971.)

Day of blood sampling	Number of correct diagnoses	Total number of diagnoses	% accuracy
19	31	42	74
20	15	17	90
21	14	14	100
22	10	11	90
23	22	25	88
24	17	18	95

(Harkness and Darling, 1971; Laing and Heap, 1971; Darling *et al.*, 1972; Heap *et al.*, 1973; Heap *et al.*, 1976); this concentration generally parallels that in blood. Whilst this approach is not applicable to maiden heifers, and will seldom be suitable for beef cattle because of problems of milk sampling in the field, it does have the advantage for the farmer of avoiding jugular blood sampling, plasma or serum preparation, and storage. Moreover, in Britain for example, the Milk Marketing Board besides running the artificial insemination service in England and Wales has also installed a semi-automated laboratory for dealing with milk progesterone assays, and offers this facility at a fee to farmers (Booth and Holdsworth, 1976). This integrated approach to cattle breeding will undoubtedly spread in the future. It seems eminently sensible that the organisation offering the insemination service should also be the one that monitors its success, since it is then in a better position to take corrective action in instances where conception has not occurred.

Milk to be assayed is collected during the fourth week, usually 21–24 days after mating or insemination. The precise nature of the milk sample may have a significant influence on the measurable concentration of progesterone and also on the storage potential of the sample, so the use of a standardised procedure is important. Either fore-milk or whole milk is collected (5 ml from each quarter) into a glass bottle containing a suitable preservative such as a 30 mg tablet of potassium dichromate (Booth and Holdsworth, 1976), and stored at 5°C until despatch to the laboratory. If fore-milk is sampled, a draw of milk is first taken to clean the teat sinus. On the other hand, whole milk from the afternoon milking now seems to be more widely used for the test and, incidentally, has a higher concentration of progesterone than morning milk (Heap *et al.*, 1976). Whilst the concentration of progesterone is also generally higher in milk than in corresponding samples of blood plasma (Harkness and Darling, 1971; Laing and Heap, 1971), it is not so much the absolute value that counts in the test but rather the comparison of milk progesterone concentration between pregnant and non-pregnant cows under the particular sampling procedures (Table IX.2). Negative assays appear to be 85–100% accurate at Day 24,

Table IX.2. To illustrate the difference in the concentrations of progesterone in milk (ng/ml) between pregnant and non-pregnant cows at various intervals after insemination. (From Heap *et al.*, 1976.)

Days after insemination	Mean concentration ± S.E.M. (no. of samples)			
	Non-pregnant	Pregnant	Δ[1]	*P* value[2]
21	15·4 ± 2·3 (63)	34·7 ± 1·9 (111)	+ 19·3 ± 3·0	<0·001
24	22·7 ± 3·3 (43)	35·0 ± 1·7 (108)	+ 12·3 ± 3·7	<0·001
28	18·8 ± 2·7 (43)	34·7 ± 1·8 (102)	+ 15·9 ± 3·3	<0·001
42	16·5 ± 3·5 (30)	33·5 ± 1·9 (93)	+ 17·0 ± 4·0	<0·001

[1] Difference between pregnant and non-pregnant values. [2] Student's *t* test.

whereas positive assays are not normally expected to be more than about 80% correct (see Heap *et al.*, 1976). False positives seem to be attributable largely to late embryonic mortality or to early foetal death.

Whilst assay of milk progesterone is a reasonably early and convenient means of pregnancy diagnosis, and even though samples can be analysed within 24 hours of receipt, this still does not normally enable non-pregnant cows to be rebred until about 42 days after first service. So the approach of measuring milk progesterone does not yet meet the requirements of an ideal pregnancy test.

Embryonic antigens

Another approach that might more closely meet the predictive requirements of a test in cattle would be a means of detecting pregnancy-specific antigens of embryonic origin. Such antigens would probably be derived from the cell layers of the trophoblast, and would need to be detectable in maternal blood. Although the idea remains attractive and much work is currently in progress examining the synthetic and secretory potential of the bovine blastocyst with regard to both steroid and protein components, a specific pregnancy test using this approach has yet to be proposed.

Pregnancy diagnosis in sheep

Whilst evidence of failure of conception in sheep may be obtained by placing a raddled vasectomised ram with the flock after breeding, this is not in itself a positive means of diagnosing pregnancy or of detecting animals that have remained in anoestrus (see Chapter XI). Nor, in many circumstances, would this approach seem more suitable than running an entire ram with the flock, since most barren ewes should be mated at the return to oestrus. Thus, leaving a ram permanently with an appropriate size of flock for two months or so after the onset of the breeding season may be the more attractive proposition in many systems of sheep production, with subsequent imposition of a test to confirm those ewes that need to receive special supervision in advanced stages of pregnancy.

Use of ultrasound

The use of a portable ultrasonic apparatus has been extensively explored by Fraser and his colleagues (Fraser and Robertson, 1968; Fraser *et al.*, 1971), and Wilson and Newton (1969), and by application of a probe to the abdomen the apparatus can detect the presence of a foetus from about the ninth week of gestation. Ultrasound—sound of very high frequency and very short wavelength—is reflected from a moving object to the transmitting source at a slightly altered frequency, the Doppler phenomenon. This enables detection of some aspect of foetal pulse (i.e. heart or umbilical cord) or, by intrapelvic examination, the characteristic sounds from the uterine arteries. The ultrasonic signals reflected from the moving object are usually amplified and analysed audibly, but they can be converted into a visual display on a small screen; with experience in positioning the probe, the instrument offers a 93% success rate (Fraser and Robertson, 1968). When the apparatus is used in conjunction with a rectal probe, an accuracy of 80% can be obtained by Day 40 of gestation (Deas, 1977). Lindahl (1966) has also described an amplitude-depth method of pregnancy detection based on the use of ultrasound for measuring the echo distance of the fluid-filled gravid uterus. The transducer probe is applied to a small shaved area of the abdomen and transmits its amplitude-depth (echo) response to a portable screen analyser. Although the accuracy of detection is 95%, the method again cannot be applied until about the 60–65th day after mating when the allantoic and amniotic membranes have become considerably swollen with fluid. In further work, Lindahl (1969) came to the conclusion that the Doppler ultrasonic method was superior to the ultrasonic echo approach in detecting pregnancy in sheep.

Radiography

Foetal radiography depends on X-ray detection of ossification after the 50th day of development (Ardran and Brown, 1964), and gives an accuracy of 90–95% by three months after breeding (Wenham and Robinson, 1972; Rizzoli *et al.*, 1976). There are no detrimental effects of the technique on either the ewe or lamb(s). But this method has not yet won favour on the grounds of the expensive equipment, even though it can be used to diagnose 400–600 ewes per day and offers the further advantages of (a) an immediate result; and (b) the identification of ewes carrying twins or triplets.

Progesterone estimation

Measurements of the concentration of progesterone in peripheral blood plasma have been made in order to detect divergence in ovarian function between pregnant and barren animals. Whilst the timing of application of this test is fairly critical, and it requires both a blood sample and laboratory facilities, it does permit a firm diagnosis (>85% accuracy) at the 17th or 18th day after mating (Robertson and Sarda, 1971; Saumande and Thimonier, 1972; Thimonier, 1973). The progesterone concentration should indicate the presence of either a functional corpus luteum or the regression of this structure preparatory to a return to oestrus.

Pregnancy-specific antigens

One attraction in the approach of attempting to detect antigens of embryonic origin in maternal blood is that an early diagnosis should be possible. Another is that after the putative antigens commence to be expressed, the timing of maternal blood sampling should be much more flexible than in the case of a progesterone estimation, where steroid secretion from a newly formed corpus luteum would complicate the diagnosis in animals entering a second oestrous cycle. A method claiming detection of embryonic antigens in sheep as early as six days after mating was described by Cerini *et al*. (1976). In essence, antisera raised in rabbits against 14-day sheep embryos were used to detect pregnancy specific antigens in maternal blood plasma by haemagglutination. This agglutination reaction was noted to occur when appropriate dilutions of anti-embryo antiserum were added to blood obtained from ewes between Days 6 and 50 of pregnancy (Cerini *et al*., 1976). However, extensive examination of the test in a number of other laboratories has so far failed to confirm the feasibility of this approach. There is still reason to hope that detection of embryonic antigens in maternal body fluids will form the basis of an early pregnancy test (see Staples *et al*., 1978). Two recent studies on sheep in France have indicated that the conceptus is indeed able to secrete both a trophoblastic protein on Days 14–16 (trophoblastin: Martal and Lacroix, 1978) and a gonadotrophin-like substance on Day 15/16 (Wintenberger-Torres, 1978). Moreover, the concentration of placental protein hormones in maternal blood may in due course enable the diagnosis of twins.

Abdominal inspection

Laparoscopy—the introduction of an endoscope into the abdominal cavity to detect a gravid uterus or maintained corpus luteum—has also been considered specifically as a means of pregnancy diagnosis in sheep, and has been found applicable as early as the 17th day after mating (Phillippo *et al*., 1971). But it requires some form of local or general anaesthesia, aseptic techniques, considerable skill and some supporting facilities. Moreover, the procedure may take 10–20 minutes per animal, depending on the amount of abdominal congestion and omental fat interfering with examination of the ovaries and uterus. Working in a research institute, Phillippo and his colleagues reported the accuracy of their diagnosis to be about 90% (see also Phillippo and Rhind, 1977), but even so laparoscopy is unlikely to find an extensive commercial application.

Abdominal palpation

A further method of diagnosis in sheep is by the recto-abdominal approach (Hulet, 1972). Since the ewe is too small for rectal palpation of the abdominal contents, a plastic rod is inserted into the rectum of the upturned animal and used to displace the gravid uterus against the abdominal wall. This is then palpated to detect a developing foetus. Pregnancy cannot be detected before about seven weeks of gestation, accuracy is low until about ten weeks, and the risk of damage to the ewe must be considered.

Pregnancy diagnosis in pigs

Until the mid-1960s, pregnancy diagnosis in pigs was usually based on the presumptive evidence of a non-return to behavioural oestrus. But during that decade, the possibility of measuring the concentration of progesterone in peripheral blood by chemical or radiochemical assay offered a positive means of diagnosing pregnancy if the test was applied, for example, about 19 or 20 days after service. In most practical situations, however, the result of the laboratory assay would not be produced in time to anticipate a return to oestrus, nor would the skills and equipment for blood sampling and processing always be available. Moreover, the problem of distinguishing between late luteal phase values of progesterone in non-pregnant animals and those derived from corpora lutea of pregnancy effectively restricts the test to a period of only a few days, since a new curve of luteal phase progesterone secretion would be established if an animal returned to oestrus and ovulated. Nonetheless, progesterone assays have been considered in some detail in experimental, as distinct from commercial, situations of pregnancy detection (Robertson and Sarda, 1971; Ellendorff *et al.*, 1976). They may also have specific clinical applications when other tests prove unsatisfactory.

Vaginal biopsy

The technique of vaginal biopsy, that is the removal of a layer of epithelial cells by some simple form of scraping from the anterior vagina, was also developed during the 1960s. The basis of the diagnosis is a change in the morphology of the epithelial cells under the influence of elevated titres of circulating progesterone. Once again, the actual diagnosis involves laboratory procedures, although in this case of a microscopical rather than a chemical or radiochemical nature. Vaginal biopsy has been reviewed in detail by Walker (1972), with descriptions of changes in epithelial morphology according to stage of the oestrous cycle or pregnancy. After the early establishment of pregnancy, the epithelium is characteristically of two or three cell layers with the nuclei arranged in rows parallel with a level basement membrane. The epithelium of "non-pregnancy" is more cellular, higher, and limited by a less regular or invaginated border (Fig. IX.3). Unfortunately, these characteristic features do not always predominate in the sample and a correct diagnosis requires experience. Even then a small proportion of samples will remain inconclusive.

Inexpensive instruments for tissue sampling have been described by Done and Heard (1968) and Morton and Rankin (1969), and their use involves little if any restraint of the animal and a negligible traumatic risk. The epithelial sample (4 × 2 mm) is subjected to standard histological procedures of fixation, staining and examination, for which phase-contrast microscopy may be advantageous; with the latter form of microscopy, unfixed but frozen specimens can be examined directly (Diehl and Day, 1973). Over and above the requirement for interpretative skills, accuracy can be improved by a standardised sampling technique and adherence to a strict laboratory routine, especially to avoid confusion among specimens from a large number of animals. When performed three weeks or more after insemination, the accuracy of the test is 90–95% where breeding dates are known (Table IX.3). False negatives are occasionally recorded, but the most frequent error is a false positive for animals in the late luteal phase of the cycle. This suggests that sampling should not be imposed at least until Day 19 or 20. Other sources of error arise from incorrect sampling (e.g. cervical tissue) and in animals with cystic ovaries. In commercial practice, the test has most often been used between one and three months after service, especially when the precise date of breeding is unknown.

Ultrasonic detection

Foetal movements or blood vessel pulsations in pigs have been detected ultrasonically on the basis of the Doppler phenomenon. In this situa-

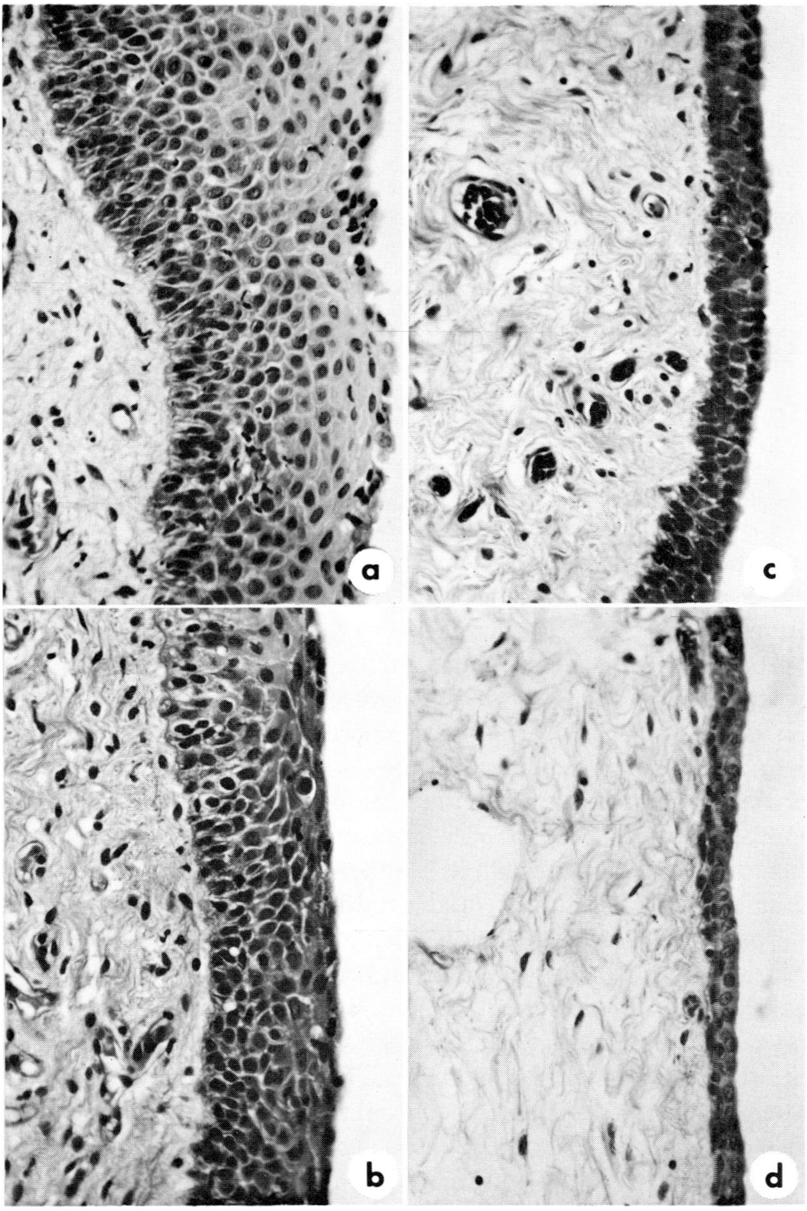

Fig. IX.3. Stained histological specimens prepared after vaginal biopsy in pigs to show the more extensive epithelium of "non-pregnancy" (a and b) and the simpler two or three cell layers of the pregnancy epithelium (c and d). (Reproduced from *Fertility and Infertility in the Domestic Animals*, Baillière Tindall, 1970. Courtesy of Professor J. A. Laing and Dr D. Mitchell.)

Table IX.3. To illustrate the accuracy of pregnancy diagnosis in pigs by the method of vaginal biopsy, as reported in a series of studies between 1963–1970. (Extracted from Walker, 1972.)

Time of tissue sampling after service (days)	Number of sows diagnosed	Accuracy of test (%)
>30	154	100
>30	192	97·4
25–93	206	93·2
31–90	162	93·2
>30	177	95·5
18–25	58	95·0
21	176	95·0
21–51	84	89·3
25	77	94·8

tion, the ultrasonic detector consists of an amplifier and a transducer which is applied as a probe to the sow's abdomen or less frequently inserted into the rectum to identify uterine arterial pulsations. This approach to diagnosing pregnancy in pigs was first described by Fraser and Robertson (1967, 1968), with the threshold of successful application being at approximately two months of gestation; but in a later report based on monitoring foetal movement and pulse with a modified apparatus, the earliest detection time was apparently reduced to one month after service with a success rate of 97% (Fraser *et al.*, 1971). This of course applies to special trials rather than to use by non-specialists under farming conditions, and experience is definitely required in interpreting the amplified sounds with accuracy.

A subsequent study using the ultrasonic principle was that of Lindahl *et al.* (1975), where the acoustical impedance of the contents of the gravid uterus provided the basis of the test, the embryonic fluids reflecting back the ultrasonic beam. The ultrasonic principle is used more objectively here because a visual reading of the echo recording is made on a small oscilloscope, with few problems of interpretation. The technique is rapid and accurate from Day 30 of gestation onwards, and simply requires a moistened transducer probe to be held in contact with the lower flank of the standing sow about 5 cm posterior to the navel and just lateral to the nipple line. An easily recognised band of echoes is obtained from a depth of 15–20 cm in pregnant sows, in contrast to echoes from a depth of only about 5 cm in non-pregnant animals (Fig.

IX.4). Confirmation of the results of ultrasonic pregnancy diagnosis by use of laboratory assays or *post mortem* inspection indicates that the technique approaches 100% accuracy for diagnosis under field conditions between 30 and 90 days of gestation. In the hands of veterinary surgeons or large scale pig producers, the accuracy of the test was >95% (Lindahl *et al.*, 1975). A modified version of Lindahl's apparatus has been produced, containing display lights instead of the oscilloscope for simplicity of diagnosis (Hansen and Christiansen, 1976).

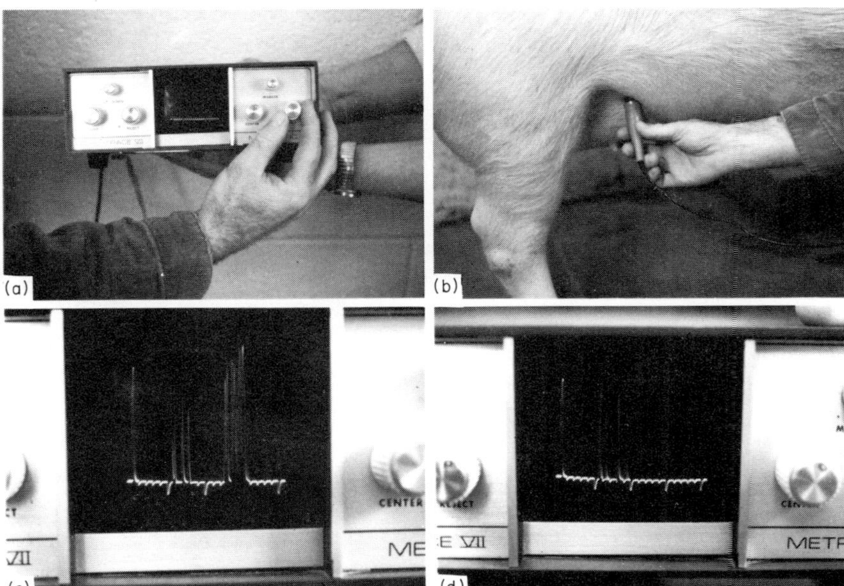

Fig. IX.4. An ultrasonic echo-trace recorder (a) and the attached transducer device (b) used for the detection of pregnancy in pigs by application to the lower flank. A characteristic band of echoes reveals a gravid uterus (c), and is clearly distinguishable from the pattern found in non-pregnant animals in which the echoes arise from the muscle, skin and fat of the body wall. (Courtesy of Professor P. J. Dziuk.)

Even though the two ultrasonic approaches just described do not achieve the objective of predicting which animals have failed to conceive and will return to service at approximately 21 days after insemination, it is worth emphasizing that there is no delay between testing and detection of pregnancy, and after purchase of the instrument there are no inherent costs in individual tests.

Rectal examination

One other method of pregnancy diagnosis in pigs that should be mentioned is that of rectal examination. Although, again, it does not meet the requirement of being applicable early in gestation, nor on physical grounds is it suitable for gilts, it may be of value in specific situations since it needs no special equipment and gives an immediate result. The objective of the examination is to detect an increase in diameter of the uterine arteries in pregnant animals, a comparison being established by palpation of the neighbouring external iliac artery (Fig. IX.5). During

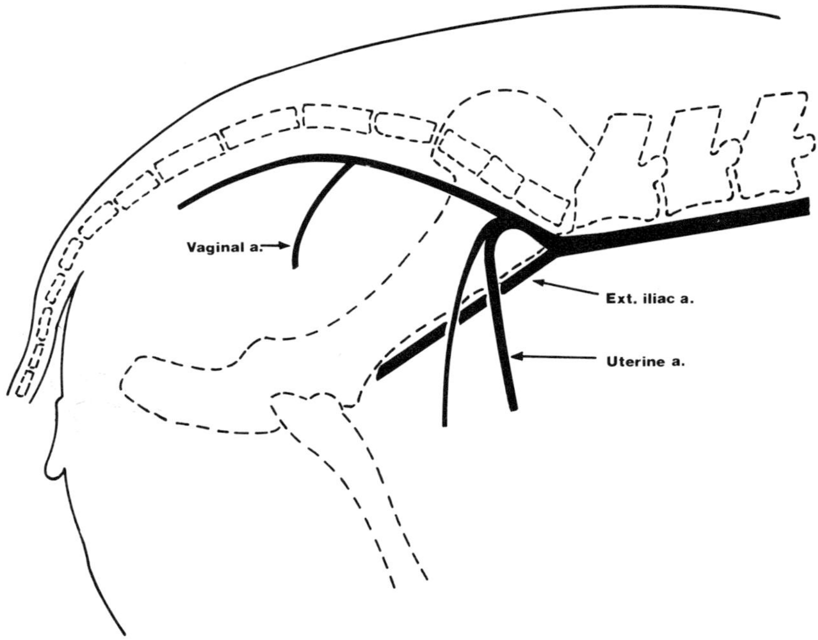

Fig. IX.5. Diagrammatic representation of the arteries associated with the reproductive tract in the sow. Digital palpation of the size and pulsation characteristics of the uterine artery can constitute a valuable aid to pregnancy diagnosis. (Courtesy of Dr M. J. Meredith.)

the examination, changes in the character of pulsation (fremitus) in the uterine arteries may be of more value than actual changes in diameter. The enlargement test is not meaningful before the sixth week of pregnancy, and even at eight weeks the test requires considerable experience (Meredith, 1976a). On the other hand, induction of fremitus in the uterine artery by means of specific digital pressure enables pregnancy detection with high accuracy from 30 days (Table IX.4). Characteristically, the induced fremitus is detected transiently during both progressive occlusion and progressive release of the vessel, but it is important to avoid pressure on the external iliac artery (Fig. IX.5) because this could also produce the phenomenon and thus lead to confusion (Meredith, 1976b).

Table IX.4. Detection of pregnancy in sows from the condition of fremitus in the uterine arteries produced in response to digital pressure. (After Meredith, 1976b.)

Stage of pregnancy (days)	Number of sows examined	Sows diagnosed as positive (i.e. pregnant)	
		Number	%
0–6	26	0	0
7–13	37	0	0
14–20	40	13	32
21–27	40	30	75
28–34	36	35	97
35–41	32	31	97
42–48	41	38	93
49–55	45	44	98
56–62	43	41	95
63–69	42	42	100
70–76	44	44	100
77–83	50	49	98
84–90	48	47	98
91–97	49	49	100
98–104	48	48	100
105–111	51	51	100
112–118	24	24	100

Other methods

For several reasons, referred to in part in the preceding discussion on methodology in sheep, the technique of laparoscopy (Wildt *et al.*, 1975) is not considered of significance as a field test for pigs, nor will radiographic detection of foetuses (Wenham *et al.*, 1969) be suggested as a method suitable for large-scale application.

Table IX.5. Summary of principal methods of pregnancy diagnosis currently applicable in large farm species. Further details of the methods are described in the text.

Species	Method	Time of application after breeding (days)	Reliability %
Cow	Rectal palpation of uterus and ovaries	>40	>90
	Blood progesterone concentration	20–22	>90
	Milk progesterone concentration	21–24	80–85[1]
Sheep	Laparoscopic inspection	17–28 onwards	90
	Blood progesterone concentration	17 or 18	>85
	Ultrasonic scanning	60–65 onwards	95
	Foetal radiography[2]	>50	95
Pig	Vaginal biopsy	20–30	90–95
	Ultrasonic detection	30–90	>95
	Fremitus induced in uterine artery	>30	>90
Horse	Rectal palpation of uterus and ovaries	>21	>85
	PMSG immunoassay	40–60	90

[1] See text for distinction between positive and negative results. [2] Twins and triplets are detectable by this method.

In most circumstances, the intensive pig producer requires an accurate method of monitoring the pregnancy status of his breeding herd, especially in situations where the animals do not have constant access to a suitable boar. As of writing, there is no simple method of diagnosing conception in the first two to two and a half weeks after mating, but application of either the vaginal biopsy technique or the ultrasonic approach can contribute to the timely identification of non-productive animals, enabling them either to be despatched from the breeding herd or alternatively to be identified, grouped and remated approximately

40–42 days after initial service. The economic decisions involved in these two courses of action do not fall within the scope of the present volume.

Pregnancy diagnosis in horses

Whilst the establishment of pregnancy in mares can be detected by exploration of the uterus and cervix *per rectum* (Day, 1940; Solomon, 1971), this is not usually performed with accuracy much earlier than 20–25 days after breeding, and even then considerable diagnostic skill is required. Application too soon after service may depress the conception rate (Voss and Pickett, 1975). Moreover, uterine palpation may not always be convenient and indeed can present hazards to the veterinary surgeon examining temperamental mares.

Progesterone assay

Progesterone concentrations in peripheral plasma have also been followed during the course of gestation in the mare (Smith, 1974; Holtan *et al.*, 1975), and could form the basis of a pregnancy test, but simpler and more attractive means of hormone estimation are available. Likewise, measurement of urinary oestrogens (Cole and Saunders, 1935; Raeside and Liptrap, 1975) is no longer considered practicable as a method of pregnancy detection.

Assay of PMSG

As in the case of women, a pregnancy test that is now widely used depends upon detection of a placental protein in a body fluid—urine in women but blood serum in mares. The endocrine product of the allanto-chorionic membranes, notably in the region of the endometrial cups, was first observed by Cole and Hart (1930) in the form of large quantities of gonadotrophin in the blood of mares at about 37–42 days of gestation, reaching a peak at about 70 days (Fig. III.1). This PMSG secretion was further characterised by Day and Rowlands (1940) and in much subsequent work (see Rowlands *et al.*, 1975). Assays for serum gonadotrophin can be either biological, in which case the ovarian or uterine response to mare serum is compared in immature rats or mice with the response of untreated control animals (Hart and Cole, 1932) or, more rapidly and conveniently, immunological where a haemagglutination-inhibition test is used. In the latter, serum from the mare under test is reacted, after extraction with acetone, with PMSG-coated formalinized erythrocytes in the presence of PMSG antiserum (Wide and Wide, 1963; Jeffcott *et al.*, 1969). Serum taken from mares after Day 40 of gestation inhibits the agglutination reaction in contrast to the influence of serum from non-pregnant mares (Fig. IX.6) and the

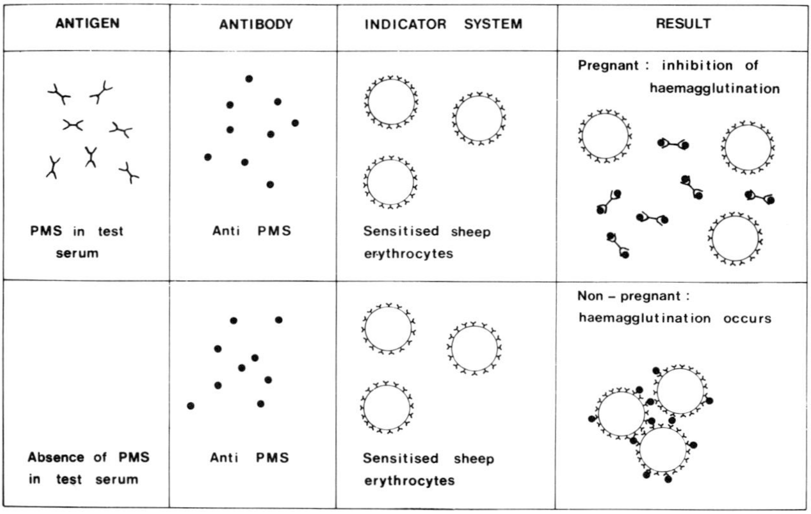

Fig. IX.6. Schematic illustration of the haemagglutination-inhibition assay for PMSG that may be used for pregnancy diagnosis in mares at 40–60 days of gestation. (Redrawn from Jeffcott *et al.*, 1969.)

assay can be read quantitatively (Wide and Wide, 1963; Allen, 1969a, b). Although the haemagglutination-inhibition assay may be more accurate than the palpation test at Days 40–60 of gestation (Parker *et al.*, 1975), embryonic or foetal loss can lead to false positives since PMSG concentrations in blood remain high for some time after embryonic death or surgical removal of the uterine contents (Allen, 1969b).

In view of the seasonal limitations to horse breeding in many countries, one quite strongly held view is that early palpation is the most attractive test, in conjunction with prostaglandin-induced luteal regression in non-cyclic mares considered not to have conceived (Irwin, 1975). In this manner, there should be a rapid return to oestrus and the opportunity to rebreed.

Surprisingly, ultrasonic methods of diagnosis (Fraser *et al.*, 1973) appear not yet to have won clinical acclaim.

Concluding remarks

Having gone in considerable detail into methods of detecting pregnancy in cattle, sheep, pigs and horses, it is worth reiterating that these approaches in themselves do nothing to improve herd or flock fertility. But in conjunction with a sound system of recording, they enable identification of animals that have mated successfully and, as a corollary, permit close supervision of those that have failed to conceive, preparatory either to reinsemination or to further investigative procedures. It is nonetheless possible to envisage, under intensive animal production systems of the future, circumstances in which females inseminated with appropriate numbers of viable spermatozoa at the optimum time relative to ovulation are presumed to have conceived. In this situation, those failing to respond positively to a pregnancy diagnosis test would *not* be further maintained and surveyed but rather eliminated straight away from the breeding herd, assuming appropriate replacement animals were available. In this hypothetical situation, the accuracy and timing of the pregnancy test become critical, for a significant loss of reproductive potential definitely occurs in Western Europe (and doubtless elsewhere) due to animals thought to be barren but in fact with established pregnancies being inadvertently sent to slaughter. A day spent in British abattoirs when mature females are being killed would emphasise all too clearly the truth of this statement.

References

Allen, W. R. (1969a). A quantitative immunological assay for pregnant mare serum gonadotrophin. *J. Endocr.* **43**, 581–591.

Allen, W. R. (1969b). Factors influencing pregnant mare serum gonadotrophin production. *Nature, London* **223**, 64–66.

Ardran, G. M. and Brown, T. H. (1964). X-ray diagnosis of pregnancy in sheep with special reference to the determination of the number of foetuses. *J. agric. Sci.* **63**, 205–207.

Aschheim, S. and Zondek, B. (1928). Schwangerschaftsdiagnose aus dem Harn (Durch Hormonnachwies). *Klin. Wschr.* **7**, 8–9.

Booth, J. M. and Holdsworth, R. J. (1976). The establishment and operation of a central laboratory for pregnancy testing in cows. *Brit. vet. J.* **132**, 518–528.

Cerini, M., Findlay, J. K. and Lawson, R. A. S. (1976). Pregnancy-specific antigens in the sheep: application to the diagnosis of pregnancy. *J. Reprod. Fert.* **46**, 65–69.

Cole, H. H. and Hart, G. H. (1930). The potency of blood serum of mares in progressive stages of pregnancy in effecting the sexual maturity of the immature rat. *Amer. J. Physiol.* **93**, 57–68.

Cole, H. H. and Saunders, F. J. (1935). The concentration of gonad-stimulating hormone in blood serum and of oestrin in the urine throughout pregnancy in the mare. *Endocrinology* **19**, 199–208.

Darling, J. A. B., Kelly, R. W., Laing, A. H. and Harkness, R. A. (1972). The isolation and identification of progesterone obtained from cows' milk during pregnancy. *J. Endocr.* **54**, 347–348.

Day, F. T. (1940). Clinical and experimental observations on reproduction in the mare. *J. agric. Sci.* **30**, 244–261.

Day, F. T. and Rowlands, I. W. (1940). The time and rate of appearance of gonadotrophin in the serum of pregnant mares. *J. Endocr.* **2**, 255–261.

Deas, D. W. (1977). Pregnancy diagnosis in the ewe by an ultrasonic rectal probe. *Vet. Rec.* **101**, 113–115.

Diehl, J. R. and Day, B. N. (1973). Utilisation of frozen sections with the vaginal biopsy technique for early pregnancy diagnosis in swine. *J. Anim. Sci.* **37**, 114–117.

Done, J. T. and Heard, T. W. (1968). Early pregnancy diagnosis in the sow by vaginal biopsy. *Vet. Rec.* **82**, 64–68.

Edgerton, L. A. and Hafs, H. D. (1973). Serum luteinizing hormone, prolactin, glucocorticoid, and progestin in dairy cows from calving to gestation. *J. Dairy Sci.* **56**, 451–458.

Ellendorff, F., Meyer, J. N. and Elsaesser, F. (1976). Prospects and problems of pregnancy and fertility diagnosis in the pig by aid of progesterone determination. *Brit. vet. J.* **132**, 543–550.

Fraser, A. F. and Robertson, J. G. (1967). The detection of foetal life in ewes and sows. *Vet. Rec.* **80**, 528–529.

Fraser, A. F. and Robertson, J. G. (1968). Pregnancy diagnosis and detection of foetal life in sheep and pigs by an ultrasonic method. *Brit. vet. J.* **124**, 239–244.

Fraser, A. F., Keith, N. W. and Hastie, H. (1973). Summarised observations on the ultrasonic detection of pregnancy and foetal life in the mare. *Vet. Rec.* **92**, 20–21.

Fraser, A. F., Nagaratnam, V. and Callicott, R. B. (1971). The comprehensive use of Doppler ultra-sound in farm animal reproduction. *Vet. Rec.* **88**, 202–205.

Hansen, L. H. and Christiansen, I. J. (1976). The accuracy of porcine pregnancy diagnosis by a newly developed ultrasonic A-scan tester. *Brit. vet. J.* **132**, 66–67.

Harkness, R. A. and Darling, J. A. B. (1971). Steroid levels in milk. *Acta endocr. Copenh. Suppl.* **155**, 92.

Hart, G. H. and Cole, H. H. (1932). A practical method for the diagnosis of pregnancy in the mare. *J. Amer. Vet. Med. Assn* **80**, 604–614.

Heap, R. B., Gwyn, M., Laing, J. A. and Walters, D. E. (1973). Pregnancy diagnosis in cows; changes in milk progesterone concentration during the oestrous cycle and pregnancy measured by a rapid radioimmunoassay. *J. agric. Sci.* **81**, 151–157.

Heap, R. B., Holdsworth, R. J., Gadsby, J. E., Laing, J. A. and Walters, D. E. (1976). Pregnancy diagnosis in the cow from milk progesterone concentration. *Brit. vet. J.* **132**, 445–464.

Holtan, D. W., Nett, T. M. and Estergreen, V. L. (1975). Plasma progestagens in pregnant mares. *J. Reprod. Fert., Suppl.* **23**, 419–424.

Hulet, C. V. (1972). A rectal-abdominal palpation technique for diagnosing pregnancy in the ewe. *J. Anim. Sci.* **35**, 814–819.

Irwin, C. F. P. (1975). Early pregnancy testing and its relationship to abortion. *J. Reprod. Fert., Suppl.* **23**, 485–488.

Jeffcott, L. B., Atherton, J. G. and Mingay, J. (1969). Equine pregnancy diagnosis. *Vet. Rec.* **84**, 80–83.

Laing, J. A. (1970) Editor. "Fertility and Infertility in the Domestic Animals", 2nd edition. Baillière Tindall and Cassell, London.

Laing, J. A. and Heap, R. B. (1971). The concentration of progesterone in the milk of cows during the reproductive cycle. *Brit. vet. J.* **127**, xix–xxii.

Lindahl, I. L. (1966). Detection of pregnancy in sheep by means of ultrasound. *Nature, London* **212**, 642–643.

Lindahl, I. L. (1969). Comparison of ultrasonic techniques for the detection of pregnancy in ewes. *J. Reprod. Fert.* **18**, 117–120.

Lindahl, I. L., Totsch, J. P., Martin, P. A. and Dziuk, P. J. (1975). Early diagnosis of pregnancy in sows by ultrasonic amplitude-depth analysis. *J. Anim. Sci.* **40**, 220–222.

Martal, J. and Lacroix, M. C. (1978). Importance d'une trophoblastine dans le contrôle endocrinien du corps jaune gestatif chez la brebis, au moment de l'implantation. *In* "Implantation de l'Oeuf" Colloque S.N.E.S.F., pp. 193–208. Masson, Paris.

Meredith, M. J. (1976a). Manual techniques of pregnancy diagnosis. *Proc. Pig Vet. Soc.* **1**, 1–4.

Meredith, M. J. (1976b). Pregnancy diagnosis in the sow by examination of the uterine arteries. *Proc. Int. Pig Vet. Soc. Congress, Ames, Iowa*, **D5**, 1–2.

Morton, D. B. and Rankin, J. E. F. (1969). The histology of the vaginal epithelium of the sow in oestrus and its use in pregnancy diagnosis. *Vet. Rec.* **84**, 658–662.

Parker, W. G., Sullivan, J. J. and Larson, L. L. (1975). Comparison of the methods of rectal palpation and haemagglutination-inhibition assay for diagnosis of pregnancy in mares. *J. Reprod. Fert., Suppl.* **23**, 489–493.

Phillippo, M. and Rhind, S. M. (1977). A new laparoscopic technique for the diagnosis of pregnancy and the estimation of foetal loss in sheep. *J. agric. Sci.* **89**, 251–252.

Phillippo, M., Swapp, G. H., Robinson, J. J. and Gill, J. C. (1971). The diagnosis of pregnancy and estimation of foetal numbers in sheep by laparoscopy. *J. Reprod. Fert.* **27**, 129–132.

Raeside, J. I. and Liptrap, R. M. (1975). Patterns of urinary oestrogen excretion in individual pregnant mares. *J. Reprod. Fert., Suppl.* **23**, 469–475.

Rizzoli, D. J., Winfield, C. G., Howard, T. J., Englund, I. K. J. and Goding, J. R. (1976). Diagnosis of multiple pregnancy in ewes on a field scale. *J. agric. Sci.* **87**, 671–677.

Robertson, H. A. and Sarda, I. R. (1971). A very early pregnancy test for mammals: its application to the cow, ewe and sow. *J. Endocr.* **49**, 407–419.

Roche, J. F., Prendiville, D. J. and Gosling, J. (1978). Synchronisation of oestrus and pregnancy diagnosis in heifers bred in autumn and winter. *Vet. Rec.* **102**, 12–14.

Rowlands, I. W., Allen, W. R. and Rossdale, P. D. (1975). Equine reproduction. *J. Reprod. Fert., Suppl.* **23**, 746 pp.

Rowson, L. E. A. and Dott, H. M. (1963). A hazard of pregnancy diagnosis in cattle: early foetal size. *Vet. Rec.* **75**, 865–866.

Saumande, J. and Thimonier, J. (1972). Utilisation d'un diagnostic précoce de gestation par estimation du taux de progestérone dans le sang périphérique de brebis. *Annls Biol. anim. Biochim. Biophys.* **12**, 661–665.

Shemesh, M., Ayalon, N. and Lindner, H. R. (1968). Early effect of conceptus on plasma progesterone level in the cow. *J. Reprod. Fert.* **15**, 161–164.

Smith, I. D. (1974). Progesterone concentrations in the peripheral plasma of the mare during pregnancy. *Res. vet. Sci.* **16**, 114–116.

Solomon, W. J. (1971). Rectal examination of the cervix and its significance in early pregnancy evaluation of the mare. *Proc. 17th A. Conv. Am. Assn Equine Prac.*, pp. 73–80.

Staples, L. D., Lawson, R. A. S. and Findlay, J. K. (1978). The occurrence of an antigen associated with pregnancy in the ewe. *Biol. Reprod.* **19**, 1076–1082.

Thimonier, J. (1973). Diagnostic précoce de la gestation par l'estimation du taux de progestérone plasmatique chez la brebis, la vache et la jument. *Rec. Méd. Vét., Alfort,* **149**, 1303–1318.

Voss, J. L. and Pickett, B. W. (1975). The effect of rectal palpation on the fertility of cyclic mares. *J. Reprod. Fert., Suppl.* **23**, 285–290.

Walker, D. (1972). Pregnancy diagnosis in pigs. *Vet. Rec.* **90**, 139–144.

Wenham, G., McDonald, I. and Elsley, F. W. H. (1969). A radiographic study of the development of the skeleton of the foetal pig. *J. agric. Sci.* **72**, 123–130.

Wenham, G. and Robinson, J. J. (1972). Radiographic pregnancy diagnosis in sheep. *J. agric. Sci.* **78**, 233–238.

Wide, M. and Wide, L. (1963). Diagnosis of pregnancy in mares by an immunological method. *Nature, London* **198**, 1017–1018.

Wildt, D. E., Morcom, C. B. and Dukelow, W. R. (1975). Laparoscopic pregnancy diagnosis and uterine fluid recovery in swine. *J. Reprod. Fert.* **44**, 301–304.

Wilson, I. A. N. and Newton, J. E. (1969). Pregnancy diagnosis in the ewe: a method of use on the farm. *Vet. Rec.* **84**, 356–358.

Wintenberger-Torres, S. (1978). Rôle actif de l'embryon avant l'implantation. *In* "Implantation de l'Oeuf", Colloque S.N.E.S.F., pp. 181–192. Masson, Paris.

Chapter X

Control of Parturition and the Incidence of Perinatal Loss

Introduction

One of the more exciting areas of investigation in reproductive physiology during the last ten years, and one in which studies in domestic animals have contributed more conspicuously than work in laboratory species, has been the process of parturition. Furthermore, the unravelling of the sequence of foetal and maternal endocrine events in late gestation, and the emergence in this regard of certain distinctions between cattle and sheep on the one hand and pigs on the other has given rise to a useful functional approach for controlling parturition in these species. The fact that gestation length and, more critically, the actual timing of the onset of birth can now be regulated fairly precisely by a semi-physiological interruption in the preparturient sequence of events could be of considerable value under intensive systems of animal production. This is not only as a means of reducing the overall period of supervision of animals at term but also for concentrating the events at birth within the working day. On clinical grounds, one could also present arguments for regulating the time of parturition in specific circumstances (see below). Nonetheless, procedures aimed at interrupting the normal course of gestation should be kept in a biological perspective, for the foetus is not normally delivered until it is adequately prepared to be expelled from the uterus and to be denied the benefits of a placental mode of nutrition. To hasten this event even by so small a margin as 48–72 hours may, on occasions, exacerbate problems of neonatal viability, although there is currently no firm evidence that this is so when the nutrition of the dam has been adequate. But because the foetus is building up its energy reserves most rapidly in the lattermost part of gestation, notably its liver and muscle glycogen, the precocious induction of parturition could have serious consequences, especially in situations where prompt suckling of the offspring cannot be ensured. As always, therefore, the attractions of intervention in any given breeding system must be balanced against potential risks to the offspring.

Rationale for inducing parturition

As already suggested, the arguments for attempting to regulate the onset of parturition are associated principally with managerial or veterinary considerations. Within the former category, the reasons concern economical use of the labour force and more efficient use of facilities that may be limited such as calving boxes or heated farrowing pens. Specialist personnel would be needed for supervising and assisting, if necessary, with the process of birth, ensuring that the respiratory passages of the new-born are clear, and that suckling is established. If calving, lambing or farrowing in groups of animals was confined to a limited and predetermined period of time, then inputs of semi-skilled labour could be reduced and attempts at cross-fostering of young made more successful. Thus, the introduction of an element of predictability into the time of parturition should be welcome.

Among the diverse veterinary considerations that would benefit from prompt induction of parturition are a bone fracture in the dam or acute disease in late pregnancy where the life of the dam or foetus is at risk. Intervention would also be warranted if complications were expected from the birth of twins, or from dairy cows in calf to large beef bulls, or if a high incidence of dystocia had previously prevailed in the herd. Metabolic disorders in the dam may also suggest induced parturition as a means of reducing the particular stress, as may the condition of hydrops of one of the foetal compartments (e.g. hydrops allantois). These are, of course, but a few examples, and further discussion of the topic can be found in appropriate veterinary texts (e.g. Arthur, 1964).

A final line of reasoning that might be offered for controlling parturition is that if a programme of synchronisation of oestrus and batch insemination has been instigated (see Chapter II), then the ultimate intervention of attempting to synchronise birth more closely than would occur spontaneously could have attractions in suitably intensive production systems. Notwithstanding these various justifications, it must be stressed that the procedure of inducing parturition of itself does nothing to improve fertility, this point being analogous to that made in the preceding chapter on diagnosis of pregnancy. However, by permitting increased supervision of treated animals and thereby reducing perinatal losses, induction of parturition should be a means of contributing towards increasing overall output.

Sequence of endocrine events at parturition

Various components of foetal stress are currently considered to underlie the initial triggering of the chain of events that culminates in expulsion of the offspring. The precise nature of the stresses upon the foetus remains rather poorly defined, but bearing in mind such factors as the rapid increase in size of the foetus and, to a lesser extent its placental membranes and fluid contents in the lattermost part of gestation, the development of active movements in the limbs and lungs of the foetus, the increasing problems of placental exchange of nutrients and waste products, it becomes clear that stress could arise from a combination of physical, respiratory and metabolic factors. More subtle endocrine interactions are doubtless also at play as the foetus endeavours to assert its own hormonal competence and somatic independence from the mother. Research in a number of laboratories is still aimed at defining a specific trigger to the events at parturition, but a point of view more widely favoured at the present time is that a combination of factors will act in unison to precipitate the time of birth: an event of such overwhelming significance to both mother and offspring would almost certainly have avoided dependence upon a single index of foetal maturity. It is within this multifactorial context that the response to a variety of foetal stresses is perhaps best interpreted.

Several lines of evidence suggest that the foetus rather than the mother predominates in initiating the endocrine sequence of events that culminates in parturition. The foetal hypothalamus monitors and responds to appropriate levels of physiological stress by releasing elevated quantities of ACTH-RH into the hypothalamo-pituitary portal bloodstream whence it acts on the foetal anterior pituitary gland to stimulate enhanced levels of ACTH release, to which the foetal adrenal cortex responds by secreting increased concentrations of corticosteroids. This pathway in the foetus appears essential for the normal initiation of the terminal events of pregnancy. Electrocoagulation of the foetal pituitary gland in sheep (Liggins *et al.*, 1967) or indeed foetal hypophysectomy (Bosc *et al.*, 1974) or decapitation in pigs (Stryker and Dziuk, 1975) leads to prolonged gestation. Likewise, the condition of an insufficiently developed foetal pituitary or adrenal cortex (Kennedy *et al.*, 1967) or surgical removal of the foetal adrenal glands (Drost and Holm, 1968) prolongs gestation in sheep and cows. The concept of a crucial involvement of the foetal adrenal glands was further extended by the observation that infusing cortisol, dexamethasone (a synthetic corticosteroid) or ACTH into foetal lambs could cause premature delivery in intact or indeed hypophysectomised animals (Liggins, 1968, 1969).

Before focusing upon the endocrine responses in the placenta and mother to this sequence of foetal events, it should be borne in mind that foetal adreno-cortical activity is also involved in the lattermost part of gestation in regulating the build-up of energy reserves in the foetus, especially in the form of liver glycogen. Thus, whilst a surge of foetal adrenal activity is necessary to precipitate the onset of parturition, an increasing level of function in the adrenal cortex has been required in late gestation to prepare the foetus metabolically for the trauma of birth and for the immediate sequel to that event of exposure to a more demanding environment. Similarly, the specific preparation of the lungs that involves coating of the alveoli with surfactant seems to be regulated in part by heightened foetal adrenal activity (see Liggins *et al.*, 1979). And, as an anatomical basis for this enhanced activity, Comline and Silver (1961) have noted that an exponential increase in foetal adrenal weight occurs only at the end of pregnancy (Fig. X.1), and Liggins (1968) has commented on the rapid growth of the adrenal cortex in foetal lambs at this time.

The response to elevated levels of foetal corticosteroid secretion that paves the way to parturition involves secretion of oestrogen and prostaglandin $F_{2\alpha}$ ($PGF_{2\alpha}$) by the placental (including uterine) tissues. An endocrine stimulus of foetal origin is thereby producing responses in the foetal and maternal placental tissues, and the primary rôle of the cortisol and $PGF_{2\alpha}$ release is to cause a decrease in progesterone secretion in the maternal circulation. In pigs, functional ovaries are essential for the maintenance of gestation (Chapter VI), and so the uterine $PGF_{2\alpha}$ acts to terminate progesterone secretion by the corpora lutea. In sheep, by contrast, the placenta becomes a principal site of progesterone synthesis for the maintenance of pregnancy after approximately Day 50 of gestation, and cortisol and/or $PGF_{2\alpha}$ therefore act primarily to reduce progesterone secretion by the syndesmo-chorial placenta (Fig. X.2) in part, at least, by its enzymatic conversion to oestrogens. The inhibitory effects of foetal cortisol and $PGF_{2\alpha}$ (Fairclough *et al.*, 1975; Edqvist *et al.*, 1976; Hunter *et al.*, 1977) are probably on ovarian

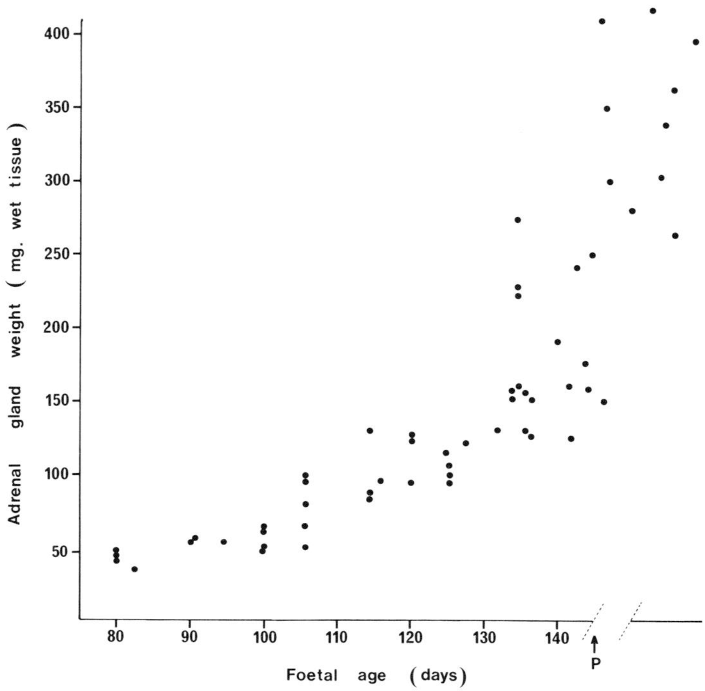

Fig. X.1. To illustrate growth of the foetal adrenal gland in sheep during the last 45 days of gestation. (Redrawn from Comline and Silver, 1961.)

progesterone secretion in cattle, even though pregnancy may be maintained in the ovariectomised cow after about Days 230–240 (see Erb *et al.*, 1968), which suggests extra-ovarian sources or reserves of progesterone in late gestation.

Approximately coincident with the reduction in systemic progesterone in the mother, there is a dramatic increase in the concentration of circulating oestrogens (Fig. X.3) which is derived principally from the foeto-placental unit rather than the ovaries, and may be metabolically dependent upon elevated corticosteroid levels. Such a relationship is certainly found in sheep where the cortisol activates placental steroidogenic enzymes that convert placental progesterone into oestrogens (Flint *et al.*, 1975). An increase in oestrogens at term is also found in the pig, and can be clearly monitored in the urinary excretion product of oestrone (Fèvre *et al.*, 1968; Fèvre, 1970); and inspection of the ovaries in sows in late gestation indicates that mature Graafian follicles are not present at that time. Moreover, ovariectomy of the sow on or after Day 108 of pregnancy neither prevents the normal course of

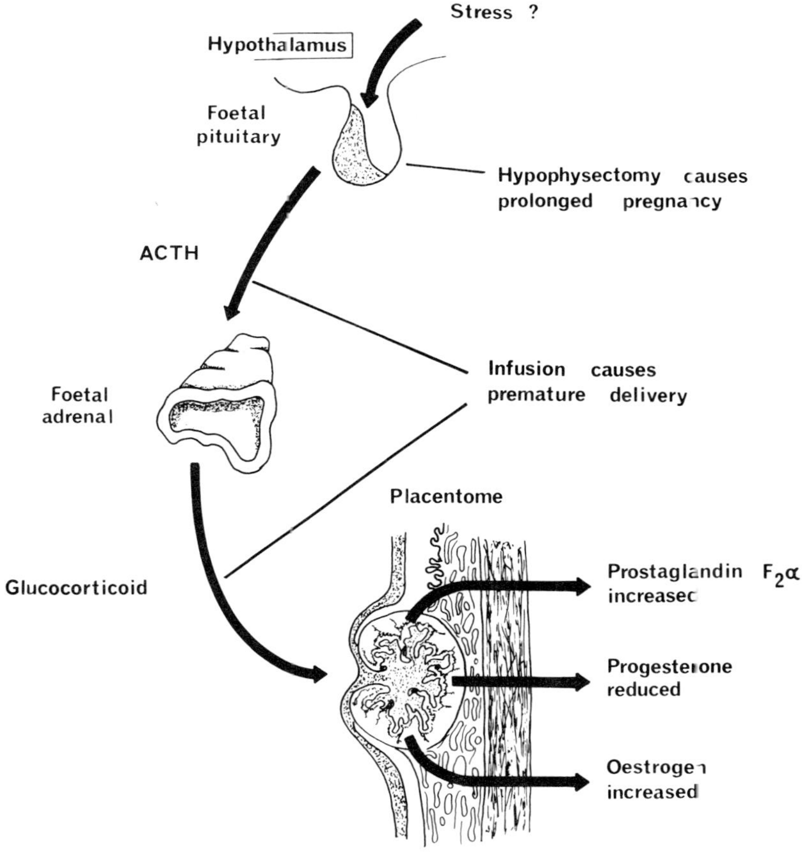

Fig. X.2. To illustrate some principal endocrine events associated with parturition in the ewe, with emphasis on the pathway between foetal ACTH secretion and placental secretion of oestrogens. The influence of experimental procedures is also shown. (Adapted from Liggins, 1969.)

parturition nor the *post partum* display of oestrus that is dependent upon the foeto-placental synthesis of oestrogen (Holness and Hunter, 1975). Although one influence of this much elevated concentration of placental oestrogens is doubtless upon the hypothalamus and pituitary gland, of more immediate concern is its influence upon the uterus and to a lesser extent the cervix and vagina. To put this fact in perspective, it is necessary to note that a classical effect of progesterone in pregnancy is to act upon the smooth muscle elements of the uterus to inhibit contractile activity, thereby providing what is commonly termed the myometrial block that enables the physical establishment and maintenance of pregnancy (Csapo, 1956, 1969). But whereas quiescence of the uterus in a contractile sense favours development of the foetus, the reverse situation is obviously essential for its expulsion, and it is in this context that withdrawal of progesterone and a marked elevation of circulating oestrogens must be viewed. In a phrase, oestrogens appear to act in very

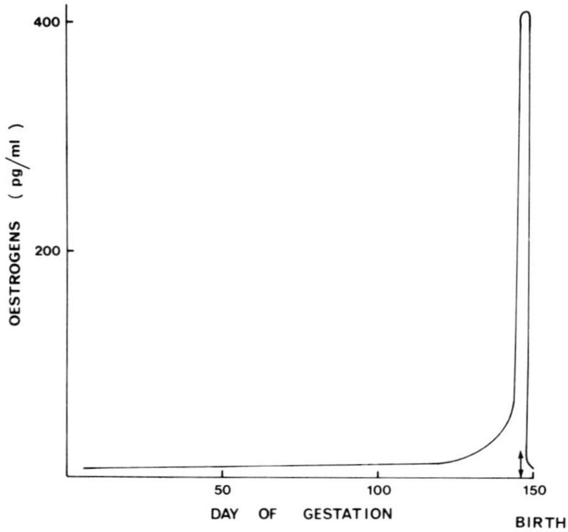

Fig. X.3. To show the dramatic increase in the plasma concentration of total unconjugated oestrogens shortly before parturition in sheep. The arrow indicates the time of parturition. (Redrawn from Challis, 1971.)

late gestation both to restore the inherent contractile activity of the myometrium and to increase its sensitivity to smooth muscle stimulating agents, thereby facilitating expulsion of the contents of the gravid uterus.

Even though the contractile potential of the uterus may have been restored by these steroid mediated events, a change in the musculature of the cervix must also occur before it can undergo dilation to permit delivery. The polypeptide hormone, relaxin, of ovarian (luteal) and perhaps also of placental sources promotes appropriate changes in the cervix and in the pelvic ligaments. Plasma relaxin concentrations may increase 50 fold over the last few days of gestation in pigs (Sherwood *et al.*, 1975). But changes in the secretions of the cervix remain principally under steroid control, mucous secretions being favoured by oestrogen predominance. Once the oestrogen and relaxin promoted effects on the reproductive tract have occurred, heightened contractions of the myometrium are found in response to oxytocin secretion by the maternal neurohypophysis, although a considerable contractile force has already developed *before* oxytocin release is detected. A local release of prostaglandins also contributes to the myometrial contractions, and recent work has suggested a possible rôle for oxytocin in stimulating release of uterine prostaglandins (Fig. X.4). In summary, therefore, the

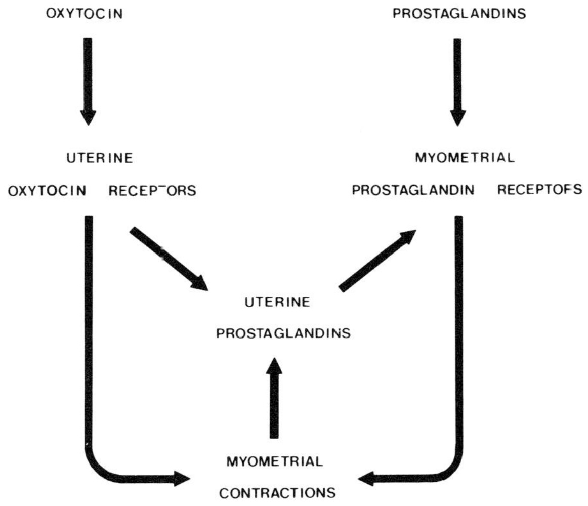

Fig. X.4. Diagrammatic representation of the stimulation by oxytocin of a uterine release of prostaglandins during parturition. Such locally released prostaglandins may reinforce the myometrial contractions promoted by oxytocin. (Redrawn after Chan, 1977.)

sequence of endocrine events in the foetus that precipitates birth is due to enhanced cortisol secretion that in turn:

(1) enables placental conversion of progesterone to oestrogens, and/or initiates uterine prostaglandin $F_{2\alpha}$ secretion, giving a reduction in progesterone titres; with
(2) uterine contractile activity being potentiated by oestrogens from the foeto-placental complex; and
(3) the foetus being expelled by powerful myometrial contractions provoked by oxytocin and/or prostaglandins.

As far as uterine contractile activity is concerned, work in sheep has indicated that labour begins about 12 hours before expulsion of the foetus (Hindson *et al.*, 1968), with clearly co-ordinated and propagated contractions of one minute duration occurring at a regular frequency of 30 per hour (Fitzpatrick and Dobson, 1979).

Control of parturition in cattle

Gestation lengths vary from 275–290 days, pregnancy usually being 5 to 10 days shorter in cattle carrying twins (Salisbury and Van Demark, 1961). Twenty-four to 48 hours before parturition, maternal plasma progesterone concentrations diminish significantly from 2·0–2·5 ng ml^{-1} to 0·7–0·9 ng ml^{-1} at term (Henricks *et al.*, 1972), although the nature and mode of action of the luteolytic stimulus require clarification. Progesterone withdrawal may constitute an important part of the mechanism initiating the onset of parturition, but very considerable increases in oestrogen concentration in peripheral blood are also found just before and at parturition; a sharp decline from concentrations of about 65 pg ml^{-1} to 4–10 pg ml^{-1} soon follows and the low values persist throughout most of the *post partum* period (Henricks *et al.*, 1972; Echternkamp and Hansel, 1973; Corah *et al.*, 1974). The concentration of individual oestrogens over this period has been presented in detail by Robertson (1974) but the question still remains as to the source of these hormones at parturition; Henricks *et al.* (1972) infer that the bulk of the increase in oestrogen before parturition is of placental origin. The relationship between these maternal blood levels of progesterone and oestrogens, and that of corticoids has yet to be critically documented, but increasing cortisol levels in the foetal calf have been noted just prior to birth (Comline *et al.*, 1974; Fairclough *et al.*, 1975; Fig. X.5) and a threefold increase in maternal serum glucocorticoids at parturition has also been reported (Edgerton and Hafs, 1973). Conversely, adrenal insufficiency in calves causes prolonged pregnancy (Holm *et al.*, 1961).

Induction of calving by administration of corticosteroids (Adams, 1969; Adams and Wagner, 1970) may well be of value in large production units, particularly if the majority of pregnancies in a herd commenced after an oestrous synchronisation programme. It may also be used in herds with a characteristically high incidence of dystocia, in situations where calving difficulties might be anticipated (as discussed above), where gestation is unduly prolonged, or where there are important seasonal constraints upon calving dates such as occur in New Zealand under a highly seasonal system of grassland agriculture (Welch *et al.*, 1973); here it would give the cows which conceived last the opportunity to be inseminated in time to calve at the optimum date the following year. Treatment involves injection of the mother with a

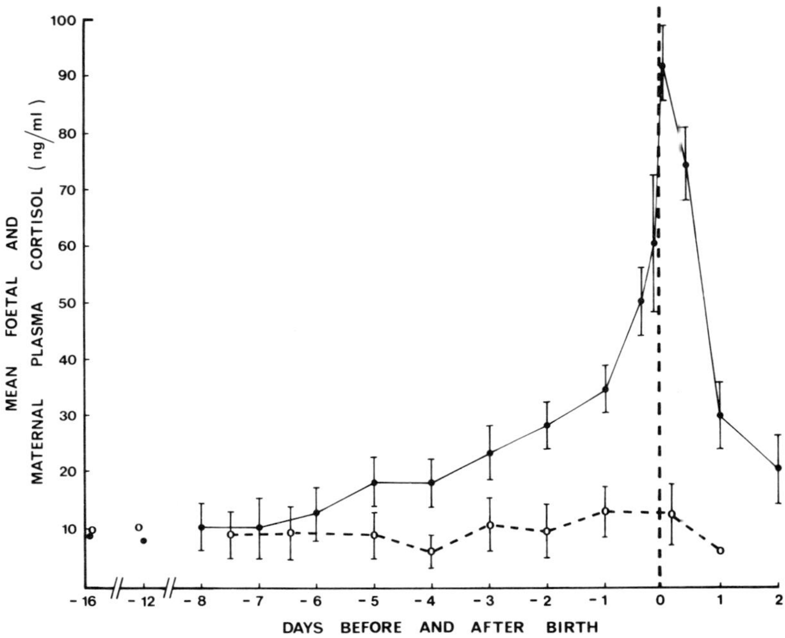

Fig. X.5. Graph to depict the increasing concentrations of cortisol in foetal plasma (solid line) and maternal plasma (dotted line) in the final days of gestation in the cow. (Redrawn from Comline *et al.*, 1974.)

preparation of a synthetic corticosteroid such as dexamethasone, betamethasone or flumethasone either in short- or long-acting form. The drug is in the free alcohol or soluble ester form when short-acting and usually induces calving within 2–4 days, whereas the long-acting versions are most frequently suspensions or insoluble esters of the synthetic steroids and provoke calving in about 2–3 weeks. When applied after approximately Days 265–270 of gestation, calving soon follows treatment with a short-acting corticosteroid (e.g. 20 mg of dexamethasone) given as a single intramuscular injection and, as a general rule, the nearer to the expected time of parturition that treatment is imposed, the more likely it is to be successful in inducing parturition. Even so, complications frequently arise in treated cattle, of which retention of the placenta and dropping of stillborn calves are the most significant (Jochle, 1971), but delayed returns to successful breeding and effects upon the milk yield have also been noted. In a study involving dexamethasone treatment of 126 cows, mostly of the Jersey breed served by a Charolais bull, Christiansen and Hansen (1974)

Table X.1. Details of the incidence of stillborn calves and retained placentae in cattle in situations of parturition induced with dexamethasone at different stages of gestation. (Adapted from Christiansen and Hansen, 1974.)

Details of calvings	Treatment to calving interval (days)	Time of treatment within the gestation period				
		< Day 277	Days 277–285	> Day 285	Information missing	Total
Number of calvings	<4	21	17	40	14	92
	>5	16	11	4	3	34
Ratio of live: stillborn calves	<4	19:5	18:1	33:7	13:1 + 1	83:14 + 1
	>5	11:5 + 1	11:0	3:1	1:2	26:8 + 1
Number of retained placentae	<4	13	5	9	4	31
	>5	1	4	1	1	7

+ 1 refers to cases of missing information concerning viability of the calf.

recorded an average of 73% of the animals calving within four days of treatment, but the incidence of stillborn calves averaged 16% and that of retained placenta 34%. A further breakdown of their results with respect to day of treatment is given in Table X.1, from which it is noted that the number of stillborn calves was increased when treatment was given before Day 277; most cases of retained placenta were also in this group.

On the other hand, use of long-acting (depot) corticosteroids may lead to calving several weeks after injection, with a lower incidence of retained placenta but a very much higher proportion of calves born dead (Welch *et al.*, 1973; O'Farrell and Crowley, 1974). Presumably, the normal sequence of *pre-partum* changes in the syndesmo-chorial placenta is better able to proceed in this protracted situation, and in most animals the membranes are shed within a few days of calving. In an extensive field trial in New Zealand (Welch *et al.*, 1973), cows were treated about 4–5 weeks before term and the induced calving occurred 10–16 days later (Table X.2). The incidence of retained placenta was 9–22%, but 17–45% of the calves were born dead (Table X.3) and most

Table X.2. The time of treatment for inducing parturition in relation to the expected time of calving and actual calving. (From Welch *et al.*, 1973).

Nature of steroid preparation	Due calving—actual calving (days)	Time of treatment to due calving (days)
Dexamethasone trimethyl-acetate	16·7	33·0
Flumethasone suspension	20·3	30·5
Triamcinolone acetonide	19·5	35·3

Table X.3. The incidence of calf mortality following induced parturition with three different steroid preparations. (From Welch *et al.*, 1973).

Nature of steroid preparation	Number of calves: Born	Number of calves: Born dead	Mortality %
Dexamethasone trimethyl-acetate (20 mg)	209	65	31
Flumethasone suspension (10 mg)	165	52	32
Triamcinolone acetonide (30 mg)	166	74	45
Dexamethasone trimethyl-acetate[1] (20 mg)	333	58	17

[1] Trial conducted in separate locality from the first three.

of these were conspicuously smaller than full-term calves. Under the specific seasonal constraints found in New Zealand, it appears that the risk of losing such a high proportion of calves may be acceptable to many farmers in order to restore late calving cows to the desired calendar of events; this is witnessed by the 150 000 cows per year in New Zealand alone presumed to have been induced to calf by treatment with corticosteroids (Welch *et al.*, 1977). Moreover, with veterinary supervision at the time of calving, and delivery at two weeks of prematurity rather than three, mortality may be reduced to less than 10% (Table X.4).

Table X.4. The influence of the stage of induced calving on the incidence of calf mortality and retained placenta. (From Welch *et al.*, 1977.)

Extent of prematurity at birth (days)	Number of animals	Incidence of calf mortality (%)	Incidence of retained placenta (%)
1–7	118	7	16
8–14	163	7	16
15–21	144	11	17
22–28	168	13	24
29–35	32	9	28
36–49	22	32	32
Total	647	9·9	18·5

Whilst short-acting corticosteroid drugs may depress milk yield, this seems not to be the case, or only marginally so, with long-acting corticosteroids, although *pre-partum* milking may be required with the latter treatment. However, use of long-acting corticosteroids after Day 260 may not significantly advance the calving date, whilst their use earlier than about Day 250 will almost certainly increase the incidence of stillbirths or neonatal mortality, and individual cows will respond unpredictably to the drug. As to the subsequent fertility of cows following treatments to induce calving, this is apparently normal and not complicated by a retarded expulsion of the placenta (Wagner *et al.*, 1974).

An overall perspective emerging from work published to date would therefore seem to suggest that induction of parturition in cattle by either form of corticosteroid may well be warranted as a managerial practice in specific circumstances, but the value of routine use of these drugs is invariably outweighed by the potential disadvantages both to the mother and to the calf. Very few production systems would tolerate a calf mortality figure of even 10%. Nor does a combined treatment with dexamethasone and the prostaglandin analogue cloprostenol 8–12 days later (Day, 1977, 1979) overcome these objections, even though the precision of the calving response can be improved and the onset of subsequent cyclic activity advanced. And calving induced with prostaglandin alone is frequently associated with a high incidence of dystocia, especially in heifers (see Henricks *et al.*, 1977).

Control of parturition in sheep

Among domestic animals, the sequence of endocrine events heralding the onset of parturition is best understood in sheep, in which species gestation lengths may commonly vary between 144–153 days. As already indicated, many individual observations both in the field and in different laboratories (see Van Rensburg, 1967; Lanman, 1968) gradually began to suggest a dominant rôle for the foetus in initiating the process of parturition. But it is perhaps most largely due to the painstaking studies of Liggins and his colleagues (Liggins *et al.*, 1972; Liggins *et al.*, 1973; Liggins, 1979) and of Thorburn *et al.* (1972, 1979), that an integrated picture of the terminal sequence of events in ovine pregnancy was constructed, and the requirement for a functional foetal hypothalamic-pituitary-adrenal axis emphasised. The activity of this axis underlies the increased level of foetal plasma cortisol that precedes birth. The involvement of uterine secretion of $PGF_{2\alpha}$ as a luteolysin during the oestrous cycle has already been discussed in Chapter VI, and now, once again, a progesterone-reducing influence of this compound is presumed to occur (see Thorburn *et al.*, 1972) following treatments to simulate foetal adreno-cortical activity. But progesterone falls near term in the spontaneous situation largely because enhanced foetal cortisol secretion activates enzymatic formation of oestrogen from placental progesterone. These changes in steroid levels in general precede a rise in prostaglandin production (see Challis *et al.*, 1979).

Because of the fact that high doses of dexamethasone (24 mg) administered to the ewe would reduce peripheral plasma concentrations of progesterone and induce parturition (Fylling, 1971), studies progressed to the stage of field-scale attempts to induce lambing with the aid of synthetic corticosteroids administered systemically. Bosc

(1972) showed that after injecting Ile-de-France ewes intramuscularly with 16 mg dexamethasone on the 144th day of gestation following a synchronised oestrus, lambing occurred at a mean of 45 hours after treatment (cf. 114 hours in the controls; Table X.5) and 75% of the

Table X.5. The mean interval between dexamethasone injection and induced lambing in ewes of the Île-de-France breed when treatment was given on the 144th day of gestation. (Adapted from Bosc, 1972.)

Dosage of dexamethasone (mg)	Number of ewes	Interval from injection to parturition Mean (hours)	± S.D.
16	19	45·0	12·4
8	20	47·3	17·3
4	21	80·6	44·0
Control	64	114·3	44·3

births in ewes receiving 8 or 16 mg of dexamethasone were concentrated within a period of 24 hours. Retention of the placenta was not observed in either treated or control ewes, nor were deleterious side effects observed after treatment; and as judged by the rate of growth of lambs, lactation was not adversely affected by the dexamethasone treatment. Joyce (1974) conducted similar trials in Ireland as a sequel to synchronisation of oestrus and synchronised breeding. Two groups of ewes of mixed breeds were injected with 12 or 16 mg of dexamethasone in solution on Day 144 of gestation, after which the mean interval to parturition was approximately 55 hours in the first group (12 mg dexamethasone) and 52 hours in the second (Table X.6). Apart from a relatively high incidence of neonatal mortality in lambs of low birth weight associated with delivery of triplets, no untoward effects of treatment were reported.

Using another form of synthetic corticosteroid, betamethasone, Penning and Gibb (1977) also induced parturition as a sequel to synchronised breeding procedures. Scottish half-bred ewes were injected 141 days after the mean mating date with 12 or 16 mg of the drug, and the mean interval between treatment and lambing fell between 42 and 45 hours during three seasons of experimentation (see Table X.7). Once again, cases of dystocia and retained placenta were not observed, and the onset of lactation appeared normal. However, the potential problem of mis-mothering during the very condensed period

Table X.6. Timing of the response to two doses of dexamethasone for inducing lambing in ewes carrying singletons, twins or triplets. Dexamethasone was injected on Day 144 of gestation. (From Joyce, 1974.)

Treatment	Number of lambs per ewe	Number of ewes treated	Injection to parturition interval Mean	± S.D. (hours)
12 mg of dexamethasone	1	14	40·9	15·3
	2	18	69·6	44·8
	3	5	43.4	25·2
TOTAL	–	37	55·2	40·6
16 mg of dexamethasone	1	10	50·1	17·1
	2	18	52·1	18·9
	3	1	63·0	–
TOTAL	–	29	51·8	20·8

of birth was highlighted, and a large number of lambing pens was therefore found to be essential. Apart from this particular problem, the advantage of being able to plan and control the date of lambing 5½ months in advance by the procedures of synchronisation of oestrus and induction of parturition would seem attractive in large-scale sophisticated systems of production. But a balanced judgement on the desirability of extensive field-scale usage of the technique must await the publication of more data, and possibly the examination of endocrine intervention at another stage in the pathway. The approach of injecting ewes with a single dose of oestradiol benzoate in an attempt to mimic the *pre-partum* surge of oestrogens in the maternal circulation (see Challis, 1971) has not been found as successful or precise as work with synthetic corticoids, added to which problems of dystocia and lamb survival were reported (Cahill *et al.*, 1976). Even so, Terqui and Delouis (1975) noted that an intramuscular injection of 15 mg oestradiol benzoate three days before expected parturition resulted in 90% of the ewes lambing within 48 hours of treatment.

Control of parturition in pigs

The sequence of endocrine events in sows during late pregnancy and parturition has been presented in some detail in a number of studies (Killian *et al.*, 1973; Molokwu and Wagner, 1973; Robertson and King,

Table X.7. The proportion of ewes lambing in each six-hour period after injection with 12 or 16 mg betamethasone some 141 days after the mean date of mating for the flock. (From Penning and Gibb, 1977.)

Year of experiment	Interval after injection until lambing (hours)														
	0–6	6–12	12–18	18–24	24–30	30–36	36–42	42–48	48–54	54–60	60–66	66–72	72–78	78–84	84+
1974	3	5	1	1		4	19	30		6	1				
1975		3	3	1	7	1	16	26	23	14	1	1	3		1
1976	1	1	2	1	2	6	18	31	22	12	2	1		1	

1974; Ash and Heap, 1975; Baldwin and Stabenfeldt, 1975; Ellendorff *et al.*, 1979; Silver *et al.*, 1979). As far as the steroid hormones are concerned, the concentration of plasma progesterone gradually diminishes over the last two weeks of gestation and rapidly over the last 24 hours (Fig. X.6) whereas the concentration of oestrogens increases very significantly over the same intervals (Fig. X.7). Changes in maternal plasma corticosteroid concentrations are apparently less dramatic, but in at least one study an increase has been shown just before parturition when there is a more marked rise in foetal cortisol levels (Silver *et al.*, 1979; Fig. X.8).

It has already been noted that functional corpora lutea are required throughout the gestation period for maintenance of pregnancy in pigs, so it is reasonable to suppose that a means of inducing luteolysis should precipitate abortion or parturition. This is, in fact, the case and the luteolytic properties of $PGF_{2\alpha}$ or potent analogues of this compound

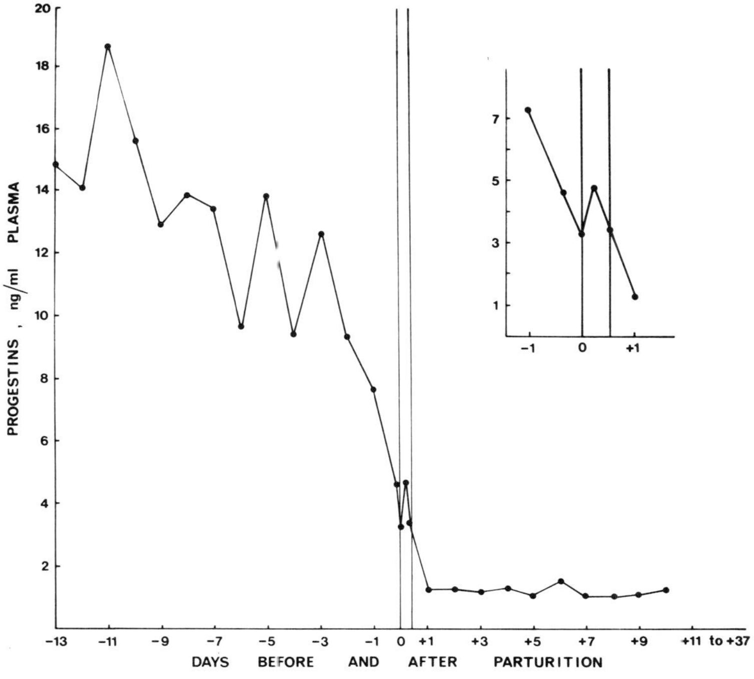

Fig. X.6. The decrease in peripheral plasma concentration of progesterone in pigs as parturition approaches, reflecting a diminishing level of secretion by the corpora lutea. The insert (upper right) presents progestin changes observed at parturition. (Redrawn from Baldwin and Stabenfeldt, 1975.)

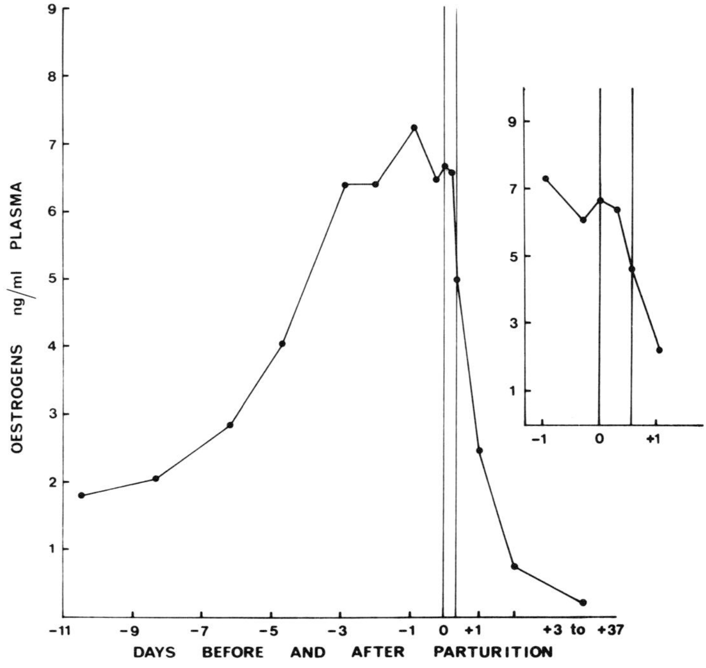

Fig. X.7. The increasing concentration of total oestrogens in the peripheral plasma of pigs as parturition approaches. The insert (upper right) presents oestrogen changes observed at parturition. (Redrawn from Baldwin and Stabenfeldt, 1975.)

can be used to achieve this end. Note, therefore, the contrast between the use of corticosteroids for induction of parturition in cattle and sheep, in part to influence the steroid synthesising activity of the placenta, and the direct use of a luteolytic agent in pigs to influence the function of the maternal ovaries. Despite the effectiveness of such a treatment, the actual mechanism initiating luteolysis at the end of pregnancy as distinct from at the end of the luteal phase in pigs still awaits a detailed description. But by analogy with the situation in sheep, and in the light of the influence of injecting ACTH into the foetus to induce parturition (Bosc, 1973), one must again consider the importance of the foetal hypophyseal-adrenal axis (Du Mesnil du Buisson, 1973). Moreover, the prolonged gestation that follows the decapitation of foetuses in late pregnancy (Stryker and Dziuk, 1975) suggests an involvement of the foetal pituitary in the normal triggering mechanism in this species. In any event, the corpora lutea must regress before parturition can commence, for induction of a new set of corpora lutea

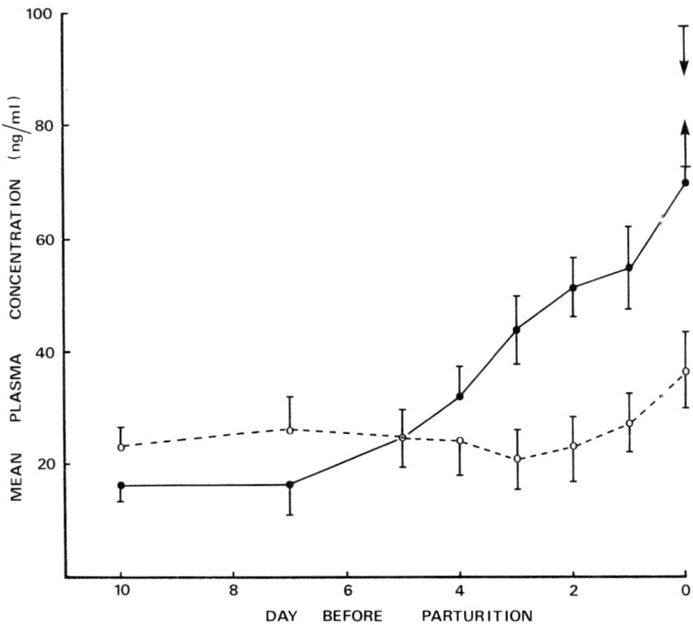

Fig. X.8. Comparison of foetal (solid line) and maternal (dotted line) plasma concentrations of corticosteroids in the last days before parturition in chronically catheterised sows. (Redrawn from Silver *et al.*, 1979.)

shortly before term prolongs gestation by about 10–12 days: this is the period that newly-formed corpora require to become sensitive to the luteolytic effects of $PGF_{2\alpha}$. Chronic studies on catheterised pigs have shown that the dramatic fall in maternal progesterone concentrations in blood plasma in the last 24 hours of gestation is accompanied by a marked rise in maternal plasma and foetal fluid prostaglandin concentrations (Silver *et al.*, 1979).

As a routine husbandry practice rather than as an aid in specific clinical situations, induction of parturition in the large domestic animals is probably most attractive in pigs, especially under intensive systems of management. Because of the relatively close approximation of gestation length to 114 or 115 days (range 111–119 days) in a majority of animals (Fig. X.9), it should be possible to apply a treatment at a well judged time before the expected day of farrowing so that birth may be advanced by only one or two days. In one of the first studies attempting to control the moment of parturition by inducing luteolysis, Ash and Heap (1973) gave intramuscular injections of the drug ICI 79 939—a

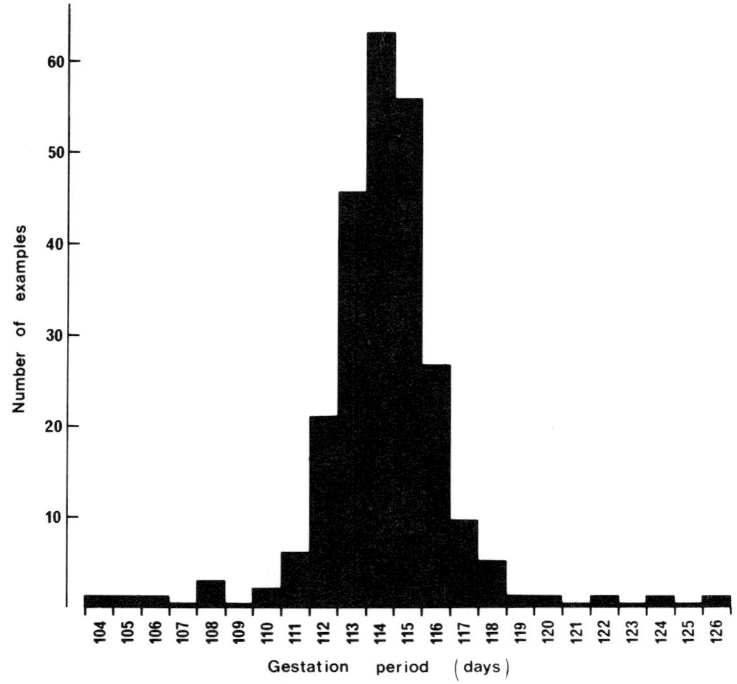

Fig. X.9. Histogram depicting the frequency distribution of gestation lengths in pigs of the Large White breed for a total of 247 pregnancies. (Redrawn from Perry, 1956.)

synthetic analogue of $PGF_{2\alpha}$ with a luteolytic potency exceeding that of the natural product. A total dose of 750 μg given during a treatment period of about 24 hours on Days 109–111 of gestation induced farrowing approximately 26 hours after the start of treatment. No troublesome side effects were observed in this small trial involving 10 animals, nor were the average birth or 3-week weaning weights of the surviving piglets significantly reduced but a surprisingly high incidence of mortality (22 of 119 piglets) was reported. In a subsequent study with preparations of the actual compound $PGF_{2\alpha}$ administered to gilts by intravenous infusion on Days 107–109 or a single intramuscular injection on Day 112, the interval from the start of treatment until birth of the first piglet was 29–30 hours (Diehl *et al.*, 1974). A steep decline in the concentration of peripheral progesterone was noted during the 30 hours after treatment.

Further extensive trials (e.g. Bosc *et al.*, 1975; Walker, 1977) with another potent analogue of $PGF_{2\alpha}$, ICI 80 996 (cloprostenol: trade name of Planate for the pig industry) have produced a simple system for controlled farrowing that is available commercially via the veterinary profession. Correct use of this analogue should not only allow effective control of the time of farrowing, but should also facilitate convenient supervision of groups of animals and enable assistance and cross-fostering of piglets where necessary. Perhaps the biggest attraction in intensive pig units is that use of the drug enables close planning of the birth schedules, thereby permitting efficient use of the expensive resources of labour and farrowing pens. As in other species, induced parturition can only be practised successfully where accurate records of the time of mating or insemination have been kept, for it is essential to treat animals at a known stage of gestation. Premature treatment may lead to a considerable reduction in the number of viable piglets born and surviving, whilst treatment too close to the time of spontaneous farrowing will fail to achieve the desired control. With respect to the mean gestation length for the herd of pigs in question, the recommendation is to give a single intramuscular injection of prostaglandin analogue (i.e. 2 ml Planate solution containing 175 μg cloprostenol) two days before the calculated date of farrowing. The majority of animals can be expected to farrow 24 ± 5 hours after the injection, and 95% will commence farrowing within 36 hours of injection; the animals that have already spontaneously commenced the sequence of events leading to farrowing constitute the exceptions. Farrowing during the working day can thus be achieved by means of injection between 1000 and 1100 hours, but of course animals can be injected at any time to suit the required delivery period and to provide batch farrowing.

A less predictable and compact period of farrowing was observed by Ehnvall *et al.* (1977) using another brand of prostaglandin analogue, but about 85% of the sows delivered within 48 hours of treatment. The question arose in this study as to whether piglet viability following induced parturition matched up to that of controls, and indeed evidence to the contrary was expressed in the earlier papers of Bosc *et al.* (1975) and Walker (1977). Once again, therefore, before drawing a general conclusion as to the desirability of controlling parturition in this species (with the shortest gestation length among the large farm animals), more data from field trials are required. Since mean birthweight depends very much on the age of the foetus and has a critical influence on viability, one improvement might be the introduction of a system where the time of ovulation and insemination was controlled (see Chapter III), thereby removing error from the calculation of foetal age.

Control of parturition in horses

The endocrinology of late pregnancy in mares has been reviewed in considerable detail in the volumes edited by Rowlands *et al*. (1975) and Rowlands and Allen (1979), and also by Pashen and Allen (1979). Although the mechanism of initiation of parturition in this species still requires much clarification, two aspects that distinguish it from the general situation in other large domestic species are that (a) the peripheral plasma concentration of (placental) progestins remains elevated until the time of delivery and (b) the concentration of plasma oestrogens is often remarkably high during the last 45 days of gestation although overall production of oestrogens probably declines after mid gestation.

Whilst horses are seldom farmed intensively in the same sense as cattle, sheep and pigs, considerable benefit might still accrue from a simple means of inducing parturition, not only for clinical purposes but also since foaling tends to be a nocturnal event (see Rossdale and Short, 1967). Although initial reports cast doubt as to whether injected corticoids would induce parturition in the mare (Drost, 1972), treatment with dexamethasone has now been used to achieve this end. For example, Alm *et al*. (1975) injected this synthetic corticoid (100 mg/day) for four days from Day 321 of gestation, and were successful in causing foaling at 328 days compared with a mean of 340 days in control animals. The birth weights of foals were significantly reduced by this treatment, but their growth rates during the first 13 weeks *post partum* were normal and so lactation was presumed to have been adequate.

Use of pure $PGF_{2\alpha}$ in mares close to term was found to be a hazardous procedure, in some manner leading to considerable foal mortality (see Alm *et al*., 1975). This lack of success led to other lines of treatment, such as sequential injections of oestrogen and oxytocin which apparently proved to be a valuable aid for inducing foaling in quite a large series of mares (Hillman, 1975). The reason for pretreatment with oestrogen in this study and in the earlier work of Purvis (1972) seems to have been to hasten cervical softening and relaxation rather than to potentiate myometrial responses to smooth muscle stimulants. More recent investigations of induced parturition, however, have involved prostaglandin analogues, and a single intramuscular injection of 250–500 μg of ICI compound 81 008 (Equimate or fluprostenol) given at term will, quite dramatically, induce foaling within one to three hours even though the cervix may not have commenced to dilate at the time of treatment (Rossdale *et al.*, 1976; Pashen and Allen, 1979). It seems possible that this potent analogue may act directly upon cervical tissues, as well as stimulating contractions of the myometrium.

Incidence of perinatal loss

Although the extent of prenatal mortality has been discussed in some detail in Chapter VI, a further major episode of reproductive loss may occur around the time of parturition and concerns apparently mature foetuses; this is best described as *perinatal* loss. This terminology takes account of foetuses dying in the last days of gestation and during the process of parturition, as well as those succumbing in the first few hours or days after birth—the latter instances being generally referred to under the heading of neonatal loss.

The dimensions of this perinatal wastage as expressed in financial terms are enormous, and concern not only the loss of offspring but also the loss of time and overheads invested in the dam during pregnancy. However, an important distinction is to be drawn between polytocous species such as the pig, where the loss falls usually on only a proportion of the litter, and essentially monotocous species such as the cow where perinatal mortality represents loss of the calf and an unfruitful investment over nine months. The nature and extent of the problem will vary very considerably between systems of animal production, but it is not the intention of these paragraphs to present estimates of losses in financial terms. Rather, the problem is best characterised in proportions, and one recent estimate suggests that the extent of the loss due to stillbirths and perinatal mortality under British conditions is approximately 7% in cattle, 15% in both lowland and upland sheep, and some 20% in pigs (Anon, 1975). Under more extensive conditions of farming than those found in Britain, the figures for cattle and sheep will often be very significantly increased, but pig production is usually a more intensive enterprise and, as such, observations on perinatal mortality may be more detailed. In this respect, Dziuk (1975) noted that the birth process in pigs is directly or indirectly responsible for more deaths than any disease or nutritional deficiency. About 7% of all normal, fully formed piglets that were alive in the uterus just before birth are stillborn. Of the piglets born alive, about 13% die within 48 hours of birth from damage suffered during birth. Together these perinatal losses account for the death of about 20% of the pig "crop". Dziuk (1975) further observed that since only a few piglets are affected in each litter, relatively little effort has been made to study and correct the problem in a scientific manner.

Causes of perinatal loss

In contrast to the situation underlying prenatal mortality, the causes of perinatal mortality are largely understood, although certain predisposing conditions such as low birth weight or the occurrence of a protracted delivery cannot readily be predicted. Nonetheless, the foetus that is to succumb is usually viable until the onset of the birth process, and with sufficient inputs of clinical and managerial attention, the extent of perinatal mortality could be considerably reduced. Some of the causes of loss and/or conditions that increase it can be listed as follows.

(1) Abnormal presentation to the cervical canal during labour.
(2) Protracted birth, placental subfunction and/or asphyxia.
(3) Dystocia due to foetal or maternal problems.
(4) Gross metabolic upset of mother before or during parturition (e.g. pregnancy toxaemia).
(5) Hypothermia. Environmental shock; and chilling of the newborn.
(6) Agalactia. Weakness, failure to suckle and obtain colostrum.
(7) Physical crushing by mother, especially of piglets.
(8) Low birth weight, exacerbating the tendencies in 5, 6 and 7.
(9) Bacterial infection, scouring, dehydration.
(10) Conditions such as anaemia, hypoglycaemia and mineral deficiencies.
(11) Mismothering or rejection, for example, of lambs.
(12) Predation, especially of lambs.

Despite this very considerable variety of factors, it should be clear that much of this loss is accessible in a way that prenatal mortality is not, and is therefore more susceptible to intervention.

Reduction of perinatal loss

Methods for reducing the loss could, of course, include veterinary supervision at parturition, indoor farrowing in a warm, draught-proof environment, provision of supplies of colostrum (frozen-stored) to boost or confer immunity where appropriate, and access to specific mineral solutions, antibiotics and general nutritional support. With the ready availability of such inputs, parturient animals and indeed the industry could be made to appear reproductively more successful. But, as always, the cost of the extra inputs might approach or exceed the value of the increased production. Hence, under systems of farming for profit, sensitive economic judgements will be called for before embarking on the exercise of providing some form of obstetrical clinics for domestic animals.

One must also ask what scope there is for research to help ameliorate the problems of perinatal loss. If birthweight is critical for survival of the newborn, and neonatal mortality increases as mean birth weight decreases, is there a straightforward treatment for increasing foetal weight and energy reserves, such as through delaying the time of parturition? The marked increase in the rate of foetal weight gain at the end of pregnancy draws attention to the potential value of a more precise regulation of gestation length, but any artificial prolongation of even one or two days has to be balanced against the risk of dystocia. If protracted birth frequently contributes to the incidence of loss, would treatment with smooth muscle stimulants such as oxytocin and prostaglandins benefit a proportion of the animals? As to possibilities after completion of parturition, attempted weaning at birth and artificial forms of rearing (such as with piglets in decks of heated stainless steel cages) will have very little appeal in most production situations.

Perhaps studies of maternal behaviour at the time of parturition, and in due course modification of the behaviour pattern, could influence the extent of perinatal loss. For example, if restlessness in sows could be controlled during the relatively lengthy period of farrowing, this should be a means of reducing crushing of poorly-orientated newborn piglets. And again, if the cues that guide the neonate to the teat could be discovered, and if these turn out in part to be of an olfactory nature, then they might be reinforced with appropriate synthetic products: possibly an aerosol spraying of the udder and the piglets' pathway to it with a chemical attractant.

But as of writing, the advent of a straightforward means for controlling the time of parturition in domestic animals—thereby enabling supervision of mother and offspring at a predetermined period—may have the most important technical contribution to make in reducing perinatal mortality. This remark clearly applies to the more intensive systems of animal production and thus to ones where the precise stage of gestation at treatment is known.

References

Adams, W. M. (1969). The elective induction of labor and parturition in cattle. *J. Am. Vet. Med. Assn.* **154**, 261–265.

Adams, W. M. and Wagner, W. C. (1970). The role of corticoids in parturition. *Biol. Reprod.* **3**, 223–228.

Alm, C. C., Sullivan, J. J. and First, N. L. (1975). The effect of a corticosteroid (dexamethasone), progesterone, oestrogen and prostaglandin $F_{2\alpha}$ on gestation length in normal and ovariectomised mares. *J. Reprod. Fert., Suppl.* **23**, 637–640.

Anon. (1975). Report of the Joint Consultative Organisation. Ministry of Agriculture, Fisheries and Food. H.M.S.O., London.

Arthur, G. H. (1964). "Wright's Veterinary Obstetrics". 3rd edition. Baillière, Tindall and Cox, London.

Ash, R. W. and Heap, R. B. (1973). The induction and synchronization of parturition in sows treated with ICI 79939, an analogue of prostaglandin $F_{2\alpha}$. *J. agric. Sci.* **81**, 365–368.

Ash, R. W. and Heap, R. B. (1975). Oestrogen, progesterone and corticosteroid concentrations in peripheral plasma of sows during pregnancy, parturition, lactation and after weaning. *J. Endocr.* **64**, 141–154.

Baldwin, D. M. and Stabenfeldt, G. H. (1975). Endocrine changes in the pig during late pregnancy, parturition and lactation. *Biol. Reprod.* **12**, 508–515.

Bosc, M. J. (1972). The induction and synchronisation of lambing with the aid of dexamethasone. *J. Reprod. Fert.* **28**, 347–357.

Bosc, M. J. (1973). Modification de la durée de gestation de la truie après administration de ACTH aux foetus. *C.r. Acad. Sci. (Paris)* **276**, 3183–3186.

Bosc, M., Du Mesnil du Buisson, F. and Locatelli, A. (1974). Mise en évidence d'un contrôle foetal de la parturition chez la truie. Interactions avec la fonction lutéale. *C.r. Acad. Sci. (Paris)* **278**, 1507–1510.

Bosc, M. J., Martinat-Botte, F. and Duchène, P. (1975). Induction de la mise bas de la truie par un analogue de la prostaglandine $F_{2\alpha}$: conséquences zootechniques. *Annls Zootech.* **24**, 661–670.

Cahill, L. P., Knee, B. W. and Lawson, R. A. S. (1976). Induction of parturition in ewes with a single injection of oestradiol benzoate. *Theriogenology* **5**, 289–294.

Challis, J. R. G. (1971). Sharp increase in free circulating oestrogens immediately before parturition in sheep. *Nature, London* **229**, 208.

Challis, J. R. G., Carson, G. D., Glickman, J. A., MacLennan, E. A., Manchester, L., Patrick, J. E. and Workewych, J. (1979). The peripartal endocrine system in sheep. *Anim. Reprod. Sci.* **2**, 29–42.

Chan, W. Y. (1977). Relationship between the uterotonic action of oxytocin and prostaglandins: oxytocic action and release of PG-activity in isolated nonpregnant and pregnant rat uteri. *Biol. Reprod.* **17**, 541–548.

Christiansen, I. J. and Hansen, L. H. (1974). Dexamethasone-induced parturition in cattle. *Brit. vet. J.* **130**, 221–228.

Comline, R. S. and Silver, M. (1961). The release of adrenaline and noradrenaline from the adrenal glands of the foetal sheep. *J. Physiol., London* **156**, 424–444.

Comline, R. S., Hall, L. W., Lavelle, R. B., Nathanielsz, P. W. and Silver, M. (1974). Parturition in the cow: endocrine changes in animals with chronically implanted catheters in the foetal and maternal circulations. *J. Endocr.* **63**, 451–472.

Corah, L. R., Quealy, A. P., Dunn, T. G. and Kaltenbach, C. C. (1974). Prepartum and postpartum levels of progesterone and oestradiol in beef heifers fed two levels of energy. *J. Anim. Sci.* **39**, 380–385.

Csapo, A. (1956). Progesterone "block". *Amer. J. Anat.* **98**, 273–291.

Csapo, A. I. (1969). The four direct regulatory factors of myometrial function. *In* "Progesterone: its Regulatory Effect on the Myometrium" (G. E. W. Wolstenholme and J. Knight, eds), pp. 13–55. Ciba Fdn Stud. Group No. 34. Churchill, London.

Day, A. M. (1977). Cloprostenol for termination of pregnancy in cattle. (a) The induction of parturition. *N.Z. vet. J.* **25**, 136–139.

Day, A. M. (1979). Induced termination of the calving season in a large dairy herd. *N.Z. vet. J.* **27**, 22–29.

Diehl, J. R., Godke, R. A., Killian, D. B. and Day, B. N. (1974). Induction of parturition in swine with prostaglandin $F_{2\alpha}$. *J. Anim. Sci.* **38**, 1229–1234.

Drost, M. (1972). Failure to induce parturition in pony mares with dexamethasone. *J. Amer. vet. Med. Ass.* **160**, 321–322.

Drost, M. and Holm, L. W. (1968). Prolonged gestation in ewes after foetal adrenalectomy. *J. Endocr.* **40**, 293–296.

Du Mesnil du Buisson, F. (1973). Facteurs lutéotropes chez la truie. *In* "Le Corps Jaune", pp. 225–237. Colloque de la Société Nationale pour l'Étude de la Stérilité et de la Fécondité. Masson and Cie, Paris.

Dziuk, P. J. (1975). Parturition in the pig. *Univ. Illinois Agric. Exp. Statn* **17**, 18–19.

Echternkamp, S. E. and Hansel, W. (1973). Concurrent changes in bovine plasma hormone levels prior to and during the first postpartum estrous cycle. *J. Anim. Sci.* **37**, 1362–1370.

Edgerton, L. A. and Hafs, H. D. (1973). Serum luteinising hormone, prolactin, glucocorticoid, and progestin in dairy cows from calving to gestation. *J. Dairy Sci.* **56**, 451–458.

Edqvist, L. E., Kindahl, H. and Stabenfeldt, G. H. (1976). On the role of prostaglandins in bovine parturition. *Proc. 8th Int. Congr. Anim. Reprod. and Artif. Insem., Krakow*, **1**, 76.

Ehnvall, R., Einarsson, S., Larsson, K. and Westerberg, L. (1977). Prostaglandin-induced parturition in swine—a field study on its accuracy after treatment with different amounts of $PGF_{2\alpha}$. *Nord. Vet.-Med.* **29**, 376–380.

Ellendorff, F., Taverne, M., Elsaesser, F., Forsling, M., Parvizi, N., Naaktgeboren, C. and Smidt, D. (1979). Endocrinology of parturition in the pig. *Anim. Reprod. Sci.* **2**, 323–334.

Erb, R. E., Gomes, W. R., Randel, R. D., Estergreen, V. L. and Frost, O. L. (1968). Effect of ovariectomy on concentration of progesterone in blood plasma and urinary estrogen excretion rate in the pregnant bovine. *J. Dairy Sci.* **51**, 420–427.

Fairclough, R. J., Hunter, J. T., Welch, R. A. S. and Payne, E. (1975). Plasma corticosteroid concentrations in the bovine foetus near term. *J. Endocr.* **65**, 139–140.

Fèvre, J. (1970). Conversion en oestrone de quelques stéroides C-19 chez la truie gestante. *Annls Biol. anim. Biochim. Biophys.* **10**, 25–35.

Fèvre, J., Léglise, P. C. and Rombauts, P. (1968). Du rôle de l'hypophyse et des ovaires dans la biosynthèse des oestrogènes au cours de la gestation chez la truie. *Annls Biol. anim. Biochim. Biophys.* **8**, 225–233.

First, N. L. (1979). Mechanisms controlling parturition in farm animals. *In* "Beltsville Symposia III on Animal Reproduction". Allanheld, Osmun and Company, Montclair, New Jersey.

Fitzpatrick, R. J. and Dobson, H. (1979). The cervix of the sheep and goat during parturition. *Anim. Reprod. Sci.* **2**, 209–224.

Flint, A. P. F., Goodson, J. D. and Turnbull, A. C. (1975). Increased concentrations of 17α, 20α-dihydroxypregn-4-en-3-one in maternal and foetal plasma near parturition in sheep. *J. Endocr.* **67**, 89–97.

Fylling, P. (1971). Premature parturition following dexamethasone administration to pregnant ewes. *Acta endocrin. (Copenhagen)* **66**, 289–295.

Henricks, D. M., Dickey, J. F., Hill, J. R. and Johnston, W. E. (1972). Plasma estrogen and progesterone levels after mating, and during late pregnancy and postpartum in cows. *Endocrinology* **90**, 1336–1342.

Henricks, D. M., Rawlings, N. C., Ellicott, A. R., Dickey, J. F. and Hill, J. R. (1977). Use of prostaglandin $F_{2\alpha}$ to induce parturition in beef heifers. *J. Anim. Sci.* **44**, 438–441.

Hillman, R. B. (1975). Induction of parturition in mares. *J. Reprod. Fert., Suppl.* **23**, 641–644.

Hindson, J.C., Schofield, B. M. and Turner, C. B. (1968). Parturient pressures in the ovine uterus. *J. Physiol., London* **195**, 19–28.

Holm, L. W., Parker, H. R. and Galligan, S. J. (1961). Adrenal insufficiency in postmature Holstein calves. *Amer. J. Obstet. Gynec.* **81**, 1000–1008.

Holness, D. H. and Hunter, R. H. F. (1975). Post-partum oestrus in the sow in relation to the concentration of plasma oestrogens. *J. Reprod. Fert.* **45**, 15–20.

Hunter, J. T., Fairclough, R. J., Peterson, A. J. and Welch, R. A. S. (1977). Foetal and maternal hormonal changes preceding normal bovine parturition. *Acta endocr.* **84**, 653–662.

Jöchle, W. (1971). Corticosteroid induced parturition in domestic animals: mechanism of action and economic importance. *Folia Vet. Lat.* **1**, 229–259.

Joyce, M. J. B. (1974). The use of dexamethasone to induce parturition in ewes synchronized for oestrus. *Irish vet. J.* **28**, 127–131.

Kennedy, P. C., Liggins, G. C. and Holm, L. W. (1967). Prolonged gestation. *In* "Comparative Aspects of Reproductive Failure" (K. Benirschke, ed.), pp. 186–193. Springer-Verlag, New York.

Killian, D. B., Garverick, H. A. and Day, B. N. (1973). Peripheral plasma progesterone and corticoid levels at parturition in the sow. *J. Anim. Sci.* **37**, 1371–1375.

Lanman, J. T. (1968). Delays during reproduction and their effects on the embryo and foetus. 3. Delayed parturition. *New Engl. J. Med.* **278**, 1092–1099.

Liggins, G. C. (1968). Premature parturition after infusion of corticotrophin or cortisol into foetal lambs. *J. Endocr.* **42**, 323–329.

Liggins, G. C. (1969). Premature delivery of foetal lambs infused with glucocorticoids. *J. Endocr.* **45**, 515–523.

Liggins, G. C. (1979). Initiation of parturition. *Brit. Med. Bull.* **35**, 145–150.

Liggins, G. C., Kennedy, P. C. and Holm, L. W. (1967). Failure of initiation of parturition after electrocoagulation of the pituitary of the foetal lamb. *Amer. J. Obstet. Gynec.* **98**, 1080–1086.

Liggins, G. C., Grieves, S. A., Kendall, J. Z. and Knox, B. S. (1972). The physiological rôles of progesterone, oestradiol-17β and prostaglandin $F_{2\alpha}$ in the control of ovine parturition. *J. Reprod. Fert., Suppl.* **16**, 85–103.

Liggins, G. C., Fairclough, R. J., Grieves, S. A., Kendall, J. Z. and Knox, B. S. (1973). The mechanism of initiation of parturition in the ewe. *Recent Prog. Horm. Res.* **29**, 111–159.

Liggins, G. C., Kitterman, J. A. and Forster, C. S. (1979). Fetal maturation related to parturition. *Anim. Reprod. Sci.* **2**, 193–207.

Molokwu, E. C. I. and Wagner, W. C. (1973). Endocrine physiology of the puerperal sow. *J. Anim. Sci.* **36**, 1158–1163.

O'Farrell, K. J. and Crowley, J. P. (1974). Some observations on the use of two corticosteroid preparations for the induction of premature calving. *Vet. Rec.* **94**, 364–366.

Pashen, R. L. and Allen, W. R. (1979). Endocrine changes after foetal gonadectomy and during normal and induced parturition in the mare. *Anim. Reprod. Sci.* **2**, 271–288.

Penning, P. D. and Gibb, M. J. (1977). The use of corticosteroid to synchronise parturition in sheep. *Vet. Rec.* **100**, 491–492.

Perry, J. S. (1956). Observations on reproduction in a pedigree herd of Large White pigs. *J. agric. Sci.* **47**, 332–343.

Purvis, A. D. (1972). Elective induction of labor and parturition in the mare. *Proc. 18th A. Conv. Am. Ass. Equine Pract.*, pp. 113–116. Cited by Hillman, R. B. (1975).

Reynolds, E. O. R. (1973). Physiological effects of reduction in lung surfactant. *Proc. R. Soc. Med.* **66**, 388–389.

Robertson, H. A. (1974). Changes in the concentration of unconjugated oestrone, oestradiol-17α and oestradiol-17β in the maternal plasma of the pregnant cow in relation to the initiation of parturition and lactation. *J. Reprod. Fert.* **36**, 1–7.

Robertson, H. A. and King, G. J. (1974). Plasma concentrations of progesterone, oestrone, oestradiol-17β and of oestrone sulphate in the pig at implantation, during pregnancy and at parturition. *J. Reprod. Fert.* **40**, 133–141.

Rossdale, P. D. and Short, R. V. (1967). The time of foaling of Thoroughbred mares. *J. Reprod. Fert.* **13**, 341–343.

Rossdale, P. D., Jeffcott, L. B. and Allen, W. R. (1976). Foaling induced by a synthetic prostaglandin analogue (fluprostenol). *Vet. Rec.* **99**, 26–28.

Rowlands, I. W. and Allen, W. R. (1979). Equine Reproduction II. *J. Reprod. Fert., Suppl.* **27**.

Rowlands, I. W., Allen, W. R. and Rossdale, P. D. (1975). Equine Reproduction I. *J. Reprod. Fert., Suppl.* **23**.

Salisbury, G. W. and Van Demark, N. L. (1961). "Physiology of Reproduction and Artificial Insemination of Cattle". W. H. Freeman and Co., San Francisco and London.

Sherwood, O. D., Chang, C. C., Bevier, G. W. and Dziuk, P. J. (1975). Radioimmunoassay of plasma relaxin levels throughout pregnancy and at parturition in the pig. *Endocrinology* **97**, 834–837.

Silver, M., Barnes, R. J., Comline, R. S., Fowden, A. L., Clover, L. and Mitchell, M. D. (1979). Prostaglandins in the foetal pig and prepartum endocrine changes in mother and foetus. *Anim. Reprod. Sci.* **2**, 305–322.

Stryker, J. L. and Dziuk, P. J. (1975). Effects of fetal decapitation on fetal development, parturition and lactation in pigs. *J. Anim. Sci.* **40**, 282–287.

Terqui, M. and Delouis, C. (1975). Les oestrogènes au cours de la gestation et de la parturition chez la Brebis. *Journées de la Recherche Ovine et Caprine, Paris*, pp. 332–341.

Thorburn, G. D. (1979). Physiology and control of parturition: reflections on the past and ideas for the future. *Anim. Reprod. Sci.* **2**, 1–27.

Thorburn, G. D., Nicol, D. H., Bassett, J. M., Shutt, D. A. and Cox, R. I. (1972). Parturition in the goat and sheep: changes in corticosteroids, progesterone, oestrogens and prostaglandin F. *J. Reprod. Fert., Suppl.* **16**, 61–84.

Van Rensburg, S. J. (1967). Gestation in sheep after foetal adrenalectomy and cortisol acetate administration. *J. Endocr.* **38**, 83–84.

Wagner, W. C., Willham, R. L. and Evans, L. E. (1974). Controlled parturition in cattle. *J. Anim. Sci.* **38**, 485–489.

Walker, N. (1977). The effects of induction of parturition in sows using an analogue of prostaglandin $F_{2\alpha}$. *J. agric. Sci.* **89**, 267–271.

Welch, R. A. S., Crawford, J. E. and Duganzich, D. M. (1977). Induced parturition with corticosteroids: a comparison of four treatments. *N.Z. vet. J.* **25**, 111–114.

Welch, R. A. S., Newling, P. and Anderson, D. (1973). Induction of parturition in cattle with corticosteroids: an analysis of field trials. *N.Z. vet. J.* **21**, 103–108.

Chapter XI

Further Limitations to Reproductive Output, and Restoration of Fertility

Introduction

Because the productivity of a breeding female may be judged in terms of the number of offspring delivered per year or per unit time, the interval from parturition until the establishment of a further pregnancy is a factor of major economic importance. But the usual sequel to parturition in the higher mammals is a phase of lactation, and during the initial stages of this process pituitary trophic secretion is directed more towards support of milk synthesis than towards resumption of cyclic ovarian activity. This situation results in a period of lactational anoestrus when the female is infertile and generally not receptive to the male (see qualification below). Another limitation to restoration of fertility concerns the condition of the reproductive tract in the *post partum* animal, particularly the uterus but also the cervix in a species such as the pig. After the distention and distortion of uterine tissues during pregnancy, and the heightened glandular development required to support the conceptus, the uterus must undergo contraction and loss of weight together with extensive regeneration of its epithelial layers. These overlapping constraints of lactational anoestrus and involution of the uterus are certainly seen as limitations to output in intensive systems of production, for they may result in a considerable period of infertility. Seasonal anoestrus in species such as sheep presents another form of limitation, but can be modified by treating ewes with gonadotrophic and/or gonadal hormones (Chapter III) or by manipulating the light régime, as well as by selecting animals with extended breeding seasons. Of the three limitations to restoration of fertility mentioned so far, involution of the uterus appears to be the most difficult to regulate by experimental intervention and seems to require a minimum timespan measured in weeks rather than in days in the large farm species (Table XI.1).

Table XI.1. Estimates of the period of time commonly required for involution of the uterus in the large farm species.

Species	Time in days
Cow	35–40
Ewe[1]	20–25
Sow	21–28
Mare	8–15

[1] Description of the situation in many breeds of sheep is complicated by the phenomenon of seasonal anoestrus (see text).

Involution of the uterus

Perhaps the simplest way to envisage the physical problem underlying involution is to consider the size of the gravid uterus at term when it is occupied not only by one or two fully developed cow or sheep foetuses, or perhaps by 13 or 14 pig foetuses, but also by the enveloping membranes and placental fluids. These all act to cause a massive distention of the uterine wall and lumen, and restoration of the organ to its cyclic condition must necessarily involve changes in connective tissue, musculature and epithelial layers. The elastic properties of the musculature permit a substantial reduction in size within hours of parturition, and myometrial contractions may remain forceful during this period and serve to displace placental débris as well as blood and residual fluids from the lumen. As mentioned in the previous chapter, the uterus comes under the influence of increasing concentrations of foeto-placental oestrogens just before birth, and such oestrogens serve to increase the tone of the uterine wall. A further factor generally held to accelerate restoration of the *post partum* uterus is the action of oxytocin elicited by the suckling reflex, whose influence will be repeated hourly, for example, by piglets, during the first days after birth. Thus, suckling activity may hasten the process of involution by causing the myometrium to contract.

The diffuse epithelio-chorial placenta of the pig and horse is relatively simple when considered macroscopically, and the demands of involution seem straightforward when compared with the situation in cattle and sheep. Here, the specialised attachment sites for the foetal cotyledons are the maternal caruncles, and involution involves substantial regression of these structures (Fig. XI.1). In the absence of shrink-

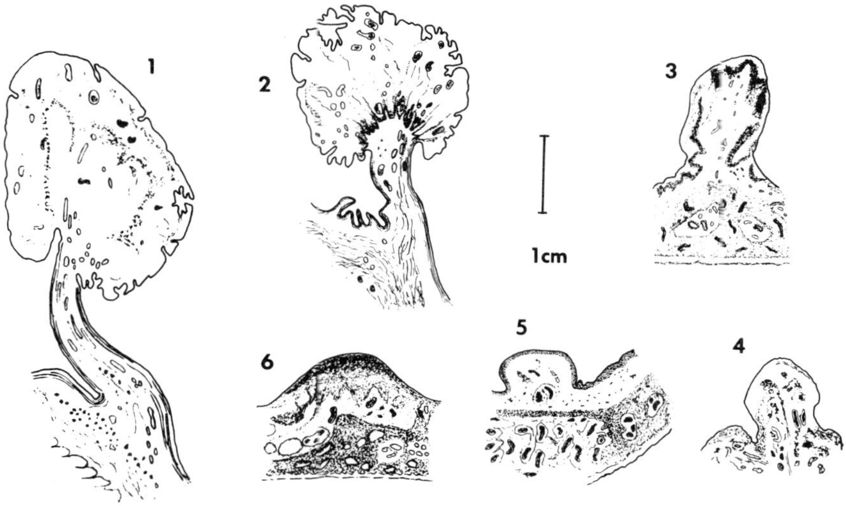

Fig. XI.1. Regression of the caruncles and restoration of a cyclic epithelium during the phase of uterine involution in cattle. The changes depicted require a period of approximately 25–30 days for completion. (Redrawn from Rasbech, 1950.)

age and modification of the caruncles, the uterus would be unable to function effectively in the processes of sperm transport and storage, if mating were to occur. Other *post partum* changes include a phase of leucocytic invasion to remove cellular débris and bacteria from the uterus, and regeneration of a secretory epithelium. Seen in the context of restoration of fertility, an animal that is to breed successfully requires a uterine environment that is not only amenable to sperm transport and survival, but also which will provide metabolic support for the developing embryo within two to three days of fertilisation. The embryo progresses through the morula and blastocyst stages to become a much enlarged conceptus before attachment is initiated, and during such growth is almost completely dependent upon the glandular secretions of the endometrium (see Chapter VI). In fact, the pattern of secretion of ovarian hormones before and after the first ovulatory oestrus causes development of a secretory endometrium, but the uterus may well be better suited to nourishing a preimplantation embryo in the *second* cycle of ovarian changes after the tissues have been exposed to the influence of a full luteal phase.

One of the reasons why the rate of involution may be limited is the period of relative ovarian inactivity found during lactational anoestrus, and thus the absence of a cycle of ovarian steroid secretions acting upon the uterine tissues. But having stressed the requirements of a phase of involution before fertility is once more established, it is worth noting that laboratory species such as the guinea pig may have a fertile heat within hours of parturition. Therefore future work attempting manipulation of the rate of involution in farm animals might, to some advantage, compare endocrine and tissue changes between species with and without a fertile *post partum* heat.

Cow

Involution of the uterus in cattle is completed within 35–40 days of parturition, although instances have been documented where the process has apparently occurred more rapidly, especially in primiparous (first calving) animals. As far as macroscopic changes are concerned, uterine size has regressed very substantially and uterine tone been restored by 15–20 days *post partum*, whilst the weight of the uterus decreases from approximately 9·0 kg at parturition to 1·0 kg at 30 days (Gier and Marion, 1968). During involution, three processes influencing the attachment sites are facilitated by vasoconstriction and myometrial activity: these are sloughing of the caruncular septum, dissolution, and detachment of the cotyledonous superficial layers (Fig. XI.1; Rasbech, 1950). A raw epithelial surface with protruding remnants of blood vessels was noted at 12 days *post partum* (Gier and Marion, 1968), at which time an increased migration of leucocytes (especially polymorphs) into these areas was conspicuous. Involutionary changes associated with the caruncles and their resurfacing with epithelium are generally completed within 25–30 days *post partum*—in the absence of a retained placenta and/or infection of the uterus (Rasbech, 1950; Gier and Marion, 1968; Wagner and Hansel, 1969).

Although the literature is somewhat equivocal, involution may be more rapid under conditions of suckling than machine milking. Nonetheless, the *frequency* of the stimulus to the teat appears to be at least as important as the *nature* of the stimulus. In these situations, the reflex release of oxytocin underlying milk ejection is undoubtedly the principal factor contributing to changes due to its simultaneous influence on myometrial contractions. Thus, Riesen *et al*. (1968) found that the stage of involution was more advanced by the thirtieth day *post partum* in 40 suckled Holsteins, compared with the situation in 35 similar cows receiving no teat stimulation at all. As to whether involution is accelerated in situations of precocious ovulation and cyclic ovarian activity, the literature is again ambiguous although one might suppose some functional relationship to have evolved. Buch *et al.* (1955) found a positive correlation between the interval to first oestrus and involution, whereas Foote and Hunter (1964) reported a non-significant relationship. But it would appear from the results of others (Marion *et al.*, 1968; Wagner and Hansel, 1969; Oxenreider and Wagner, 1971) that cyclic ovarian activity is not essential for involution and that conditions following ovariectomy may actually hasten the rate of involution.

Sheep

Information on the period required for uterine involution in sheep is less detailed than in cattle and pigs, doubtless because most animals are seasonal breeders and the rate of involution is therefore not a factor limiting fertility. However, a period of 20–24 days for the *post partum* changes is cited by Hafez and Jainudeen (1974), with the qualifying remark that an additional 10–12 days are required before the uterus will permit a further establishment of pregnancy. Robertson (1977) observed that in sheep lambing during the breeding season and not permitted to suckle, ovulation will generally have occurred by Day 20 *post partum* whilst conception is usually not found until about Day 35, i.e. at the next ovulation. Bearing in mind that sheep also develop specialised maternal caruncles which enlarge very considerably during gestation (Fig. VI.8), a period of three to four weeks for full regression of these structures and regeneration of a "cyclic" epithelium would seem a reasonable estimate. Necrosis of the caruncles is apparently involved, followed by liquefaction and exfoliation of the necrotic tips (Terrill, 1974).

Pig

Because of the less complex form of the epithelio-chorial placenta in pigs, involution in this species when viewed macroscopically is more concerned with a reduction in uterine size and weight than with conformational changes in the epithelium. As judged from comparative studies, the uterus may take three to four weeks to be fully restored to its *pre-partum* weight (Palmer *et al.*, 1965a), but three-quarters of this loss in weight is achieved within 8–10 days of farrowing (Fig. XI.2). Changes in the endometrium have also reached an advanced stage

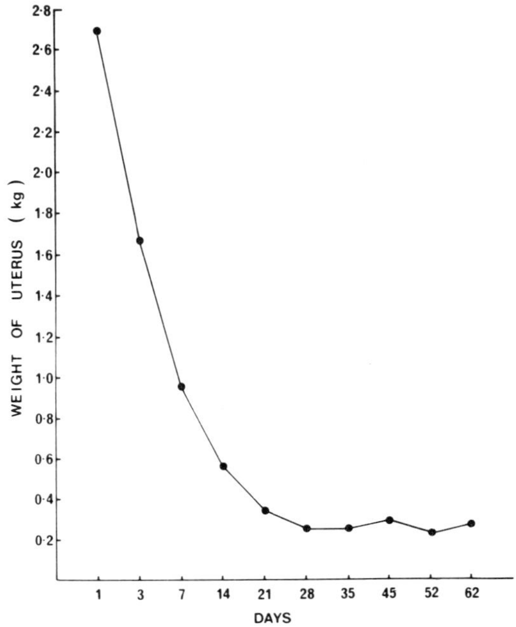

Fig. XI.2. Graphical illustration of the reduction in uterine weight in the *post partum* sow during the phase of uterine involution. (Redrawn from Palmer *et al.*, 1965a.)

within a week of birth, with a reduction in the oedematous condition and a concomitant degeneration of the endometrial glands of pregnancy. Despite some further atrophy and destruction of cells in the myometrium, regenerative changes are distinguishable in the cells of the endometrium and increase during the second week after farrowing (Palmer *et al.*, 1965b; Graves *et al.*, 1967). The ability of the uterus to support developing blastocysts would seem to be restored during the third week, though clearly this can be tested only where cyclic ovarian events have been initiated. The uterine epithelium is fully regenerated by the twenty-first day (Palmer *et al.*, 1965b) and involution is considered complete within 21–28 days of farrowing (Palmer *et al.*, 1965a). Although involutional changes focus attention on the uterus, the cervix may also require substantial regeneration after the events of parturition before it can function normally at mating for both intromission and "locking" of the penis and retention of the ejaculate in the uterus.

Horse

Following the compact period of delivery in the mare, the epitheliochorial placenta is usually released within an hour and involution proceeds rapidly in the interval preceding the foal heat some 5–15 (commonly 8–13) days later. Whilst it is uncertain whether involution is complete by this time, the final restoration of uterine condition may not be critical in a chronological sense because of the lengthy preimplantation phase of the developing embryo (see Chapter VI). Even so, conception rates are usually much improved when breeding occurs subsequent to the first *post partum* heat.

Lactational anoestrus

As an immediate sequel to parturition, a major reproductive function of the pituitary gland is to support milk synthesis and release by the mammary tissues. Whereas oxytocin is involved in the process of release, milk synthesis is regulated by hormones of the lactogenic complex. These include adreno-corticotrophic hormone (ACTH) and thyroid stimulating hormone (TSH), but prolactin can be viewed as a key hypophyseal component. Much recent research has suggested that there is an inverse relationship between pituitary prolactin and gonadotrophin secretion in the first weeks after delivery in cattle, sheep and pigs, and that this relationship underlies the condition of lactational anoestrus. Thus, when prolactin is being actively secreted in response to inputs from the suckling stimulus, gonadotrophic activity is much diminished if not in complete abeyance, presumably due to a dearth of hypothalamic releasing factor (Gn-RH) secretion. In this way, the earliest and heaviest phase of lactation is seen to inhibit resumption of cyclic ovarian activity, and may be interpreted as a mechanism to prevent establishment of a further pregnancy whilst the heaviest demands of the newborn are being met. As the offspring grow and develop, so the intensity of the suckling stimulus diminishes, and the reduction in frequency of suckling apparently permits sufficient secretion of gonadotrophic hormones for resumption of follicular maturation and cyclic ovarian activity. The reduced metabolic demand on the mother may also be involved.

One interpretation of these events would be that afferent inputs to the brain arising from teat stimulation in the *post partum* animal lead to an antagonism of two hypothalamic neurosecretions, prolactin-inhibiting factor and Gn-RH. Artificial manipulation of the frequency of suckling in pigs, by restricting access of the piglets to the sow, has been demonstrated to shorten the period of lactational anoestrus and to accelerate the return of the sow to a fertile oestrus (Smith, 1961; Crighton, 1970). Likewise, use of a drug based upon an ergot alkaloid derivative (bromo-ergocryptine) that blocks prolactin synthesis and/or secretion has also been shown to facilitate resumption of ovarian activity in *post partum* animals (Kann and Martinet, 1975). Another line of evidence associated with this interplay between prolactin and gonadotrophin predominance concerns the stimulus of suckling versus that of the milking machine. The complex suckling activities of the calf might well be viewed as a more positive form of afferent input than that of the milking cluster suspended from the teats, and indeed a shorter period of anoestrus in machine milked rather than suckled cows has been amply documented (Wiltbank and Cook, 1958; Wagner and Hansel, 1969). But despite these lines of evidence, the precise nature of the relationship between prolactin and gonadotrophin secretion in the *post partum* animal is little understood: prolactin may be exerting inhibitory effects directly at the hypothalamic and/or ovarian levels (see Kann *et al.*, 1978; Lamming, 1978).

Cow

There is a slow morphological regression of the non functional corpus luteum following parturition in cattle, and gradual growth and development of several prominent Graafian follicles resulting in increasing oestrogen titres. Most breeds of dairy cow under Western systems of agriculture return to oestrus 30–60 days after calving, except perhaps in the depth of winter, whereas beef cows suckling their calves show oestrus 3–4 weeks later. But cows that are milked four times a day show a protracted anoestrus when compared with those milked twice a day (Casida, 1968). The level of nutrition also has an important influence on the resumption of cyclic ovarian activity, especially in beef cows, for low energy intake invariably extends the period of anoestrus (Wiltbank, 1970).

One approach to accelerating the onset of *post partum* oestrus has been to impose a progestagen-implant treatment (Chupin *et al.*, 1975) or a short-term vaginal coil treatment commencing four weeks or more after calving (Roche, 1976). As can be seen from Table XI.2, there is a strong suggestion that such a treatment advances the time of first oestrus in animals whose ovaries were previously regarded as inactive by means of rectal palpation. Progestagen treatments have also been followed by a PMSG injection at the time of sponge removal (Mulvehill and Sreenan, 1977), but the use of gonadotrophin or synthetic releasing hormones alone to stimulate *post partum* ovarian activity in nursing cows (Britt, 1975; Webb and Haynes, 1976; Webb *et al.*, 1977) has so far remained a purely experimental approach.

Table XI.2. Influence of the *post partum* interval in dairy cows on the oestrous synchronisation response following a 12-day progesterone treatment by means of intra-vaginal coils. (Taken from Roche, 1976.)

	Post-partum interval in days	No. of animals treated	Observed in Oestrus No.	Observed in Oestrus %
Control Cows				
	10–30	35	6	17
	31–45	29	15	52
	>45	14	14	100
Treated Cows				
No prior oestrus	10–30	22	10	45
	>30	13	12	92
Prior oestrus	>30	34	33	97

[1] Animals observed in oestrus 2 to 4 days after removal of the coils.

Sheep

The period of lactational anoestrus in sheep is obscured in many breeds by seasonal anoestrus (see below). However, where lambing occurs during the breeding season, the length of the *post partum* anoestrus is related to the degree of mammary stimulation. In ewes suckling their offspring, lactational anoestrus lasts for 5–7 weeks or more, but most ewes will show oestrus within two weeks of their lambs being weaned. Again, there is evidence that the release of prolactin is in some manner antagonising gonadotrophin secretion and/or the ovarian response during early lactation, whereas denervation of the mammary gland or pharmacological suppression of prolactin secretion in suckling ewes is followed by an early restoration of cyclical ovarian activity (Kann and Martinet, 1975).

Pig

Although the corpora lutea of pregnancy regress rapidly, the heat that may occur 3–5 days after farrowing is anovulatory, being derived from the peak of foeto-placental oestrogens rather than an ovarian source of these steroids (Fèvre *et al.*, 1968; Holness and Hunter, 1975). If the young are removed at birth and lactation is not permitted, ovulatory cycles may resume within 8–16 days (Polge, 1972). But under normal (*ad lib*) conditions of suckling, oestrus and ovulation are suppressed during the active phase of lactation, that is for the first 5–6 weeks, although oestrus will follow within 4–7 days of weaning the young (Burger, 1952)—a period corresponding to the growth and final maturation of the Graafian follicles. As a general statement, the longer that the process of lactation has been in force, the shorter and less variable the subsequent interval after weaning until restoration of oestrus (see Self and Grummer, 1958). But by reducing the *frequency* of the suckling stimulus at four weeks by preventing access of the piglets to the sow for 12 hours daily during a period of 3 days, oestrus and ovulation may resume spontaneously (Smith, 1961), although this response is only reliable at three weeks when a single injection of PMSG is given at the end of the separation treatment (Crighton, 1970). A more straightforward approach to inducing ovarian activity and reestablishing a pregnancy might be simply to inject gonadotrophic hormones, but the earlier work of Cole and Hughes (1946) and Heitman and Cole (1956) suggested that ovulation was not a predictable response in normally lactating sows before the fortieth day *post partum*. By contrast, Epstein and Kadmon (1969) reported that a single injection of PMSG on the twenty-eighth day of lactation would induce oestrus and ovulation, but conception rate was invariably lower than at spontaneous oestrus. In an attempt to minimise the period of lactational anoestrus, various systems of very early weaning of piglets have been examined and the endocrine implications of this approach reviewed by Polge (1972). The penalties associated with very early weaning are not insignificant: the earlier the interval *post partum* that weaning is practised, the longer the delay in returning to oestrus, the lower the conception rate and the smaller the litter size.

Horse

The period of *post partum* anoestrus in mares may be only a matter of days, and can hardly be viewed as a true lactational anoestrus. The foal heat has already been referred to, and occurs in most animals within about two weeks of parturition, even though the foal is actively suckling.

Seasonal anoestrus

Whilst mature cattle and pigs are polyoestrous under climatic conditions where developed farming is more commonly found, and when nutritional and environmental constraints are not severe, breeding activity is nonetheless lower in winter months than in summer. In cattle, in particular, this is reflected in shorter heat periods (i.e. 6–12 hours) and in ovulation without oestrous behaviour (i.e. silent heat), and in practice usually means a greater delay *post partum* before conception for autumn calving animals than for those having spring calves. In some contrast, many breeds of sheep at the higher latitudes exhibit a seasonality in their reproductive performance. The period of seasonal anoestrus may extend for 4–5 months or more and is found as a sequel to the period of lambing, and thus overrides any expression of lactational anoestrus. But it may still act as a limitation to reproductive output, preventing the possibility of spontaneous establishment of a further pregnancy until resumption of the breeding season. Much classical experimental work has demonstrated that seasonal anoestrus is brought about principally by a change in the pattern of daylength (Yeates, 1949; Hart, 1950; Hafez, 1952), whose effects are eventually mediated through modified hypothalamic activity and reflected in a diminished and/or altered frequency of secretion of gonadotrophic hormones. Artificial régimes of lighting in intensively housed sheep can therefore be imposed to stimulate enhanced pituitary secretion which in turn will cause ovarian activity, oestrous cyclicity and the opportunity to establish pregnancy if suitable samples of semen are available (see Mauléon and Rougeot, 1962; Thibault *et al.*, 1966).

A more attractive approach to reducing the influence of seasonal anoestrus on reproductive output in sheep has been to select or import breeds of sheep showing extended seasons of ovarian activity—breeds such as the Dorset Horn or Finnish Landrace which at many latitudes may yield two crops of lambs per year (see Land and McLelland, 1971). A third approach to mitigating the influence of seasonal anoestrus, already referred to in Chapter III, has been to use injections of gonadotrophic hormones to stimulate ovarian activity, such treatment usually being imposed after a period of exposure to depot progestagens administered by means of intra-vaginal sponges (e.g. Gordon, 1975a, b; Thimonier *et al.*, 1975). Whilst hormonal therapy can be used successfully to extend the breeding season, responses have been consistently poor when applied during deep anoestrus. However, a treatment that uses the combined influence of progestagens and artificial lighting (Robinson *et al.*, 1975) may permit imposition of a continuous breeding cycle, especially when applied to breeds such as the Finnish Landrace, Dorset Horn or their crosses.

Restoration of fertility

After the phase of inhibition of breeding activity during lactational and/or seasonal anoestrus, and when involution of the uterus is completed, restoration of fertility is signalled by a return to oestrous cycles, receptivity to the male and subsequent conception. Once cyclic ovarian activity is expressed, involution of the uterus is unlikely to remain a factor limiting fertility but as already indicated, conception rates may be improved, especially in cattle, if animals are mated or inseminated at their second period of oestrus after lactational or seasonal ovarian inactivity (see Whitmore *et al.*, 1974). Fertility in cows bred within the first two months of calving is usually lower than normal (Fig. XI.3) and, as a general rule, the longer the interval from calving to first oestrus, the greater the conception rate obtained (Table XI.3). The ovulation rate itself is not depressed once spontaneous cyclic ovarian activity has been resumed, in fact the converse is the case during the first few pregnancies in a polytocous species as can be judged from the increasing litter size with parity in pigs (Perry, 1956; French *et al.*, 1979). However, in situations of precocious weaning of the offspring when an attempt is made artificially to curtail the period of lactation anoestrus, there is frequently some depression in the number of eggs released during the first oestrus. The underlying mechanism is thought to be associated in part with the reciprocal relationship between gonadotrophin and prolactin activities in the *post partum* animal.

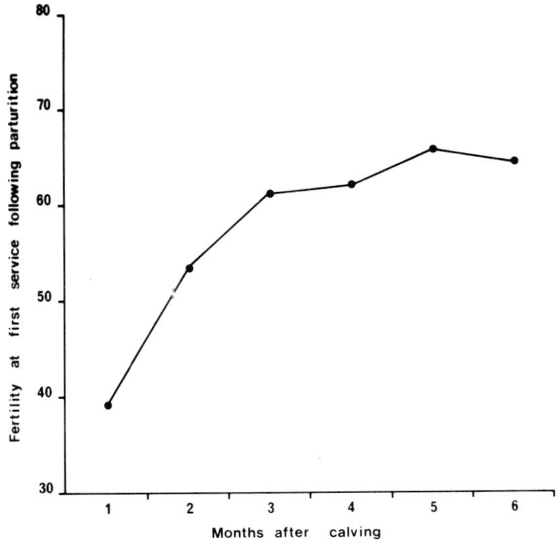

Fig. XI.3. Representation of the improved conception rate generally found in dairy cattle as the interval from calving to first service increases. (Redrawn from Thibault and Levasseur, Fig 4–7, in *Reproduction in Farm Animals*, ed. E. S. E. Hafez, 3rd edition, 1974, Lea and Febiger.)

Table XI.3. To illustrate the increase in the conception rate in dairy cattle with the number of oestrous cycles elapsing since parturition. (Taken from Whitmore *et al*., 1974.)

Order number of the oestrous period at first insemination	No. of animals inseminated	Animals conceiving %
First	17	35·3
Second	28	50·0
Third	67	73·1
Fourth	61	72·1
Fifth	7	100·0

Decline in fertility with age

Whilst the scope of this book does not extend to considerations of infectious diseases, nutritional deficiencies or the influence of temperature stress, fertility may also decline with age and with increasing parity beyond a certain number of pregnancies. In perhaps the majority of instances, cattle, sheep and pigs may not be retained in the breeding herd long enough for this condition to be exhibited, for culling is principally on account of reproductive disorders whether suspected or real. However, reproductive decline has been clearly documented in laboratory animals after repeated pregnancies and some of the underlying factors have been examined systematically (Jones and Krohn, 1960, 1961; Biggers *et al.*, 1962; Adams, 1964). Although the shedding of potentially defective oocytes (Henderson and Edwards, 1968) may contribute to diminished performance with advancing parity, Biggers *et al.* (1962), Biggers (1969) and Adams (1970) considered that defects in the uterus are, in the first instance, mainly responsible for reduced fertility with advancing age. In rabbits, mice and hamsters, the increased embryonic death underlying smaller litter size in old females occurs *after* implantation (Biggers, 1969), and would seem to implicate the uterus rather than the embryo. Use of the embryo transfer technique to instil embryos from young mothers into the uterus of an ageing recipient (Table XI.4) indicates unequivocally that the uterus of aged

Table XI.4. To emphasise the influence of deleterious changes in the reproductive system of ageing females. Fertilised eggs were transferred between animals of markedly different ages, and observations made on those surviving to birth. (After Adams, 1970.)

Nature of transfer	Rabbit	Hamster	Mouse
	Eggs surviving to term (%)		
Young to old	1·5	8·3	14
Young to young	50·0	49·2	48
Old to young	12·5	4·5	54

laboratory animals does constitute an unfavourable milieu for development (Biggers, 1969). A condition of the endometrium termed cystic glandular hyperplasia was particularly prominent in ageing rabbits (Adams, 1970), and others have invoked increased deposition of collagen in the ageing tract as a cause underlying embryonic death. But changes in the uterine vascular bed and impairment of capillary blood flow may be at least as important, and still await critical examination.

Cow

In large domestic species, there seems to be as yet insufficient evidence to enable reproductive decline to be attributed on the one hand to ovarian limitations (e.g. ovulation rate and/or oocyte quality) and on the other to uterine subfunction. But Erickson *et al.* (1976) did not find a significant difference in germ cell numbers between fertile and infertile ageing cows (Table XI.5). Although cattle may remain capable of

Table XI.5. The relationship between ovarian follicular populations and the reproductive status of a group of cows aged 14 to 15 years. (Adapted from Erickson *et al.*, 1976.)

Reproductive status of animals	Number of animals	Classification of ovarian follicles: Primordial (No. X10^3)	Growing (No.)	Healthy antral (No.)	Atretic antral (No.)
Fertile	69	25±3[1]	268±19	41±3	26±2
Infertile	78	23±3	204±15	40±3	27±2

[1] Mean ± standard error.

reproduction to a considerable age (see Krohn, 1964), the calving interval tends to increase after five or six pregnancies, usually leading to elimination of such animals from the herd. The increasing interval could reflect the influence of a large number of factors, but there is evidence that failure or anomalies of fertilisation or preimplantation death of the embryo may underlie many instances of repeat breeding (Table XI.6). Specific evidence derived from large scale artificial

Table XI.6. The extent of early embryonic loss in cattle classified as "repeat breeders" and in a group of first service heifers. (Adapted from Casida, 1961.)

Proportion of animals with:	Normal first service heifer %	Repeat breeder: Heifer (%)	Repeat breeder: Cow (%)
Abnormalities of reproductive organs	2·7	13·5	6·0
Fertilisation failure or early zygote death	21·7	35·3	39·3
Death of embryos before 30 days	16·0	24·8	32·5
Apparently normal embryos at 30–34 days	59·6	26·4	22·2

insemination programmes in dairy cows indicates that fertility increases slightly up to 3–4 years of age (largely due to selection against heifers with anatomical or endocrine abnormalities), and then declines in cows of 6–7 years and older (Tanabe and Salisbury, 1946; Boyd and Reed, 1961; Spalding *et al.*, 1975); but fertility in beef cattle may not decline until 9–10 years of age. Detailed studies on a population of cows in the phase of reproductive decline are clearly needed but will prove expensive, and the findings may not discourage farmers from eliminating such animals from the breeding herd.

Sheep

Because of the extensive conditions under which sheep are traditionally bred, with the rams in attendance for the duration of several oestrous cycles, accurate information concerning the onset of impaired fertility with age is difficult to obtain. Ewes failing to produce a lamb are usually drafted from the breeding flock. However, Laing (1970) suggests that the maximum birth rate is reached at about 5–6 years of age, after which there is a gradual decline.

Pigs

Several reports indicate that litter size in pigs declines with maternal age especially after the fourth to sixth pregnancy, although the rate of decline may be considerably less than in rodents (see Biggers *et al.*, 1962). Nonetheless, the general pattern of an increase in litter size for about the first four pregnancies and then a gradual decrease is widely accepted (Pomeroy, 1960; Strang, 1970). Once again, the loss could represent ovarian shortcomings, specific problems associated with fertilisation or implantation, or embryonic or foetal loss due to uterine subfunction. Evidence from Perry (1954) indicated that ovulation rate increases with age (Table XI.7) whilst litter size in due course decreases, so if the delicate assumption is made that the oocytes shed do not decline in quality with maternal age, then the inference here is a deleterious influence of the ageing genital tract.

Table XI.7. To illustrate the increasing number of corpora lutea (and hence eggs ovulated) in relation to the number of previous litters in pigs. (Modified from Perry, 1954.)

	Number of previous litters							
	0 (gilts)	1	2	3	4	5	6	7 and over
No. of animals	27	14	20	17	12	6	8	16
Mean ovulation rate	13·6	18·6	18·8	21·8	21·9	19·8	22·6	24·4
Standard error ±	0·4	0·6	0·9	1·3	1·7	1·2	0·8	1·6

Reproductive failure

If once again the influence of infectious disease and nutritional deficiencies is excluded, and also the specialised conditions of intersexes and hermaphrodites (Biggers and McFeely, 1966; Short, 1969; Breeuwsma, 1970), failure to breed or to maintain a pregnancy may be due to the following categories of problem.

(1) Anatomical aberrations of the female genital tract, including those associated with freemartinism.

(2) A short luteal phase (i.e. inadequate progesterone secretion) or, alternatively, a persistent corpus luteum.

(3) Development of cystic ovaries comprised of follicular or luteal cysts.

(4) Anovulatory oestrus or silent heats (i.e. ovulation without oestrus).

(5) Deleterious effects of phyto-oestrogens, ergot contamination, etc.

These and other disorders underlying infertility—as distinct from sterility—are described in detail by Laing (1970) and to a lesser extent by Nalbandov (1976). Whereas gross anatomical aberrations of the female genital tract are nearly always manifested at the time of birth, the conditions referred to in categories 2, 3 and 4 above tend to increase with age. A further generalisation would be that whilst infertility due to microorganisms can usually be detected quite rapidly and treated clinically, endocrine and internal anatomical disorders require a systematic and more costly examination. Such an examination, moreover, may not always afford a satisfactory explanation of the infertility or diagnose a condition amenable to treatment.

Anatomical aberrations

The incidence of anatomically abnormal genital tracts is influenced genetically and may increase with the degree of inbreeding, especially in pigs, but in contrast to the incidence of ovarian problems (see below), the gross abnormalities do not increase with age. Anatomical abnormalities may take many forms, as is well illustrated in Nalbandov (1976, p. 308), but perhaps most frequently they represent an excessive degree of fusion in the Mullerian duct system. This may give rise to conditions such as blind uterine horns, occluded portions of the Fallopian tubes or uterus, or missing segments of these ducts. The well documented condition of "White heifer disease" of cattle involves anomalous development of the Mullerian ducts, is inherited, and is thought to be due to a recessive gene associated with that for white coat colour (Laing, 1957).

Also to be considered under the heading of anatomical aberrations of the female genital tract is the condition of freemartinism, which is found most commonly in cattle (Lillie, 1916, 1917) but has been noted in sheep and goats (Short, 1969, 1970). There are three essential prerequisites for the freemartin condition to arise: that there should be at least two ovulations giving rise to dizygotic twins, that the embryos should be of different genetic sex, and that placental fusion leading to anastomosis of allanto-chorionic blood vessels between the two embryos should occur. As a result of the last condition, the female foetus co-twinned to a male usually (>90% in cattle) shows suppression of ovarian development and abnormalities of the reproductive tract. These are associated with fusion and/or inhibition of development of portions of the Mullerian ducts, expressed commonly as a vulva terminating in a short vagina sealed off from the rest of the tract.

The most frequently offered explanation for the freemartin has been both an androgenic and an antifeminising influence of the precociously developing male gonads upon the Mullerian ducts and gonads of the female co-twin. But transplacental migration of cells rather than substances has been proposed (Fechheimer *et al.*, 1963) and reported between the freemartin and its male co-twin (Ohno *et al.*, 1962). Colonisation of the female with cells from the male foetus might therefore underlie certain instances of disruption and partial sex reversal of the female gonad. In this manner, secretions from rudimentary testicular components (the medullary cords) within the disorganised gonad could be the principal deleterious influence on the Mullerian ducts (Short, 1969). Although a form of germ cell chimaerism may thus be involved in producing freemartins, other more subtle components such as sensitivity of the XX gonad to H-Y antigen (Ohno *et al.*, 1976) may also underlie the aetiology of this condition.

Short luteal phase; persistent corpora lutea

Whilst problems of a short or inadequate luteal phase certainly arise in domestic animals, especially in cattle, the underlying causes remain largely speculative since the condition is frequently diagnosed retrospectively and expressed in an oestrous cycle significantly shorter than normal. But the bovine corpus luteum is known to be susceptible to breakdown during the first five or six days of development, and this has been shown experimentally, for example, by intravenous (Armstrong and Hansel, 1959) or subcutaneous (Donaldson *et al.*, 1970) administration of oxytocin. Inadequate luteotrophic support of corpora lutea induced in cattle and sheep in response to Gn-RH injection leads to subfunction of these structures (see Chapter III), perhaps here due to a dearth of prolactin secretion. Endocrine imbalances or irregularities would seem to underlie most cases of luteal insufficiency.

Persistence of the corpus luteum, on the other hand, is often associated with infection of the uterus which may have arisen due to retention of placental remnants from the previous pregnancy. Infection of the uterus giving rise to pus in the lumen (pyometritis) interferes with the normal functioning of the luteolytic mechanism, as has been demonstrated experimentally in sheep by inoculating the uterine lumen with *Corynebacterium pyogenes* or *Vibrio foetus* to cause pyometritis (Coudert and Short, 1966). Persistent corpora lutea are best enucleated manually *per rectum* in cattle, following which the increasing oestrogen titres may be sufficient to promote conditions for removing the infection, due to the infiltration of polymorphonuclear leucocytes. (Development of uterine infection is facilitated under progesterone dominance—Rowson *et al.* (1953).) Treatment of persistent corpora lutea has more recently involved regression with a luteolytic dose of a prostaglandin analogue injected intramuscularly.

Cystic ovaries

Abnormal development of follicles, especially in the form of cysts, frequently gives rise to intense and persistent sexual behaviour expressed as the condition of nymphomania. However, in a proportion of animals with such follicles, anoestrus rather than persistent oestrus is noted. The granulosa cell layers of cystic follicles degenerate, but the follicles can continue to enlarge and secrete oestrogens and probably also androgens from the thecal cells. Another form of ovarian cyst is a luteinised follicle, in which layers of luteal cells usually surround a fluid-filled or haemorrhagic cavity. Both forms of cyst are thought to arise from an endocrine imbalance, often during the period of lactation, and both forms tend to persist and require clinical intervention rather than regressing spontaneously. Whilst the incidence of their formation may be influenced genetically, and the condition is prominent in pigs (Perry and Pomeroy, 1956), cystic ovarian problems may also increase with age and therefore underlie instances of reproductive failure. Experiments in cattle have suggested that cystic follicles may respond more predictably to treatment with Gn-RH than with LH or HCG (Kittok *et al.*, 1973; Britt, 1975), but if the follicle is showing extensive morphological degeneration, an ovulatory or luteinisation response even to endogenous gonadotrophins is difficult to envisage (see Chapter III).

Anovulatory oestrus or silent heats

Dissociation between behavioural oestrus and the ensuing process of ovulation may occur in certain circumstances, leading to anovulatory oestrus or, alternatively, to "silent" heat. A differing threshold of oestrogen feedback from the maturing Graafian follicle(s) is apparently required to trigger on the one hand oestrous behaviour and on the other the preovulatory surge of gonadotrophin. The problem of silent heat is more common and has already been referred to in the section on seasonal anoestrus. In practice, animals, especially cows, are undergoing ovarian cycles after ovulation but not showing oestrous behaviour and thus not being mated. This situation is not associated with permanent infertility, but may frequently occur at the first *post partum* ovulation or at the onset of the breeding season in sheep. The absence of an influence of progesterone from a regressing corpus luteum on the hypothalamic behaviour centres may in part explain the failure to show oestrus.

Deleterious effects of phyto-oestrogens, etc.

Ingestion of unsuitable feedstuffs can certainly compromise fertility, and continued access to the material may be associated with prolonged infertility, but the general topic is far too extensive to be embarked upon here. Suffice it to repeat (see Chapter III) that grazing of plants containing oestrogenic substances which may be found in pasture grasses and specific strains of clover can cause endocrine disturbances and temporary infertility. Feedstuffs containing grain contaminated with ergot, *Claviceps purpurea*, can also rapidly terminate a pregnancy (Nordskog and Clark, 1945) and be followed by a period of infertility.

Non-specific causes of infertility

In contrast to the influence of some of the disorders discussed above, a significant proportion of cows culled from the breeding herd as supposedly infertile and subsequently examined at veterinary investigation centres will show oestrous cycles and conceive (Dawson, 1977). This exhibition of fertility is seen without rendering any specific clinical treatment apart from rectal palpation of the ovaries and possibly also jugular blood sampling. It is debatable whether veterinary investigation on the farm would have relieved the problem, but the result found upon moving animals is open to several interpretations: for example, removing the animal from some form of local stress or, alternatively, directly stimulating follicular maturation, oestrous behaviour and ovulation. The latter possibility is well documented in pigs where transport of gilts on the threshold of puberty, or mature animals in an anoestrous state, provokes oestrous behaviour in many animals some 4–5 days later (Du Mesnil du Buisson and Signoret, 1962; Signoret, 1970). This interval corresponds closely to that required for follicular maturation. Nalbandov (1976) has also discussed the problem of pigs that mate repeatedly but fail to conceive in commercial situations, yet do so readily when moved to an experimental farm. These observations therefore lead to the conclusion that a considerable, though variable, proportion of animals despatched for slaughter on the grounds of infertility are simply showing temporary abeyance of successful breeding activity.

In the light of these remarks, the importance of accurate records for calculating breeding efficiency can no longer be denied, and with the convenience of modern methods of data handling (see Esslemont and Eddy, 1977), inspection of such records would seem essential before formulating breeding or culling strategies.

References

Adams, C. E. (1964). The influence of advanced maternal age on embryo survival in the rabbit. *Proc. 5th Int. Congr. Anim. Reprod. Artif. Insem., Trento*, **2**, 305–308.

Adams, C. E. (1970). Ageing and reproduction in the female mammal with particular reference to the rabbit. *J. Reprod. Fert., Suppl.* **12**, 1–16.

Armstrong, D. T. and Hansel, W. (1959). Alteration of the bovine estrous cycle with oxytocin. *J. Dairy Sci.* **42**, 533–542.

Biggers, J. D. (1969). Problems concerning the uterine causes of embryonic death, with special reference to the effects of ageing of the uterus. *J. Reprod. Fert., Suppl.* **8**, 27–43.

Biggers, J. D. and McFeely, R. A. (1966). Intersexuality in domestic mammals. *Advances Rep. Physiol.* **1**, 29–59.

Biggers, J. D., Finn, C. A. and McLaren, A. (1962). Long-term reproductive performance of female mice. II. Variation of litter size with parity. *J. Reprod. Fert.* **3**, 313–330.

Boyd, H. and Reed, H. C. B. (1961). Investigations into the incidence and causes of infertility in dairy cattle—fertility variations. *Brit. vet. J.* **117**, 18–35.

Breeuwsma, A. J. (1970). Studies on intersexuality in pigs. Dissertation, Drukkerij Bronder-Offset N. V., Rotterdam.

Britt, J. H. (1975). Ovulation and endocrine response after LH-RH in domestic animals. *Annls Biol. anim. Biochim. Biophys.* **15**, 221–231.

Buch, N. C., Tyler, W. J. and Casida, L. E. (1955). Postpartum estrus and involution of the uterus in an experimental herd of Holstein-Friesian cows. *J. Dairy Sci.* **38**, 73–79.

Burger, J. F. (1952). Sex physiology of pigs. *Onderstepoort J. vet. Res., Suppl.* **2**.

Casida, L. E. (1961). Present status of the repeat-breeder cow problem. *J. Dairy Sci.* **44**, 2323–2329.

Casida, L. E. (1968). The postpartum cow—a résumé. *Univ. of Wisconsin Res. Bull.* **270**, 48–54.

Chupin, D., Pelot, J. and Thimonier, J. (1975). The control of reproduction in the nursing cow with a progestagen short-term treatment. *Annls Biol. anim. Biochim. Biophys.* **15**, 263–271.

Cole, H. H. and Hughes, E. H. (1946). Induction of estrus in lactating sows with equine gonadotrophin. *J. Anim. Sci.* **5**, 25–29.

Coudert, S. P. and Short, R. V. (1966). Prolongation of the functional life of the corpus luteum in sheep with experimental uterine infections. *J. Reprod. Fert.* **12**, 579–582.

Crighton, D. B. (1970). Induction of pregnancy during lactation in the sow. *J. Reprod. Fert.* **22**, 223–231.

Dawson, F. L. M. (1977). Reproductive potential in female cattle discarded as infertile. *J. Reprod. Fert.* **51**, 53–56.

Donaldson, L. E., Bassett, J. M. and Thorburn, G. D. (1970). Peripheral plasma progesterone concentration of cows during puberty, oestrous cycles, pregnancy and lactation, and the effects of undernutrition or exogenous oxytocin on progesterone concentration. *J. Endocr.* **48**, 599–614.

Du Mesnil du Buisson, F. and Signoret, J. P. (1962). Influences de facteurs externes sur le déclenchement de la puberté chez la truie. *Annls Zootech.* **11**, 53–59.

Epstein, H. and Kadmon, S. (1969). Physiological and economic feasibility of hormonally induced ovulation in lactating Large White sows. *J. agric. Sci.* **72**, 365–370.

Erickson, B. H., Reynolds, R. A. and Murphree, R. L. (1976). Ovarian

characteristics and reproductive performance of the aged cow. *Biol. Reprod.* **15**, 555–560.
Esslemont, R. J. and Eddy, R. G. (1977). The control of cattle fertility; the use of computerised records. *Brit. vet. J.* **133**, 346–355.
Fechheimer, N. S., Herschler, M. S. and Gilmore, L. O. (1963). Sex chromosome mosaicism in unlike sexed cattle twins. *In* "Genetics Today" (S. J. Geerts, ed.), p. 265. Pergamon Press, Oxford.
Fèvre, J., Léglise, P. C. and Rombauts, P. (1968). Du rôle de l'hypophyse et des ovaires dans la biosynthèse des oestrogènes au cours de la gestation chez la truie. *Annls Biol. anim. Biochim. Biophys.* **8**, 225–233.
Foote, W. D. and Hunter, J. E. (1964). Post-partum intervals of beef cows treated with progesterone and estrogen. *J. Anim. Sci.* **23**, 517–520.
French, L. R., Rutledge, J. J. and First, N. L. (1979). Effect of age and parity on litter size in pigs. *J. Reprod. Fert.* **57**, 59–60.
Gier, H. T. and Marion, G. B. (1968). Uterus of the cow after parturition: involutional changes. *Amer. J. vet. Res.* **29**, 83–96.
Gordon, I. (1975a). The use of progestagens in sheep bred by natural and artificial insemination. *Annls Biol. anim. Biochim. Biophys.* **15**, 303–315.
Gordon, I. (1975b). Hormonal control of reproduction in sheep. *Proc. Brit. Soc. Anim. Prod.* **4**, 79–93.
Graves, W. E., Lauderdale, J. W., Kirkpatrick, R. L., First, N. L. and Casida, L. E. (1967). Tissue changes in the involuting uterus of the postpartum sow. *J. Anim. Sci.* **26**, 365–371.
Hafez, E. S. E. (1952). Studies on the breeding season and reproduction of the ewe. *J. agric. Sci.* **42**, 189–265.
Hafez, E. S. E. and Jainudeen, M. R. (1974). Gestation, prenatal physiology and parturition. *In* "Reproduction in Farm Animals", 3rd edition (E. S. E. Hafez, ed.), Chapter 8. Lea and Febiger, Philadelphia.
Hart, D. S. (1950). Photoperiodicity in Suffolk sheep. *J. agric. Sci.* **40**, 143–149.
Heitman, H., Jr and Cole, H. H. (1956). Further studies in the induction of estrus in lactating sows with equine gonadotrophin. *J. Anim. Sci.* **15**, 970–977.
Henderson, S. A. and Edwards, R. G. (1968). Chiasma frequency and maternal age in mammals. *Nature, London* **218**, 22–28.
Holness, D. H. and Hunter, R. H. F. (1975). Post-partum oestrus in the sow in relation to the concentration of plasma oestrogens. *J. Reprod. Fert.* **45**, 15–20.
Jones, E. C. and Krohn, P. L. (1960). The effect of unilateral ovariectomy on the reproductive lifespan of mice. *J. Endocr.* **20**, 129–134.
Jones, E. C. and Krohn, P. L. (1961). The relationships between age, numbers of oocytes and fertility in virgin and multiparous mice. *J. Endocr.* **21**, 469–495.
Kann, G. and Martinet, J. (1975). Prolactin levels and duration of *postpartum* anoestrus in lactating ewes. *Nature, London* **257**, 63–64.
Kann, G., Martinet, J. and Schirar, A. (1978). Hypothalamic-pituitary control during lactation in sheep. *In* "Control of Ovulation" (D. B. Crighton,

G. R. Foxcroft, N. B. Haynes and G. E. Lamming, eds), pp. 319–333. Butterworths, London and Boston.

Kittok, R. J., Britt, J. H. and Convey, E. M. (1973). Endocrine response after Gn RH in luteal phase cows and cows with ovarian follicular cysts. *J. Anim. Sci.* **37**, 985–989.

Krohn, P. L. (1964). The reproductive lifespan. *Proc. 5th Int. Congr. Anim. Reprod. Artif. Insem., Trento,* **3**, 23–55.

Laing, J. A. (1957). Female fertility. *In* "Progress in the Physiology of Farm Animals' (J. Hammond, ed.), pp. 760–792. Butterworths Scientific Publications, London.

Laing, J. A. (1970) Editor. "Fertility and Infertility in the Domestic Animals", 2nd edition. Baillière Tindall and Cassell, London.

Lamming, G. E. (1978). Reproduction during lactation. *In* "Control of Ovulation" (D. B. Crighton, G. R. Foxcroft, N. B. Haynes and G. E. Lamming, eds), pp. 335–353. Butterworths, London and Boston.

Land, R. B. and McLelland, T. H. (1971). The performance of Finn-Dorset sheep allowed to mate four times in two years. *Anim. Prod.* **13**, 637–642.

Lillie, F. R. (1916). The theory of the free-martin. *Science, N.Y.* **43**, 611–613.

Lillie, F. R. (1917). The free-martin; a study of the action of sex hormones in the foetal life of cattle. *J. exp. Zool.* **23**, 371–452.

Marion, G. B., Norwood, J. S. and Gier, H. T. (1968). Uterus of the cow after parturition: factors affecting regression. *Amer. J. vet. Res.* **29**, 71–75.

Mauléon, P. and Rougeot, J. (1962). Régulation des saisons sexuelles chez des brebis de races différentes au moyen de divers rhythmes lumeneux. *Annls Biol. anim. Biochim. Biophys.* **2**, 209–222.

Mulvehill, P. and Sreenan, J. M. (1977). Improvement of fertility in post-partum beef cows by treatment with PMSG and progestagen. *J. Reprod. Fert.* **50**, 323–325.

Nalbandov, A. V. (1976). "Reproductive Physiology of Mammals and Birds", 3rd edition. W. H. Freeman and Company, San Francisco.

Nordskog, A. W. and Clark, R. T. (1945). Ergotism in pregnant sows, female rats and guinea pigs. *Amer. J. vet. Res.* **6**, 107–116.

Ohno, S. (1969). The problem of the bovine freemartin. *J. Reprod. Fert., Suppl.* **7**, 53–61.

Ohno, S., Trujillo, J., Stenius, C., Christian, L. C. and Teplitz, R. L. (1962). Possible germ cell chimeras among newborn dizygotic twin calves (*Bos taurus*). *Cytogenetics* **1**, 258–265.

Ohno, S., Christian, L. C. Wachtel, S. S. and Koo, G. C. (1976). Hormone-like role of H-Y antigen in bovine freemartin gonad. *Nature, London* **261**, 597–599.

Oxenreider, S. L. and Wagner, W. C. (1971). Effect of lactation and energy intake on postpartum ovarian activity in the cow. *J. Anim. Sci.* **33**, 1026–1031.

Palmer, W. M., Teague, H. S. and Venzke, W. G. (1965a). Macroscopic observations on the reproductive tract of the sow during lactation and early postweaning. *J. Anim. Sci.* **24**, 541–545.

Palmer, W. M., Teague, H. S. and Venzke, W. G. (1965b). Histological changes in the reproductive tract of the sow during lactation and early postweaning. *J. Anim. Sci.* **24**, 1117–1125.

Perry, J. S. (1954). Fecundity and embryonic mortality in pigs. *J. Embryol. exp. Morph.* **2**, 308–322.

Perry, J. S. (1956). Observations on reproduction in a pedigree herd of Large White pigs. *J. agric. Sci.* **47**, 332–343.

Perry, J. S. and Pomeroy, R. W. (1956). Abnormalities of the reproductive tract of the sow. *J. agric. Sci.* **47**, 238–248.

Polge, C. (1972). Reproductive physiology in the pig with special reference to early weaning. Proc. Brit. Soc. Anim. Prod., 5–18.

Pomeroy, R. W. (1960). Infertility and neonatal mortality in the sow. I. Lifetime performance and reasons for disposal of sows. *J. agric. Sci.* **54**, 1–17.

Rasbech, N. O. (1950). The normal involution of the uterus of the cow. *Nord. Vet.-Med.* **2**, 655–687.

Riesen, J. W., Saiduddin, S., Tyler, W. J. and Casida, L. E. (1968). Relation of postpartum interval to corpus luteum development, pituitary prolactin activity, and uterine involution in dairy cows. *Univ. of Wisconsin Res. Bull.* **270**, 27–41.

Robertson, H. A. (1977). Reproduction in the ewe and the goat. *In* "Reproduction in Domestic Animals" 3rd edition (H. H. Cole and P. T. Cupps, eds), Chapter 18. Academic Press, New York and London.

Robinson, J. J., Fraser, C. and McHattie, I. (1975). The use of progestagens and photoperiodism in improving the reproductive rate of the ewe. *Annls Biol. anim. Biochim. Biophys.* **15**, 345–352.

Roche, J. F. (1976). Synchronisation of oestrus in cattle. *World Rev. Anim. Prod.* **12**, 79–88.

Rowson, L. E. A., Lamming, G. E. and Fry, R. M. (1953). The relationship between ovarian hormones and uterine infection. *Vet. Rec.* **65**, 335–340.

Self, H. L. and Grummer, R. H. (1958). The rate and economy of pig gains and the reproductive behaviour in sows when litters are weaned at 10 days, 21 days or 56 days of age. *J. Anim. Sci.* **17**, 862–868.

Short, R. V. (1969). An introduction to some of the problems of intersexuality. *J. Reprod. Fert., Suppl.* **7**, 1–8.

Short, R. V. (1970). The bovine freemartin: a new look at an old problem. *Phil. Trans. Roy. Soc. Lond. B*, **259**, 141–147.

Signoret, J. P. (1970). Reproductive behaviour of pigs. *J. Reprod. Fert., Suppl.* **11**, 105–117.

Smith, D. M. (1961). The effect of daily separation of sows from their litters upon milk yield, creep intake, and energetic efficiency. *N.Z. Jl agric. Res.* **4**, 232–245.

Spalding, R. W., Everett, R. W. and Foote, R. H. (1975). Fertility in New York artificially inseminated Holstein herds in dairy herd improvement. *J. Dairy Sci.* **58**, 718–723.

Strang, G. S. (1970). Litter productivity in Large White pigs. I. The relative importance of some sources of variation. *Anim. Prod.* **12**, 225–233.

Tanabe, T. and Salisbury, G. W. (1946). The influence of age on breeding efficiency of dairy cattle in artificial insemination. *J. Dairy Sci.* **29**, 337–344.

Terrill, C. E. (1974). Sheep. *In* "Reproduction in Farm Animals" 3rd edition (E. S. E. Hafez, ed.), Chapter 12. Lea and Febiger, Philadelphia.

Thibault, C., Courot, M., Martinet, L., Mauléon, P., Du Mesnil du Buisson, F., Ortavant, R., Pelletier, J. and Signoret, J. P. (1966). Regulation of breeding season and estrous cycles by light and external stimuli in some mammals. *J. Anim. Sci.* **25**, Suppl., 119–142.

Thimonier, J., Cognie, Y., Cornu, C., Schneberger, J. and Vernusse, G. (1975). Intensive lamb production. *Annls Biol. anim. Biochim. Biophys.* **15**, 365–367.

Wagner, W. C. and Hansel, W. (1969). Reproductive physiology of the *post partum* cow. I. Clinical and histological findings. *J. Reprod. Fert.* **18**, 493–500.

Webb, R. and Haynes, N. B. (1976). Induction of ovulation in the post-partum nursing cow. *Proc. 8th Int. Congr. Anim. Reprod. Artif. Insem., Krakow*, **3**, 518–519.

Webb, R., Lamming, G. E., Haynes, N. B., Hafs, H. D. and Manns, J. G. (1977). Response of cyclic and post-partum suckled cows to injections of synthetic LH-RH. *J. Reprod. Fert.* **50**, 203–210.

Whitmore, H. L., Tyler, W. J. and Casida, L. E. (1974). Effects of early postpartum breeding in dairy cattle. *J. Anim. Sci.* **38**, 339–346.

Wiltbank, J. N. (1970). Research needs in beef cattle reproduction. *J. Anim. Sci.* **31**, 755–762.

Wiltbank, J. N. and Cook, A. C. (1958). The comparative reproductive performance of nursed and milked cows. *J. Anim. Sci.* **17**, 640–648.

Yeates, N. T. M. (1949). The breeding season of sheep with particular reference to its modification by artificial means using light. *J. agric. Sci.* **39**, 1–43.

Chapter XII

Future Developments: Hormonal, Cellular and Nuclear Manipulations

Introduction

This final chapter is intended to be concise, since speculation about future trends is relatively easy to offer whereas implementation of many of the techniques outlined in the preceding chapters may be more important to animal productivity. Even so, this book would not be complete without reference, for example, to methods that might be used to select animals of superior fertility or to control the sex of the offspring. The application of synthetic pheromones requires brief mention and some evaluation of the prospects for propagating desirable genotypes by the technique of cloning is also relevant. But many of the exciting new developments in molecular biology and immunology that may one day contribute fruitfully to mammalian reproductive studies have not been mentioned, even though discussion of such advances certainly has a place in the teaching curriculum.

Fertility ranking

Males used for breeding through artificial insemination services have been selected after careful evaluation of their fertility in a wide range of herds, but within such a population of males it is usually possible to demonstrate a hierarchy in terms of the fertilising ability of their spermatozoa. An apparently superior fertility of one male over another is best revealed in situations of "competitive fertilisation", that is when the spermatozoa of at least two males are introduced into the female tract at closely coincident times. Heterospermic insemination (Beatty *et al.*, 1969, 1976) and double mating (Dziuk, 1970; Martin and Dziuk, 1977) are both techniques that achieve this purpose, and they have been referred to in Chapter VI. A third method that is objective and deserves further attention, especially in polytocous species, involves

X-irradiation of one of the samples of semen as a means of identifying the parental origin of the spermatozoon that has been successful at fertilisation (Bedford and Hunter, 1968).

In essence, one of the semen samples to be evaluated is exposed to 6000–10 000 rads in order to "mark" the spermatozoa (Bedford and Overstreet, 1972). As judged from extensive experiments in rabbits, this treatment does not influence the ability of such spermatozoa to penetrate the egg investments and complete syngamy, but subsequent cleavage of the egg is disrupted due to the effects of irradiation damage in the sperm nucleus. Retarded cleavage and/or fragmentation of the egg(s) provides a ready identification of the successful male (Bedford and Hunter, 1968; Bedford and Overstreet, 1972), whereas tests involving competitive mating or heterospermic insemination cannot lead to such identification until the stage of foetal development or term. Artificial insemination programmes that have already achieved high conception rates and are therefore well advanced on the curve of diminishing returns might consider a further ranking of their stud males using the irradiation technique.

A predictive means for ranking females on the basis of their potential fertility would also be valuable, but performance rather than prediction has been the conventional approach so far. Nonetheless, with the ready availability of sensitive methods of hormone estimation on samples of peripheral blood, geneticists and animal breeders working together have drawn attention to the possibility of physiological selection for reproductive traits. A thoughtful paper by Land (1973) suggested that fertility might not be sex-limited in the sense that performance traits hitherto regarded as peculiar to the female might be reflected in the pituitary and gonadal activity of closely-related males. In other words, many components of reproductive performance may originate from the same genes—autosomal genes—and thus expression of these genes should be measurable in both sexes. Therefore, it might be possible to select females on the basis of reproductive traits shown by their male relatives, and indeed to select a male's potential daughters from measurements of hormone concentration or the response to hormone treatment even, for example, in young ram lambs. Progress has been made in this field in terms of developing experimental protocols and in relating hormone measurements to breed differences in fertility (reviewed by Land, 1977). However, a serious snag in collecting and evaluating the endocrine data arises from the episodic or pulsatile nature of gonadotrophin secretion and the sampling frequency required to plot realistic curves and to establish significant differences between breeds or strains. A balanced assessment of this approach to reproductive selection suggests that it is still very much at the experimental stage, and to date has been more fruitful in yielding basic physiological information than in promoting selection programmes.

Pheromones

In his Cambridge inaugural lecture in 1961, Parkes made reference to the neurohumoral effects of social stimuli and, in particular, of odours that permit communication between animals. He drew a distinction between substances secreted internally for the integration of the individual (hormones) and those secreted externally by individuals to integrate a group or population of animals (pheromones); a contrast was therefore drawn between endocrinology and exocrinology (Parkes, 1962). Whitten (1966) has defined pheromones as substances or mixtures of substances which are produced to the exterior by one animal and may be received by a second individual of the same species, in which they produce one or more specific reactions. Subsequent reviews have elaborated on the properties of pheromones in mammals (Bruce, 1970; Whitten and Champlin, 1973, 1978), and a number of specific conditions have been described in which male pheromonal activity has a dominant influence on female reproductive physiology. These include the Whitten effect in which the distribution of oestrous cycles in large groups of mice can be synchronised in 40–50% of the animals by the introduction of a mature male, and the Bruce effect in which odours of a "strange" male (i.e. differing in strain from the stud male) can cause termination of preimplantation stages of pregnancy in groups of mice. A principal source of the odorous substances in rodents seems to be in the urine of males (Dominic, 1966), and their production depends on male sex hormone activity.

In the large farm species, an influence of the male upon female reproductive activity has been repeatedly documented in sheep and pigs, where presentation of a mature male to appropriate groups of females stimulates the onset of oestrous behaviour. Thus, introducing a ram to a group of anoestrous ewes in the late summer may hasten the onset of the breeding season (Watson and Radford, 1960; Robertson, 1977). Similarly, penning a boar with a group of gilts on the verge of puberty will provoke the onset of oestrus in many animals a few days later (Brooks and Cole, 1970). This particular effect can be mimicked by introducing prepuberal gilts into a pen recently vacated by a boar, so a pheromonal stimulus may again be involved even though the act of moving the gilts may complicate interpretation of the oestrous response (see Du Mesnil du Buisson and Signoret, 1962). But assuming that pheromones are involved in the "boar effect" and that the appropriate odoriferous compounds can be synthesised, it is not difficult to envisage treatments for provoking and/or synchronising the onset of the first oestrus.

At the time of writing, the only specific treatment of farm animals involving synthetic pheromones is to assist the detection of oestrus in pigs. The characteristic behavioural response of gilts and sows in oestrus (lordosis) is most readily elicited in the presence of male odours, although the response is further reinforced by the sight and sound of the boar (see Signoret, 1970). Whereas the preputial gland is of importance, a major source of the characteristic male odours is from the sub-maxillary salivary glands. These glands act as secondary sex tissues in the sense that they are influenced by testicular androgens, and the compound that has pheromonal activity is apparently 5α-androst-16-en-3-one or a close derivative (see Patterson, 1968; Booth *et al.*, 1973). Compounds of this nature have been synthesised and produced in aerosol form, and are used for detection of oestrus under conditions of artificial insemination (Melrose *et al.*, 1971; Reed *et al.*, 1974).

Apart from the possible use of pheromones for triggering oestrous activity in seasonal breeders and for synchronising oestrus at puberty or *post partum*, compounds with pheromonal properties may one day also be used for facilitating cross-fostering and for reducing aggressive behaviour under conditions of intensive housing. In these latter situations, olfactory cues have a key influence on behaviour, and clarification of the molecules involved might lead the way to valuable treatments. In the first instance, however, co-operative studies between animal behaviourists and chemists would seem to be needed.

Cellular and nuclear manipulations

In a physical or mechanical sense, the micromanipulation of mammalian eggs and embryos has remained a specialised and skilful field during the last 10 to 15 years, although studies on the much larger amphibian eggs have been extensive and achieved prominence (see

Gurdon, 1962, 1968). Separation of blastomeres in 2- or 4-cell eggs is perhaps the most straightforward manoeuvre in this category, but piercing the zona pellucida and displacing individual blastomeres in a controlled manner still requires the use of a micromanipulator. Because cells in the very young embryo are totipotent and have not commenced irreversible differentiation, splitting of 2- or 4-cell eggs into constituent blastomeres has been considered a means of yielding identical (monozygotic) twins or quadruplets, thereby propagating embryos of desirable strains. In practice, the technique usually leads to (a) damage of one or more blastomeres and (b) problems of survival of blastomeres that are not enclosed within a zona pellucida. Even so, separation of blastomeres in 2-, 4- or 8-cell rabbit eggs has been achieved with intact single cells subsequently developing into viable young (Moore *et al.*, 1968). In a small series of experiments, similar manipulations have been performed on pig embryos (Moore *et al.*, 1969) and mechanical division of sheep blastocysts has also been reported (Trounson and Moore, 1974). However, until a simpler and more successful means of separating and replacing blastomeres in "foster" zonae is developed, or of culturing divided blastocysts during a period of repair, these approaches will not significantly increase rates of propagation and will probably be restricted to the research laboratory.

Formation of chimaeras

The word chimaera refers to a composite animal in which different cell populations are derived from more than one zygote, or the union of more than two gametes, and primary chimaeras are those in which the genetically different cell populations have co-existed from a very early stage of embryogenesis, or from fertilisation itself (see McLaren, 1976). Chimaeras have been constituted in the laboratory by fusing the embryos of mice (Tarkowski, 1961) or by introducing somatic cells from one individual into the blastocoele of a second embryo (Gardner, 1968). The technique of chimaera formation is an elegant way of studying problems of differentiation and de-differentiation by examining the fate of cells introduced in the above manner at different stages of development; it may also yield valuable information on malignancy and on possible ways of controlling the growth of tumours. The literature dealing with observations on mammalian chimaeras has been reviewed in the monograph of McLaren (1976). As far as applications in farm animals are concerned, formation of chimaeras has been suggested as a means of incorporating desirable traits from one breed into another, in effect producing new strains or breeds of animal. Experimentation with this objective has yet to be reported for cattle, sheep or pigs. Nonetheless, chimaeric offspring have been formed in sheep by embryo fusion during a period of culture (Pighills *et al.*, 1968) and by injecting disaggregated blastomeres from young embryos into sheep morulae: offspring of tetraparental origin were demonstrated (Tucker *et al.*, 1974).

Nuclear transplants and cloning

The work of Gurdon (1962, 1968) with amphibia has shown that it is possible to introduce the nucleus of a differentiated somatic cell into the cytoplasm of an unfertilised egg whose own nucleus has been removed, and thereby to obtain normal development. Studies in mammals have centred on the mouse egg, and nuclei obtained from the disrupted cells of young morulae have been inserted into single-cell zygotes. In this situation, introduction of a single nucleus leads to the formation of a tetraploid embryo. Development to the blastocyst stage has occurred following this manipulation (Modlinski, 1978), and refinement of the technique may enable formation of diploid embryos. Experiments involving nuclear transplantation in farm animals, if successful, could perhaps permit the development of clones of identical individuals and might therefore be viewed as desirable in a production context (i.e. clones of high-yielding dairy cows). Once again, however, considerable technical advances will be needed before cloning can be agriculturally useful. Furthermore, it must be recalled that all the manipulations discussed in this section require the ultimate step of egg or embryo transplantation before they can be fruitful.

Predetermination or regulation of sex

Before discussing methods that have been applied in an attempt to control the sex of mammalian offspring, it is necessary to make reference to the primary, secondary or tertiary sex ratio of males to females in a polytocous species or population of animals. The primary sex ratio is that expressed at fertilisation, according to whether the egg has been penetrated by a spermatozoon carrying an X- or a Y-chromosome with respect to the sex chromosomes. The secondary sex ratio is the proportion of males to females found at some point during gestation or at term and, due to instances of prenatal loss in polytocous species, may differ considerably from that found at fertilisation. The tertiary sex ratio is that expressed at puberty when the population under consideration is competent to breed.

Control of the primary sex ratio would be possible if one could separate the two distinct types of spermatozoa within an ejaculate, one type bearing an X-chromosome, the other a Y-chromosome. Coupled with the techniques of artificial insemination or *in vitro* fertilisation, this should then enable predetermination of the sex of the offspring according to whether a population of X- or Y-bearing spermatozoa was inseminated. In a less flexible sense, control of the primary sex ratio would also be effected by techniques that led to parthenogenetic development of embryos. Here, of course, only X-chromosomes would be under consideration since the mammalian female is the homogametic sex, so there is not really the question of choice of sex. Whilst much experimentation has been pursued in mammals in relation to techniques of sperm separation and parthenogenesis, it is important to state firmly at the outset that there is currently no prospect of controlling the sex of farm animals by either of these approaches.

Sperm separation

Spermatozoa are haploid cells and methods of separation of X- and Y-bearing sperm in the ejaculate must necessarily assume that there is expression of the haploid genome. But the extent to which there may be haploid gene expression is still a controversial question, so approaches to separating the two kinds of spermatozoa have been empirical rather than rational. However, since the Y-chromosome is smaller than the X-chromosome, a possible sperm separation has been suggested on the basis of mass, for some phenotypic difference between the two types of spermatozoa might be expected. Clear cut differences in mass on the basis of the sex chromosome constitution seem highly improbable. One conclusive demonstration of a difference between sperm cells has been that of fluorescent "F-bodies" in Y-bearing human spermatozoa (Barlow and Vosa, 1970; Evans, 1972; Beatty, 1977). Physico-chemical methods attempting separation of X- and Y-bearing spermatozoa have included sedimentation, counterstream centrifugation or electrophoresis of the cells under a variety of conditions, and the use of systems that depend on differential migration of spermatozoa in colloidal media. Although the sex ratio of offspring has on occasions been substantially modified from the theoretical ratio of 50/50 males to females, the change has not been repeatable and predictable when strict precautions concerning experimental design have been enforced. The literature has been reviewed in detail by Courot (1975), who was unable to conclude his paper on an optimistic note. So whilst it would be rash to rule out the possibility of a chance observation in the laboratory leading to a satisfactory technique of separating X- and Y-bearing spermatozoa, a perspective that has been recently offered is that . . . "the tantalising goal of artificial regulation of the primary sex ratio remains a mirage on the horizon" (Short, 1979). This statement takes into account the immunological approach of trying to eliminate Y-bearing spermatozoa in an ejaculate by treatment with H-Y antibody (see Ohno and Wachtel, 1978).

Sex identification in embryos

A characteristic of the somatic cells of genetically female embryos is that one of the two X-chromosomes becomes inactivated, and in this form gives rise to a lump of chromatin—the sex chromatin or Barr body—that can be recognised in the nucleus upon staining. Accordingly, determination of the sex of embryos should be possible by screening constituent cells for sex chromatin, and in practice this can be achieved in blastocysts that are held temporarily in culture whilst the procedures are completed (Gardner and Edwards, 1968). Since the technique involves removing a small group of cells from the developing embryo, this is best done when extensive cell multiplication has already occurred and generally involves excision of a small portion of trophoblast using a micromanipulator, and then culture of the embryos during a period of recovery. Staining of the excised cells enables prediction of the sex of individual blastocysts, and the accuracy of the technique is established by transferring the blastocysts to suitable recipient animals. A high level of success has been achieved by this approach in laboratory animals (Gardner and Edwards, 1968), but the technique of staining trophoblast cells for sex chromatin is not readily adaptable for farm animals due to granular elements in the nucleus. On the other hand, the tissue sample can be processed for chromosome analysis (karyotyping), and this approach has been used to predict the sex of a calf after removing a fragment of a 4 mm long conceptus and analysing the sex chromosomes (Mitchell, 1977). But even with this advance, it is difficult to imagine such techniques being applied extensively in production situations, although they might be used to expedite multiplication of selected female lines.

Other predictive methods

Once the placental membranes have developed, other approaches to predicting as distinct from controlling the sex of the offspring in monotocous species include amniocentesis and endocrine estimations in placental fluids. In the latter case, measurements of testosterone in allantoic fluid have enabled prediction of foetal sex (Bongso *et al.*, 1976). Amniocentesis involves puncture of the amniotic membrane and removal of a small volume of fluid in which the cells are then prepared for chromosomal analysis. Whilst amniocentesis will certainly reveal the sex chromosome constitution of the foetus, its principal application is for genetic screening in man in situations where anomalies of the chromosome complement such as in Downe's syndrome might be suspected. Once again, neither amniocentesis nor steroid measurements in allantoic fluid seem likely to have a rôle in conventional systems of animal production, not least since regulation of sex using these approaches would involve aborting foetuses of the unwanted sex.

Opioid peptides and ovarian responses

Brief reference has already been made in Chapter I to the opioid pentapeptide molecules termed enkephalins and endorphins that seem to be involved in regulating hypothalamic and pituitary function. For example, endorphins in the pituitary gland are specifically localised in the intermediate and anterior lobes, and there is evidence that β-endorphin and ACTH are synthesised by pituitary cells from a common precursor polypeptide (see Simantov, 1979). Likewise, there is evidence to suggest a direct inhibitory rôle of glucocorticoids on the synthesis of endorphins by the pituitary cells, and it is possible that a similar feedback system may operate from other endocrine glands. Of particular relevance here is the observation that fluctuations in the content of pituitary endorphins occur during the oestrous cycle of rats (Simantov, 1979).

Endogenous opiates, released either by the median eminence into the hypothalamo-hypophyseal portal system or by the pituitary into the general circulation, may also be involved in the physiological regulation of prolactin secretion (see Enjalbert *et al.*, 1979). Accordingly, if endorphins or enkephalins could be used to manipulate secretion of pituitary hormones, and perhaps specifically to modify the interplay between prolactin and gonadotrophin secretion, then a treatment with these compounds might be developed to regulate ovarian function. This, however, is definitely looking to the future, for the physiological actions of these peptide molecules are only just beginning to be explored in a reproductive context, and there is no direct evidence *in vivo* that they can be used to modify secretion of gonadotrophic hormones.

References

Barlow, P. and Vosa, C. G. (1970). The Y chromosome in human spermatozoa. *Nature, London 226*, 961–962.

Beatty, R. A. (1977). F-bodies as Y chromosome markers in mature human sperm heads: a quantitative approach. *Cytogenet. Cell Genet.* **18**, 33–49.

Beatty, R. A., Bennett, G. H., Hall, J. G., Hancock, J. L. and Stewart, D. L. (1969). An experiment with heterospermic insemination in cattle. *J. Reprod. Fert.* **19**, 491–502.

Beatty, R. A., Stewart, D. L., Spooner, R. L. and Hancock, J. L. (1976). Evaluation by the heterospermic insemination technique of the differential effect of freezing at −196°C on fertility of individual bull semen. *J. Reprod. Fert.* **47**, 377–379.

Bedford, J. M. and Hunter, R. H. F. (1968). The influence of X-irradiation of rabbit spermatozoa on fertilisation and early cleavage. *J. Reprod. Fert.* **17**, 49–57.

Bedford, J. M. and Overstreet, J. W. (1972). A method for objective evaluation of the fertilising ability of spermatozoa irrespective of genetic character. *J. Reprod. Fert.* **31**, 407–414.

Bongso, T. A., Basrur, P. K. and Younglai, E. V. (1976). Prediction of fetal sex in cattle by testosterone levels in allantoic fluid. *J. Reprod. Fert.* **46**, 441–442.

Booth, W. D., Hay, M. F. and Dott, H. M. (1973). Sexual dimorphism in the submaxillary gland of the pig. *J. Reprod. Fert.* **33**, 163–166.

Brooks, P. H. and Cole, D. J. A. (1970). The effect of the presence of a boar on the attainment of puberty in gilts. *J. Reprod. Fert.* **23**, 435–440.

Bruce, H. M. (1970). Pheromones. *Brit. med. Bull.* **26**, 10–13.

Courot, M. (1975). Le choix du sexe est-il possible chez les mammifères? *In* "La Fécondation", Colloque S.N.E.S.F., pp. 113–130. Masson, Paris.

Dominic, C. J. (1966). Observations on the reproductive pheromones of mice. I. Source. *J. Reprod. Fert.* **11**, 407–414.

Du Mesnil du Buisson, F. and Signoret, J. P. (1962). Influences de facteurs externes sur le déclenchement de la puberté chez la truie. *Annls Zootech.* **11**, 53–59.

Dziuk, P. J. (1970). Estimation of optimum time for insemination of gilts and ewes by double-mating at certain times relative to ovulation. *J. Reprod. Fert.* **22**, 277–282.

Enjalbert, A., Ruberg, M., Arancibia, S., Priam, M. and Kordon, C. (1979). Endogenous opiates block dopamine inhibition of prolactin secretion *in vitro*. *Nature, London* **280**, 595–597.

Evans, H. J. (1972). Properties of human X and Y sperm. *In* "The Genetics of the Spermatozoon" (R. A. Beatty and S. Gluecksohn-Waelsch, eds), pp. 144–157. Proc. Int. Symp., Edinburgh, 1971. Departments of Genetics of the University of Edinburgh and of the Albert Einstein College of Medicine, Edinburgh and New York.

Gardner, R. L. (1968). Mouse chimaeras obtained by the injection of cells into the blastocyst. *Nature, London* **220**, 596–597.

Gardner, R. L. and Edwards, R. G. (1968). Control of the sex ratio at full term in the rabbit by transferring sexed blastocysts. *Nature, London* **218**, 346–348.

Gurdon, J. B. (1962). The developmental capacity of nuclei taken from intestinal epithelium cells of feeding tadpoles. *J. Embryol. exp. Morph.* **10**, 622–640.

Gurdon, J. B. (1968). Transplanted nuclei and cell differentiation. *Sci. Am.* **219**, 24–35.

Land, R. B. (1973). The expression of female sex-limited characters in the male. *Nature, London* **241**, 208–209.

Land, R. B. (1977). Genetic variation and improvement. *In* "Reproduction in Domestic Animals" 3rd edition (H. H. Cole and P. T. Cupps, eds), pp. 577–604. Academic Press, New York and London.

McLaren, A. (1976). "Mammalian Chimaeras". Cambridge University Press, London, New York and Melbourne.

Martin, P. A. and Dziuk, P. J. (1977). Assessment of relative fertility of males (cockerels and boars) by competitive mating. *J. Reprod. Fert.* **49**, 323–329.

Melrose, D. R., Reed, H. C. B. and Patterson, R. L. S. (1971). Androgen steroids associated with boar odour as an aid to the detection of oestrus in pig artificial insemination. *Brit. vet. J.* **127**, 497–502.

Mitchell, D. (1977). Sexing of embryos. *In* "Embryo Transfer in Farm Animals" (K. J. Betteridge, ed.), pp. 26–27. Canada Dept. of Agriculture, Monograph 16.

Modlinski, J. A. (1978). Transfer of embryonic nuclei to fertilised mouse eggs and development of tetraploid blastocysts. *Nature, London* **273**, 466–467.

Moore, N. W., Adams, C. E. and Rowson, L. E. A. (1968). Developmental potential of single blastomeres of the rabbit egg. *J. Reprod. Fert.* **17**, 527–531.

Moore, N. W., Polge, C. and Rowson, L. E. A. (1969). The survival of single blastomeres of pig eggs transferred to recipient gilts. *Aust. J. biol. Sci.* **22**, 979–982.

Ohno, S. and Wachtel, S. S. (1978). On the selective elimination of Y-bearing spermatozoa. *Immunogenetics* **7**, 13–16.

Parkes, A. S. (1962). Prospect and retrospect in the physiology of reproduction. *Brit. med. J.* **ii**, 71–75.

Patterson, R. L. S. (1968). Identification of 3α-hydroxy-5α-androst-16-ene as the musk odour component of boar submaxillary salivary gland and its relationship to the sex odour taint in pork meat. *J. Sci. Fd Agric.* **19**, 434–438.

Pighills, E., Hancock, J. L. and Hall, J. G. (1968). Attempted induction of chimaerism in sheep. *J. Reprod. Fert.* **17**, 543–547.

Reed, H. C. B., Melrose, D. R. and Patterson, R. L. S. (1974). Androgen steroids as an aid to the detection of oestrus in pig artificial insemination. *Brit. vet. J.* **130**, 61–67.

Robertson, H. A. (1977). Reproduction in the ewe and the goat. *In* "Reproduction in Domestic Animals" 3rd edition. (H. H. Cole and P. T. Cupps, eds), pp. 475–498. Academic Press, New York and London.

Short, R. V. (1979). Sex determination and differentiation. *Brit. med. Bull.* **35**, 121–127.

Signoret, J. P. (1970). Reproductive behaviour of pigs. *J. Reprod. Fert., Suppl.* **11**, 105–117.

Simantov, R. (1979). Glucocorticoids inhibit endorphin synthesis by pituitary cells. *Nature, London* **280**, 684–685.

Tarkowski, A. K. (1961). Mouse chimaeras developed from fused eggs. *Nature, London* **190**, 857–860.

Trounson, A. O. and Moore, N. W. (1974). Attempts to produce identical offspring in the sheep by mechanical division of the ovum. *Aust. J. biol. Sci.* **27**, 505–510.

Tucker, E. M., Moor, R. M. and Rowson, L. E. A. (1974). Tetraparental sheep chimaeras induced by blastomere transplantation. *Immunology* **26**, 613–621.

Watson, R. H. and Radford, H. M. (1960). The influence of rams on onset of oestrus in Merino ewes in the spring. *Aust. J. agric. Res.* **11**, 65–71.

Whitten, W. K. (1966). Pheromones and mammalian reproduction. *Advances Reprod. Physiol.* **1**, 155–177.

Whitten, W. K. and Champlin, A. K. (1973). The role of olfaction in mammalian reproduction. *In* "Handbook of Physiology", Section 7, Endocrinology II (R. O. Greep and E. B. Astwood, eds), pp. 109–141. American Physiological Society, Washington.

Whitten, W. K. and Champlin, A. K. (1978). Pheromones, estrus, ovulation, and mating. *In* "Methods in Mammalian Reproduction" (J. C. Daniel, Jr., ed.), pp. 403–417. Academic Press, New York and London.

Subject Index